Advances in Intelligent Systems and Computing

Volume 389

Series editor

Janusz Kacprzyk, Polish Academy of Sciences, Warsaw, Poland
e-mail: kacprzyk@ibspan.waw.pl

About this Series

The series "Advances in Intelligent Systems and Computing" contains publications on theory, applications, and design methods of Intelligent Systems and Intelligent Computing. Virtually all disciplines such as engineering, natural sciences, computer and information science, ICT, economics, business, e-commerce, environment, healthcare, life science are covered. The list of topics spans all the areas of modern intelligent systems and computing.

The publications within "Advances in Intelligent Systems and Computing" are primarily textbooks and proceedings of important conferences, symposia and congresses. They cover significant recent developments in the field, both of a foundational and applicable character. An important characteristic feature of the series is the short publication time and world-wide distribution. This permits a rapid and broad dissemination of research results.

More information about this series at http://www.springer.com/series/11156

Ryszard S. Choraś
Editor

Image Processing and Communications Challenges 7

Editor
Ryszard S. Choraś
Institute of Telecommunications
University of Technology and Life Sciences
Bydgoszcz
Poland

ISSN 2194-5357 ISSN 2194-5365 (electronic)
Advances in Intelligent Systems and Computing
ISBN 978-3-319-23813-5 ISBN 978-3-319-23814-2 (eBook)
DOI 10.1007/978-3-319-23814-2

Library of Congress Control Number: 2015948782

Springer Cham Heidelberg New York Dordrecht London

Printed on acid-free paper

Springer International Publishing AG Switzerland is part of Springer Science+Business Media
(www.springer.com)

Preface

This book contain papers accepted for IP&C 2015, the International Conference on Image Processing and Communications, held at UTP University of Science and Technology, Bydgoszcz, Poland, September 9–11, 2015. This was the eighth edition in the IP&C series of annual conferences. The idea of organizing these conferences was to facilitate exchange between researchers and scientists in the broad fields of image processing and communications, addressing recent advances in theory, methodology, and applications.

At this IP&C conference, a wide variety of topics are covered by a relatively small selection of papers presented in single-track sessions. This year, there were 62 manuscripts submitted to the conference. Each paper was carefully reviewed by scientists and researchers in IP&C areas. Of these papers, 34 were accepted for presentation.

The book should be of interest to a large group of researchers, engineers, and practitioners in image processing and communication. I hope that, by reading this book, researchers in both academia and industry with an interest in this area will be encouraged and facilitated to pursue it further.

We acknowledge the hard work and dedication of many people. We thank the authors who have contributed their work. We are grateful for the help, support, and patience of the Springer publishing team.

Bydgoszcz
September 2015

Ryszard S. Choraś

Contents

Part I
Image Processing

Classifier Selection Uses Decision Profiles in Binary Classification Task

Paulina Baczyńska and Robert Burduk

Abstract The dynamic selection of classifiers plays an important role in the creation of an ensemble of classifiers. The paper presents the dynamic selection of a posteriori probability function based on the analysis of the decision profiles. The idea of the dynamic selection is exemplified with the binary classification task. In addition, a number of experiments have been carried out on ten benchmark data sets.

Keywords Ensemble pruning methods · Classifiers selection · Multiple classifier system

1 Introduction

Classification is one of the important steps in pattern recognition, which belongs to machine learning fields [1]. The classification task can be accomplished by a single classifier or by a team of classifiers. In the literature, the use of the multiple classifiers for a decision problem is known as the multiple classifier systems (MCS) or an ensemble of classifiers EoC [4, 10]. The construction of MSC consists of three phases: generation, selection and integration [2]. In the second phase, which is discussed in this paper, one or a subset of the base classifiers is selected to make the final decision which it is to assign an object to the class label.

The output of an individual classifier can be divided into three types [17].

- The abstract level—the classifier ψ assigns the unique label j to a given input x.
- The rank level—in this case for each input x, each classifier produces an integer rank array. Each element within this array corresponds to one of the defined class labels. The array is usually sorted and the label at the top being the first choice.

P. Baczyńska · R. Burduk (✉)
Department of Systems and Computer Networks, Wroclaw University of Technology, Wybrzeze Wyspianskiego 27, 50-370 Wroclaw, Poland
e-mail: robert.burduk@pwr.edu.pl

R.S. Choraś (ed.), *Image Processing and Communications Challenges 7*,
Advances in Intelligent Systems and Computing 389,
DOI 10.1007/978-3-319-23814-2_1

- The measurement level—the output of a classifier is represented by a measurement value that addresses the degree of assigning the class label to the given output x. An example of such a representation of the output is a posteriori probability returned by Bayes classifier.

According to these three types of outputs of the base classifier, various problems of the combination function of classifier' outputs are considered. The problems studied in [18, 24] belong to the abstract level. The combining outputs for the rank level are presented in [12] and problems studied in [15, 16] belong to the last level.

The selection of classifiers is one of the important problems in the creation of EoC [14, 23]. This task is related to the choice of a set of classifiers from all the available pool of classifiers. Here you can distinguish between the static or dynamic selection [20]. In the static classifier selection one set of classifiers is selected to create an EoC. This EoC is used in the classification of all the objects from the testing set. The main problem in this case is to find a pertinent objective function for selecting the classifiers. One of the best objective functions for the abstract level of classifier' outputs is the simple majority voting error [22]. In the dynamic classifier selection for each unknown sample a specific subset of classifiers is selected [3]. It means that we are selecting different EoCs for different objects from the testing set. In this type of the classifier selection, the classifier is chosen and assigned to the sample based on different features [25] or different decision regions [6, 13].

In this work we propose the dynamic selection of a posteriori probability functions (PPFs). In detail we propose the new method based on the analysis of decision profiles. In particular we use the average and standard deviation of the PPFs from the correct prediction of the base classifiers.

The text is organized as follows: after this introduction, in Sect. 2 the idea of EoC is presented. Section 3 contains the description of the proposed dynamic selection of PPFs. The experimental results on ten benchmark data sets are presented in Sect. 4. Finally, conclusions from the experiments and future research proposals are presented.

2 Ensemble of Classifiers

Let us assume that we possess K of different classifiers $\Psi_1, \Psi_2, \ldots, \Psi_K$. Such a set of classifiers, which is constructed on the basis of the same learning sample is called an ensemble of classifiers or a combining classifier. However, any of Ψ_i classifiers is described as a component or base classifier. As a rule K is assumed to be an odd number and each of Ψ_i classifiers makes an independent decision. As a result, of all the classifiers' action, their K responses are obtained. Having at the disposal a set of base classifiers one should determine the procedure of making the ultimate decision regarding the allocation of the object to the given class. It implies that the output information from all K component classifiers is applied to make the ultimate decision.

In this work we consider the situation when each base classifier returns the estimation of a posteriori probability. This means that the output of all the base classifiers is at the measurement level. Let us denote a posteriori probability estimation by $\hat{p}_k(i|x)$, $k = 1, 2, \ldots, K$, $i = 1, 2, \ldots, M$, where M is the number of the class labels. One of the possible methods for such outputs is the linear combination method. This method makes use of the linear function like Sum, Prod or Mean for the combination of the outputs. In the sum method the score of the group of classifiers is based on the application of the following sums:

$$s_i(x) = \sum_{k=1}^{K} \hat{p}_k(i|x), \qquad i = 1, 2, \ldots, M. \tag{1}$$

The final decision of the group of classifiers is made following the maximum rule and is presented accordingly, depending on the sum method (1):

$$\Psi_S(x) = \arg\max_i s_i(x). \tag{2}$$

3 Proposal Selection Methods of a Posteriori Probability Estimations

3.1 Algorithm with the Use of Average

Being given K classifiers from the initial pool of classifiers we select a posteriori probability functions (PPFs) returned by this pool. The selected PPFs are integrated and are used to built the ensemble. The final decision is made on the basis of the dynamically selected PPFs. It means that the selection is performed for each new object (from the testing sets).

Now we present a novel algorithm for the selection of PPFs. For K base classifier their outputs are arranged in the decision profile:

$$DP(x) = \begin{bmatrix} \hat{p}_1(1|x) & \hat{p}_1(2|x) \\ \vdots & \vdots \\ \hat{p}_K(1|x) & \hat{p}_K(2|x) \end{bmatrix}. \tag{3}$$

for the case of the binary classification task.

During learning of the base classifiers we obtain m decision profiles, where m is the number of objects from the learning set. In the first stage of the proposed selection of PPFs algorithm we calculate the decision scheme according to the formula:

$$DS_{S-M} = \begin{bmatrix} \widehat{\overline{ds}}_{11} & \widehat{\overline{ds}}_{12} \\ \vdots & \vdots \\ \widehat{\overline{ds}}_{K1} & \widehat{\overline{ds}}_{K2} \end{bmatrix}, \quad (4)$$

where

$$\widehat{\overline{ds}}_{k\omega} = \frac{\sum_{n=1}^{m} I(\Psi_k(x_n) = \omega_n)\, \hat{\overline{p}}_k(\omega_n|x_n)}{\sum_{n=1}^{m} I(\Psi_k(x_n) = \omega_n)}. \quad (5)$$

The $\widehat{\overline{ds}}_{k\omega}$ is calculated only from those PPFs for which the classifier k did not make an error.

The above decision scheme is used in the selection of PPFs from the decision profile for the new object according to the formula:

$$\text{if } \hat{\overline{p}}_k(\omega|x) < \widehat{\overline{ds}}_{k\omega} \text{ then } \hat{\overline{p}}_k(\omega|x) = null, k = 1, \ldots, K, \ \omega = 1, 2. \quad (6)$$

The obtained decision profile, designated as DP_{S-M}, for the new object contains the selected PPFs. Based on the DP_{S-M} we can use the various algorithms for the integration of PPFs. In experimental studies we use the sum method (2) to make the final decision by the ensemble classifier after the selection of PPFs. The algorithm using this method is denoted as Ψ_{S-M}.

3.2 Algorithm with the Use of Average and Standard Deviation

The condition described in (5) concerns the use of averages form PPFs. Now propose a different decision profile which is calculated using mean and standard deviation of PPFs. This decision profile DP_{S-MS} is calculated as follows:

$$DS_{S-MS} = \begin{bmatrix} \widehat{ds}_{11} & \widehat{ds}_{12} \\ \vdots & \vdots \\ \widehat{ds}_{K1} & \widehat{ds}_{K2} \end{bmatrix}, \quad (7)$$

where

$$\widehat{ds}_{k\omega} = \widehat{\overline{ds}}_{k\omega} + \sqrt{\frac{\sum_{n=1}^{m}(I(\Psi_k(x_n) = \omega_n) \ - \widehat{\overline{ds}}_{k\omega})^2}{m-1}}. \quad (8)$$

Now, by substituting in condition (6) values from the decision profile DS_{S-MS} we obtain the new algorithm Ψ_{S-M} which uses average and standard deviation of PPFs.

In the experimental studies we also use the proposed algorithms in which the normalization of PPFs was performed. The normalization refers to the columns of the decision profile $DP(x)$. The algorithms with the normalization are labeled as Ψ^N_{S-MS} and Ψ^N_{S-M}.

4 Experimental Studies

In the experiments 7 base classifiers were used. One of them (labeled as Ψ_1) used the decision trees algorithms, with the splitting rule based on entropy, the number of branches equal 2 and the depth of the precision tree having at most 6 levels. Two of them work according to $k - NN$ rule where k parameter is equal to 3 or 5 and are labeled as Ψ_2 and Ψ_3 respectively. The classifier labeled as Ψ_4 is the rule induction classifier. This classifier uses a tree-based modeling in which a tree is run and models for which the purity is at or above a specified threshold are removed from the training data set and placed to the side. The fifth classifier Ψ_5 uses Support Vector Machines models with Decomposed Quadratic Programming estimation method. The sixth classifier Ψ_6 uses least squares regression model to predict the class label. The last of the base classifiers Ψ_7 is a multilayer perception model with 3 hidden units. In the experiments SAS Enterprise Miner environment has been used.

The aim of the experiments was to compare the proposed selection method algorithms with the base classifiers. In the experiential research we used 10 benchmark binary data sets. The eight benchmark data sets come from the UCI repository [8]. The last two are generated randomly—they are the so called Banana and Higleyman sets. The numbers of attributes, examples and ration in the classes are presented in Table 1.

The results containing the classification accuracy and mean rank positions produced by the Friedman test [23] are presented in Tables 2 and 3. The studies did not include the impact of the feature selection process on the quality of classifications. Therefore, the feature selection process [11, 21] was not performed. The presented results are obtained via 10-fold-cross-validation method.

Table 2 shows the results of the classification for the proposed selection methods of PPFs with and without normalization. Additionally, the mean ranks obtained by the Friedman test were presented. To compare the results the post-hoc Nemenyi test

Table 1 Description of data sets selected for the experiments

Data set	Example	Attribute	Ration (0/1)
Banana	400	2	1.0
Blood	748	5	3.2
Breast cancer wisconsin	699	10	1.9
Haberman	306	3	0.2
Highleyman	400	2	1.0
Ionosphere	351	34	1.8
Indian liver patient	583	10	0.4
Mammographic mass	961	6	1.2
Parkinson	197	23	0.3
Pima Indians diabetes	768	8	1.9

Table 2 Classification accuracy and mean rank positions for the proposed selection algorithms produced by the Friedman test

Data set	Ψ^N_{S-MS}	Ψ^N_{S-M}	Ψ_{S-MS}	Ψ_{S-M}
Banana	0.928	0.967	0.889	0.970
Blood	0.775	0.779	0.795	0.784
Cancer	0.942	0.955	0.841	0.949
Haberman	0.726	0.749	0.664	0.749
Higle	0.838	0.923	0.793	0.930
Ion	0.867	0.907	0.787	0.892
Liver	0.657	0.700	0.556	0.676
Mam.	0.831	0.827	0.771	0.836
Park.	0.856	0.913	0.882	0.908
Pima	0.773	0.761	0.734	0.748
Mean rank	2.9	1.7	3.6	1.7

Table 3 Classification accuracy and mean rank positions for the base classifiers ($\Psi_1, \ldots, \Psi_7$) and proposed selection Ψ_{S-M} algorithm produced by the Friedman test

Data set	Ψ_1	Ψ_2	Ψ_3	Ψ_4	Ψ_5	Ψ_6	Ψ_7	Ψ_{S-M}
Banana	0.953	0.987	0.984	0.942	0.859	0.859	0.989	0.970
Blood	0.780	0.747	0.742	0.757	0.760	0.776	0.788	0.784
Cancer	0.929	0.962	0.956	0.940	0.960	0.945	0.948	0.949
Haberman	0.716	0.723	0.739	0.720	0.733	0.742	0.723	0.749
Higle	0.913	0.925	0.920	0.920	0.828	0.788	0.940	0.930
Ion	0.884	0.841	0.844	0.878	0.878	0.864	0.887	0.892
Liver	0.687	0.669	0.657	0.676	0.714	0.711	0.690	0.676
Mam.	0.818	0.799	0.788	0.830	0.800	0.798	0.810	0.836
Park.	0.892	0.867	0.851	0.903	0.877	0.882	0.872	0.908
Pima	0.739	0.722	0.724	0.727	0.768	0.777	0.736	0.748
Mean rank	4.7	5.4	5.9	4.9	4.2	4.7	3.3	2.4

was used. The critical difference for this test at $p = 0.05$ is equal $CD = 0.086$. The results clearly show that the proposed algorithms which use only means of PPFs in the process of creating the decision scheme are better than other proposed algorithms. This is because the post-hoc test detects significant differences between the algorithms, for example between Ψ_{S-MS} and Ψ_{S-M}.

Table 3 shows the results of the classification for all base classifiers and one of the proposed algorithms Ψ_{S-M} which was chosen on the basis of previous results. The critical difference for this set of algorithms at $p = 0.05$ is equal $CD = 3.32$. It means, that the mean ranks of the proposed algorithm Ψ_{S-M} and base classifier Ψ_3 are significantly different. For the other pair of classifiers this property does not occur. It should be noted that the proposed selection algorithm has the lowest average rank which is a promising result.

5 Conclusion

The paper presents new selection algorithms of PPFs. The selection process the decision profiles are used. After this process the integration of the outputs of base classifier is performed in accordance with the sum rule. This means that a classifier ensemble is created wherein not all PPFs are used to make the final decision. The paper presents two approaches. In one of them only an average of PPFs from the decision profile is used. In the second case under consideration average and standard deviation of PPFs from the decision profile are used.

In the paper several experiments on benchmark data sets were carried out. The obtained results for algorithms which utilize only averages of PPFs from the decision profiles are promising. This is due to statistical difference in mean ranks for the certain algorithms. Additionally, the paper presents the selection methods of PPFs which can work in the parallel and distributed environment or can be applied in various practical tasks involving multiple elementary classification tasks [5, 7, 9]. The distributed computing approaches enable an efficient use of the classification systems with ensemble methods [19].

In the future works we plan to discuss the impact of use of the standard deviation, presented in the Sect. 3.2 with an additional parameter. In addition, the further experiments should apply not only to the binary classification task.

Acknowledgments This work was supported by the Polish National Science Center under the grant no. DEC-2013/09/B/ST6/02264 and by the statutory funds of the Department of Systems and Computer Networks, Wroclaw University of Technology.

References

1. Bishop, C.M.: Pattern Recognition and Machine Learning (Information Science and Statistics). Springer-Verlag New York Inc, Secaucus (2006)
2. Britto, A.S., Sabourin, R., Oliveira, L.E.: Dynamic selection of classifiers—a comprehensive review. Pattern Recogn. **47**(11), 3665–3680 (2014)
3. Cavalin, P.R., Sabourin, R., Suen, C.Y.: Dynamic selection approaches for multiple classifier systems. Neural Comput. Appl. **22**(3–4), 673–688 (2013)
4. Cyganek, B.: One-class support vector ensembles for image segmentation and classification. J. Math. Imaging Vision **42**(2–3), 103–117 (2012)
5. Cyganek, B., Woźniak, M.: Vehicle logo recognition with an ensemble of classifiers. In: Intelligent Information and Database Systems, Lecture Notes in Computer Science, vol. 8398, pp. 117–126. Springer (2014)
6. Didaci, L., Giacinto, G., Roli, F., Marcialis, G.L.: A study on the performances of dynamic classifier selection based on local accuracy estimation. Pattern Recogn. **38**, 2188–2191 (2005)
7. Forczmański, P., Łabędź, P.: Recognition of occluded faces based on multi-subspace classification. In: Computer Information Systems and Industrial Management, Lecture Notes in Computer Science, vol. 8104, pp. 148–157. Springer (2013)
8. Frank, A., Asuncion, A.: UCI machine learning repository (2010). http://archive.ics.uci.edu/ml

9. Frejlichowski, D.: An algorithm for the automatic analysis of characters located on car license plates. In: Image Analysis and Recognition, Lecture Notes in Computer Science, vol. 7950, pp. 774–781. Springer (2013)
10. Giacinto, G., Roli, F.: An approach to the automatic design of multiple classifier systems. Pattern Recogn. Lett. **22**, 25–33 (2001)
11. Guyon, I., Elisseeff, A.: An introduction to variable and feature selection. J. Mach. Learn. Res. **3**, 1157–1182 (2003)
12. Ho, T.K., Hull, J.J., Srihari, S.N.: Decision combination in multiple classifier systems. IEEE Trans. Pattern Anal. Mach. Intell. **16**(1), 66–75 (1994)
13. Jackowski, K., Woźniak, M.: Method of classifier selection using the genetic approach. Expert Syst. **27**(2), 114–128 (2010)
14. Jackowski, K., Krawczyk, B., Woźniak, M.: Improved adaptive splitting and selection: the hybrid training method of a classifier based on a feature space partitioning. Int. J. Neural Syst. **24**(03) (2014)
15. Kittler, J., Alkoot, F.M.: Sum versus vote fusion in multiple classifier systems. IEEE Trans. Pattern Anal. Mach. Intell. **25**(1), 110–115 (2003)
16. Kuncheva, L.I.: A theoretical study on six classifier fusion strategies. IEEE Trans. Pattern Anal. Mach. Intell. **24**(2), 281–286 (2002)
17. Kuncheva, L.I.: Combining Pattern Classifiers: Methods and Algorithms. Wiley, New York (2004)
18. Lam, L., Suen, C.Y.: Application of majority voting to pattern recognition: an analysis of its behavior and performance. IEEE Trans. Syst. Man Cybern. Part A **27**(5), 553–568 (1997)
19. Przewoźniczek, M., Walkowiak, K., Woźniak, M.: Optimizing distributed computing systems for k-nearest neighbours classifiers-evolutionary approach. Logic J. IGPL 357–372 (2010)
20. Ranawana, R., Palade, V.: Multi-classifier systems: review and a roadmap for developers. Int. J. Hybrid Intell. Syst. **3**(1), 35–61 (2006)
21. Rejer, I.: Genetic algorithms in eeg feature selection for the classification of movements of the left and right hand. In: Proceedings of the 8th International Conference on Computer Recognition Systems CORES 2013. Advances in Intelligent Systems and Computing, vol. 226, pp. 579–589. Springer (2013)
22. Ruta, D., Gabrys, B.: Classifier selection for majority voting. Inf. Fusion **6**(1), 63–81 (2005)
23. Smętek, M., Trawiński, B.: Selection of heterogeneous fuzzy model ensembles using self-adaptive genetic algorithms. New Gener. Comput. **29**(3), 309–327 (2011)
24. Suen, C.Y., Legault, R., Nadal, C.P., Cheriet, M., Lam, L.: Building a new generation of handwriting recognition systems. Pattern Recogn. Lett. **14**(4), 303–315 (1993)
25. Woloszyński, T., Kurzyński, M.: A probabilistic model of classifier competence for dynamic ensemble selection. Pattern Recogn. **44**(10–11), 2656–2668 (2011)

2DHMM-Based Face Recognition Method

Janusz Bobulski

Abstract So far many methods of recognizing the face arose, each has the merits and demerits. Among these methods are methods based on Hidden Markov models, and their advantage is the high efficiency. However, the traditional HMM uses one-dimensional data, which is not a good solution for image processing, because the images are two-dimensional. Transforming the image in a one-dimensional feature vector, we remove some of the information that can be used for identification. The article presents the full ergodic 2D-HMM and applied for face identification.

Keywords Biometrics · Face recognition · Two dimensional Hidden Markov models

1 Introduction

Face recognition has great potentials in many applications dealing with uncooperative subjects, in which the full power of face recognition being a passive biometric technique can be implemented and utilised. Face recognition has been an active area of research in image processing and computer vision due to its extensive range of prospective applications relating to biometrics, information security, video surveillance, law enforcement, identity authentication, smart cards, and access control systems. The benefits of facial recognition are that it is not intrusive, can be done from a distance even without the user being aware they are being scanned. A number of approaches have been developed for extracting features from still images. Turk and Pentland [15] proposed Eigenfaces employed Karhunen-Loeve Transform (KLT) [6]. Others popular methods for face features extraction are wavelet transform [9]. The most popular method of face identification is Principal Component Analysis (PCA) [10]. PCA is an unsupervised learning method, which treats samples of the

J. Bobulski (✉)
Czestochowa University of Technology, Institute of Computer and Information Science,
Dabrowskiego Street 73, 42-200 Czestochowa, Poland
e-mail: januszb@icis.pcz.pl

R.S. Choraś (ed.), *Image Processing and Communications Challenges 7*,
Advances in Intelligent Systems and Computing 389,
DOI 10.1007/978-3-319-23814-2_2

different classes in the same way. Fisherfaces proposed by Belhumeour and Hespanha [2] is a supervised learning method using the category information associated with each sample to extract the most discriminatory features. Other popular methods use Wavelet Transform [8], Hidden Markov Models [14] or characteristic points [12]. Previous methods which based on HMM processed one-dimensional data. This is not a problem in application such as speech recognition, because feature vectors are only one dimension. 1D HMM is unpractical in image processing, because the images are two-dimensional. When we convert an image from 2D to 1D, we lose some information. So, if we process two-dimensional data, we should apply two-dimensional HMM, and this 2D HMM should works with 2D data. One of solutions is pseudo 2D HMM [7, 16], which is extension of classic 1D HMM. There are super-states hiding linear one-dimensional hidden Markov models. So, we have 1D model with 1D data in practise. Article [17] presents analytic solution and proof of correctness two-dimensional HMM, which is similar to MRF [11, 13], and works with one-dimensional data. Additional it can be apply only for left-right type of HMM. This article presents real solution for 2D problem in HMM. There is shown true 2D HMM which processes 2D data. Similar to 1D HMM, the most important thing for 2D HMMs is also to solve two basic problems, namely probability evolution and parameters estimation. Moreover the presented algorithms are regarding ergodic models, rather than of type "left-right" [17]. In this paper we focus on the face recognition method from single digital images with two dimensional hidden Markov models (2D-HMM) base on the feature extraction with wavelets. The paper is organized as follows: first we preprocessing procedure to get the pure face image; next we describe feature extraction; then we present the problem two dimensional data in HMM; afterwards, experiments are performed on the facial database with different experimental conditions. Finally, conclusions are given in last section.

2 Propose Method

2.1 Pre-processing Procedure

Pre-processing procedure is the most important step for face recognition. The ideal output of processing is to obtain face images, which have normalized intensity, equal size and containing whole face in vertical pose. Moreover this procedure should also eliminate the effect of illumination and lighting. The database [5] used in the experiment provides the majority of these conditions. The pre-processing procedure of our system performs the following steps in converting image to a normalized face image for feature extraction: (1) locating and cropping the face region using a rectangle according to face shape; (2) selecting face area; (3) scales image in that way, that distance between the inner corners of the eyes is equal 120 pixels. A detailed description of the face normalization procedure can be found in [4]. The

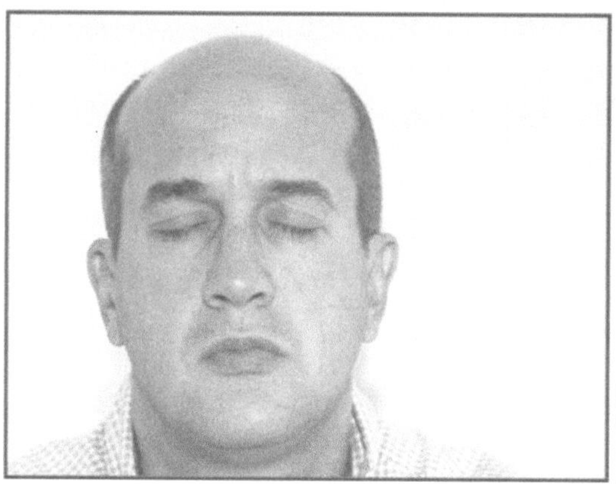
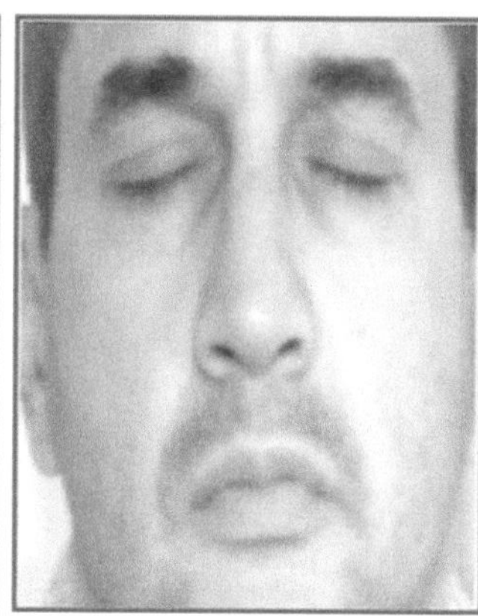

Fig. 1 The effect of pre-processing procedure (image from [5])

points coordinates of the inner corners of the eyes is obtained from database. The effect of pre-processing procedure is shown in Fig. 1.

2.2 Features Extraction

One of the parts persons identification systems is features extraction, and this process is very important because effectiveness of system depend of it. The features extraction has to get out information from a signal (image), which will be base for person identification. The separation of useful information from face is very important, because this data will be use to identification and should describing clearly the face. One of the popular technique for features extraction is Wavelet Transform (WT) [8]. One major advantage afforded by wavelets is the ability to perform local analysis—that is, to analyse a localized area of a larger signal. In wavelet analysis, we often speak about approximations and details. The approximations are the high-scale, low-frequency components of the signal. The details are the low-scale, high-frequency components. Using 2D WT, the face image is decomposed into four subimages via the high-pass and low-pass filtering. The image is decomposed along column direction into subimages to high-pass frequency band H and low-pass frequency band L. Assuming that the input image is a matrix of $m \times n$ pixels, the resulting subimages become $m/2 \times n$ matrices. At second step the images H and L are decomposed along row vector direction and respectively produce the high and low frequency band HH and HL for H, and LH and LL for L. The four output images become the matrices of $m/2 \times n/2$ pixels. Low frequency subimage LL possesses high energy, and is a smallest copy of original images. The remaining subimages LH, HL, and HH respectively extract the changing components in horizontal, vertical, and diagonal direction. The very important aspect of features extraction with WT is suitable choice of wavelet function [1]. The choice should adapt shape of wavelet to individual case and take into consideration the properties of the signal or image. The bad choice of wavelet will cause problems of analysis and identification processed signal. In order to point the best wavelet function was made the experiment. The best result achieved with function db10 from among accessible function.

2.3 2D HMM

HMM is a double stochastic process with underlying stochastic process that is not observable (hidden), but can be observed through another set of stochastic processes that produce a sequence of observation. HMM is the good tools and solution for process one-dimensional data. Unfortunately, this is unpractical in image processing, because the images are two-dimensional. When we convert an image from 2D to 1D, we lose some information. So, if we process two-dimensional data, we should apply two-dimensional HMM, and this 2D HMM should works with 2D data [3].

The statistical parameters of the 2D model (Figs. 2 and 3):

- The number of states of the model N^2
- The number of data streams $k_1 \times k_2 = K$
- The number of symbols M
- The transition probabilities of the underlying Markov chain, $A = \{a_{ijl}\}, 1 \leq i, j \leq N, 1 \leq l \leq N^2$, where a_{ij} is the probability of transition from state ij to state l

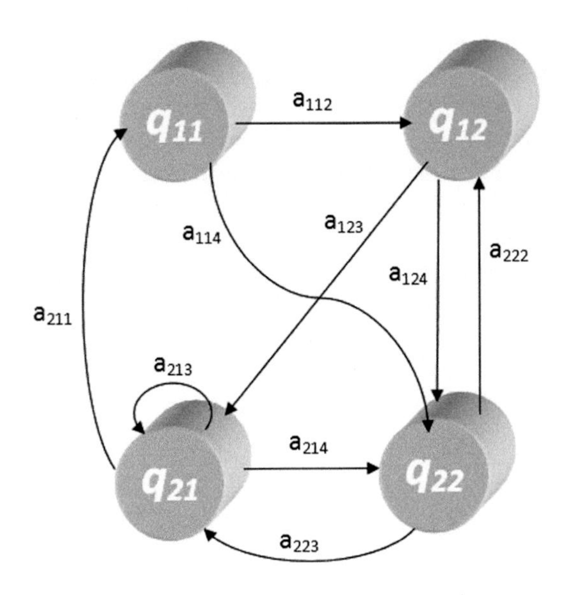

Fig. 2 Two-dimensional ergodic HMM

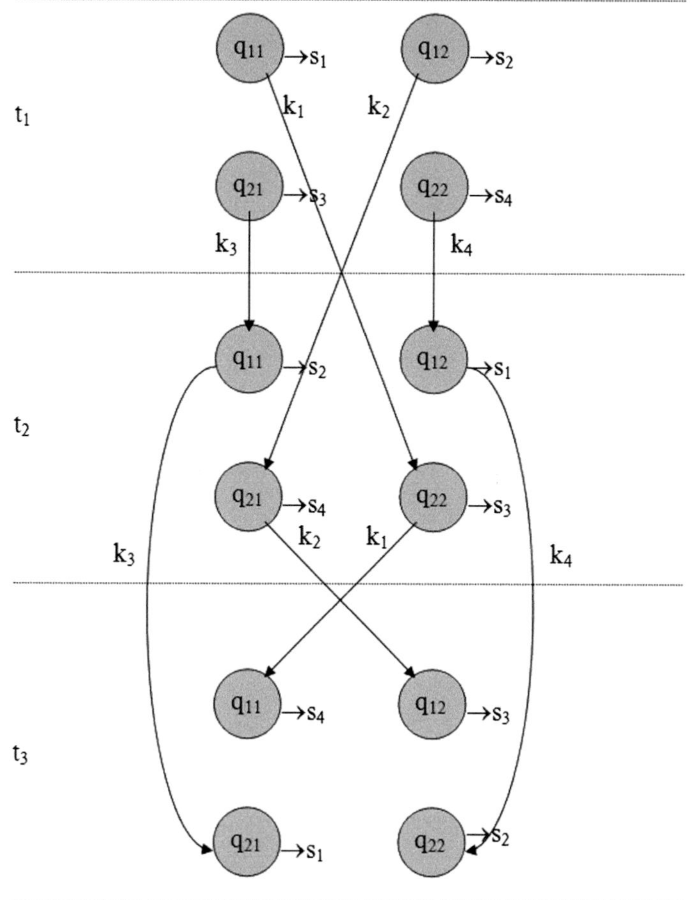

Fig. 3 The idea of 2DHMM

- The observation probabilities, $B = \{b_{ijm}\}, 1 \leq i, j \leq N, 1 \leq m \leq M$ which represents the probability of generate the m_{th} symbol in the ij_{th} state.
- The initial probability, $\Pi = \{\pi_{ijk}\}, 1 \leq i, j \leq N, 1 \leq k \leq K$.
- Observation sequence $O = \{o_t\}, 1 \leq t \leq T, o_t$ is square matrix simply observation with size $k_1 \times k_2 = K$

There are two fundamental problems of interest that must be solved for HMM to be useful in face recognition applications. These problems are the following:

1. Given observation $O = (o_1, o_2, \ldots, o_T)$ and model $\lambda = (A, B, \Pi)$, efficiently compute $P(O|\lambda)$
2. Given observation $O = (o_1, o_2, \ldots, o_T)$, estimate model parameters $\lambda = (A, B, \Pi)$ that maximize $P(O|\lambda)$

Solution to Problem 1

The modified forward algorithm

- Define forward variable $\alpha_t(i, j, k)$ as:

$$\alpha_t(i, j, k) = P(o_1, o_2, \ldots, o_t, q_t = ij|\lambda) \tag{1}$$

- $\alpha_t(i, j, k)$ is the probability of observing the partial sequence $(o_1, o_2, \ldots, o_t)$ such that the state q_t is i, j for each k_{th} stream of data
- Induction

1. Initialization:

$$\alpha_1(i, j, k) = \pi_{ijk} b_{ij}(o_1) \tag{2}$$

2. Induction:

$$\alpha_{t+1}(i, j, k) = \left[\sum_{l=1}^{N} \alpha_t(i, j, k) a_{ijl}\right] b_{ij}(o_{t+1}) \tag{3}$$

3. Termination:

$$P(O|\lambda) = \sum_{t=1}^{T} \sum_{k=1}^{K} \alpha_T(i, j, k) \tag{4}$$

Solution to Problem 2

The modified parameters re-estimation algorithm

- Define $\xi(i, j, l)$ as the probability of being in state ij at time t and in state l at time $t + 1$ for each k_{th} stream of data

$$\begin{aligned} \xi_t(i, j, l) &= \frac{\alpha_t(i, j, k) a_{ijl} b_{ij}(o_{t+1}) \beta_{t+1}(i, j, k)}{P(O|\lambda)} \\ &= \frac{\alpha_t(i, j, k) a_{ij} b_{ij}(o_{t+1}) \beta_{t+1}(i, j, k)}{\sum_{k=1}^{K} \sum_{l=1}^{N^2} \alpha_t(i, j, k) a_{ijl} b_{ij}(o_{t+1}) \beta_{t+1}(i, j, k)} \end{aligned} \tag{5}$$

- Define $\gamma(i, j)$ as the probability of being in state i, j at time t, given observation sequence.

$$\gamma_t(i, j) = \sum_{l=1}^{N^2} \xi_t(i, j, l) \tag{6}$$

- $\sum_{t=1}^{T} \gamma_t(i, j)$ is the expected number of times state i, j is visited
- $\sum_{t=1}^{T-1} \xi_t(i, j, l)$ is the expected number of transition from state ij to l

  Update rules:

- $\bar{\pi_i} jk$ = expected frequency in state i, j at time $(t = 1) = \gamma_1(i, j)$

Table 1 Comparison of recognition rate

Method	Number of faces	Correctly	Incorrectly	Recognition rate [%]
PCA	60	56	4	94
1D HMM	60	54	6	90
2D HMM	60	55	5	92

- $\bar{a}_{ij}$ = (expected number of transition from state i, j to state l)/(expected number of transitions from state i, j):

$$\bar{a}_{ijl} = \frac{\sum_t \xi_t(i, j, l)}{\sum_t \gamma_t(i, j)} \tag{7}$$

- $\bar{b_{ij}}(k)$ = (expected number of times in state j and observing symbol k)/(expected number of times in state j):

$$\bar{b}_{ij}(k) = \frac{\sum_{t, o_t = k} \gamma_t(i, j)}{\sum_t \gamma_t(i, j)} \tag{8}$$

3 Experiment

The image database *UMB-DB* was used for experiments. The University of Milano Bicocca 3D face database is a collection of multimodel (3D + 2D colour images) facial acquisitions. The database is available to universities and research centers interested in face detection or face recognition. They recorded 1473 images of 143 subjects (98 male, 45 female). The images show the faces in variable condition, lighting, rotation and size [5].

We chose three sets of 60 persons each, in order to verify the method, and for each individual chose three images for learning and three for testing. The 2D HMM implemented with parameters $N = 4$; $N^2 = 16$; $K = 16$; $M = 25$. Wavelet transform was chosen as features extraction technique, and db10 as wavelet function. Table 1 presents the average results of experiments.

4 Conclusion

The obtained results are satisfactory in comparison to other method and proposed method may be the alternative solution to the others. Recognition rate of the method is 92 %, which is better than 1D HMM. In comparison to PCA the result is worse but our method is faster. In this paper, the new conception of face recognition with two-dimensional hidden Markov models is presented. We show solutions of principle problems for ergodic 2D HMM, which may be applied for 2D data. Presented method allows for faster face processing and recognition because they do not have to change the two-dimensional input data in the image form into a one dimensional data. Therefore, we do not lose the information contained in the image.

References

1. Antoniadis, A., Oppenheim, G.: Wavelets and Statistics, Lecture Notes in Statistics, vol. 103. Springer, New York (1995)
2. Belhumeour, P.N., Hespanha, J.P., Kriegman, D.J.: Eigenfaces versus fisherfaces: recognition using class specific linear projection. IEEE Trans. Pattern Anal. Mach. Intell. **19**, 711–720 (1997)
3. Bobulski, J., Adrjanowicz, L.: Two-dimensional hidden markov models for pattern recognition. Artificial Intelligence and Soft Computing, Lecture Notes in Computer Science, vol. 7894, pp. 515–523 (2013)
4. Bobulski, J., Kubanek, M.: Person identification system using an identikit picture of the suspect. Opt. Appl. **42**(4), 865–873 (2012)
5. Colombo, A., Cusano, C., Schettini, R.: UMB-DB: a database of partially occluded 3D faces. In: Proceedings of the ICCV 2011 Workshops, pp. 2113–2119 (2011)
6. Duda, R.O., Hart, P.E., Stork, D.G.: Pattern Classification. Wiley, New York (2001)
7. Eickeler, S., Mller, S., Rigoll, G.: High performance face recognition using pseudo 2-D Hidden Markov models. In: European Control Conference (1999). http://citeseer.ist.psu.edu
8. Garcia, C., Zikos, G., Tziritas, G.: Wavelet pocket analysis for face recognition. Image Vis. Comput. **18**, 289–297 (2000)
9. Kozuani, A.Z., He, F., Sammut, K.: Wavelet packet face representation and recognition. In: Proc. IEEE Conf. Syst. Man Cybern. **2**, 1614–1619 (1997)
10. Kirby, M., Sirovich, L.: Application of the karhunen-loeve procedure for the characterization of human faces. IEEE Trans. Pattern Anal. Mach. Intell. **12**(1), 103–108 (1990)
11. Kindermann, R., Snell, J.L.: Markov Random Fields and Their Applications. American Mathematical Society, Providence (1980)
12. Kubanek, M.: Automatic Methods for Determining the Characteristic Points in Face Image. Lecture Notes in Artificial Intelligence, vol. 6114, Part I, 523–530 (2010)
13. Li, J., Najmi, A., Gray, R.M.: Image classification by a two dimensional Hidden Markov model. IEEE Trans. Signal Process. **48**, 517–533 (2000)
14. Samaria, F., Young, S.: HMM-based architecture for face identification. Image Vis. Comput. **12**(8), 537–543 (1994)
15. Turk, M., Pentland, A.: Eigenfaces for recognition. J. Cogn. Neuro-science **3**, 71–86 (1991)
16. Vitoantonio Bevilacqua, V., Cariello, L., Carro, G., Daleno, D., Mastronardi, G.: A face recognition system based on Pseudo 2D HMM applied to neural network coefficients. Soft. Comput. **12**(7), 615–621 (2008)
17. Yujian, L.: An analytic solution for estimating two-dimensional hidden Markov models. Appl. Math. Comput. **185**, 810–822 (2007)

Corner Detection Based on Directional Gradients—Comparative Study

Arkadiusz Cacko, Marcin Iwanowski, Grzegorz Sarwas and Michal Swiercz

Abstract The paper presents a comparative study of image corner detectors based on directional gradient and computation of cornerness function, the maxima of which indicates the position of corner points. Various combinations of different directional gradients and formulas defining the cornerness function are investigated. In particular, gradients computed not only along principle axes of the image coordinates system but also along two diagonals are considered. The experiments conducted show that some of combinations perform better than classic Harris and KLT corner detectors.

Keywords Corner detection · Feature points · Local features

1 Introduction

Corner detectors are very popular tools used for the description of visual scene. They are widely used to characterise the content of a digital image by pointing at details of the visual scene. They are also used as feature points, based on which the descriptors are extracted.

In this paper we focus on classical solutions which generic principle of feature points detection using corners is based on the three-step scheme. At first the image gradients are computed. Next, the gradients are combined together in order to obtain

A. Cacko (✉) · M. Iwanowski · M. Swiercz
Institute of Control and Industrial Electronics, Warsaw University of Technology, ul.Koszykowa 75, 00-662 Warszawa, Poland
e-mail: arkadiusz.cacko@ee.pw.edu.pl

M. Iwanowski
e-mail: iwanowski@ee.pw.edu.pl

M. Swiercz
e-mail: kubaswiercz@gmail.com

G. Sarwas · M. Swiercz
Lingaro, ul. Puławska 99a, 02-595 Warszawa, Poland
e-mail: grzegorz.sarwas@lingaro.com

R.S. Choraś (ed.), *Image Processing and Communications Challenges 7*,
Advances in Intelligent Systems and Computing 389,
DOI 10.1007/978-3-319-23814-2_3

a cornerness function that describes, for each image pixel, the possibility of finding the corner in the particular image point. Finally, from the cornerness function, a list of feature points is extracted. We investigate in this paper the application of various diagonal gradients as well as various formulas of cornerness function in view of their efficiency and invariance to basic image transformations.

Based on this scheme several popular corner detectors has been proposed [2, 3]. Some improvements of this idea was introduced in [8, 9] where authors proposed has more accurate algorithm for corner and edge detections using different quadratic polynomial to approximate the variation around pixels. In [4, 5] authors propose Harris-like scale and affine invariant detectors using local extrema over scale of normalized derivatives.

However it is possible to propose other combinations of particular methods that also allow to detect corners within the image [1, 6]. The main goal of the research described in these paper have been comparing various combinations of directional gradients and defining the cornerness function. It will be shown that some of those combinations surpasses the classical methods in terms of the accuracy of pointing at meaningful image details as well as invariance to some image transformations.

The paper is divided as follows: In Sect. 2 we describe classic corner detector methods. Then in Sect. 3 we propose some modifications of original version which have been tested. Results of these tests are described in Sect. 4. Section 5 concludes the paper.

2 Classic Corner Detectors

Harris corner detector [2] is commonly used for finding of feature points. It is an improvement of Moravec's corner detector [7] based on the convolution operator with simple linear gradient masks:

$$H = \begin{bmatrix} -1 \; \underline{0} \; 1 \end{bmatrix}, V = \begin{bmatrix} 1 \; \underline{0} \; -1 \end{bmatrix}^T . \tag{1}$$

The above masks are used to get two directional linear gradients using convolution operator (denoted as $\otimes$): $g_H = I \otimes H$, $g_V = I \otimes V$. The directional response obtained (Eq. 1) is used to construct the covariance matrix:

$$M = \begin{bmatrix} I_{HH} & I_{HV} \\ I_{HV} & I_{VV} \end{bmatrix}, \tag{2}$$

where:

$$I_{HH} = g_H^2 \otimes W \ , \ I_{HV} = (g_H \cdot g_V) \otimes W \ , \ I_{VV} = g_V^2 \otimes W, \tag{3}$$

are second order moments convoluted using Gaussian mask W. The final cornerness function is defined by:

$$r = det(M) - k \cdot trace(M)^2 = \lambda_1\lambda_2 - k(\lambda_1 + \lambda_2)^2, \tag{4}$$

where λ_1 and λ_2 are eigenvalues of matrix M. Parameter k is chosen manually, often set up as $k = 0.04$.

In order to obtain the list of corner pixels, the regional maxima of cornerness function r are computed. These maxima may suffer from two defects. Their values could be insignificantly low. It means that either the corner described by this maximum is poorly visible or it is not a real corner of a visual scene, but a kind of artefact being the result of e.g. noise. In order to solve this problem, the cornerness function is thresholded at given level t, and only maxima that are higher than t are considered. The second problem is the fact that single maximum may consist of several connected pixels. To solve it, the reduction of connected set of pixels into single pixel is usually performed, by means of e.g. thinning.

There exist many modifications of original Harris algorithm. One of them is Kanade-Lucas-Tomasi detector—KLT [3] where the authors proposed to modify cornerness function, so that it is computed as minimum of eigenvalues of matrix M: $r = \min(\lambda_1, \lambda_2)$. Other approximation was proposed in [8, 9] where authors show other way to compute cornerness function.

3 Modifications of Classic Scheme

Classic detectors consist of three principal steps, that refer to computation of gradient, cornerness function and finding points. The gradients used in first two of these stages may be replaced by some other ones. In our research we investigated some of alternatives in view of the quality of corner detection. We assumed that directional gradients, apart from vertical and horizontal versions, may also be formulated in two diagonal variants. The choice of diagonal gradients was motivated by the assumption that vertical and horizontal gradients may not be enough to achieve high quality of corner detection. Considering the diagonal variants may increase accuracy of finding local corner points. Using both normal and diagonal gradients provide us with four different edge responses.

In case of the second issue we investigated gradient masks covering larger than two-pixel only (comparing to described by the Eq. 1) neighborhood. Larger (in terms of the number of pixels belonging to) neighborhoods allow to consider wider context in which the corner pixel may be detected.

In our research we examined various directional gradients as well as methods of combining obtained directional responses into single cornerness function.

The following gradient masks have been investigated:

1. Classic masks:

$$H_1 = \begin{bmatrix} -1 \ \underline{0} \ 1 \end{bmatrix}, \ V_1 = \begin{bmatrix} 1 \ \underline{0} \ -1 \end{bmatrix}^T, \ D_1 = \begin{bmatrix} 0 & 0 & 1 \\ 0 & \underline{0} & 0 \\ -1 & 0 & 0 \end{bmatrix}, \ D_1' = \begin{bmatrix} 1 & 0 & 0 \\ 0 & \underline{0} & 0 \\ 0 & 0 & -1 \end{bmatrix}. \tag{5}$$

2. Prewitt masks:

$$H_2 = \begin{bmatrix} -1 & 0 & 1 \\ -1 & \underline{0} & 1 \\ -1 & 0 & 1 \end{bmatrix}, \ V_2 = \begin{bmatrix} 1 & 1 & 1 \\ 0 & \underline{0} & 0 \\ -1 & -1 & -1 \end{bmatrix}, \ D_2 = \begin{bmatrix} -1 & -1 & 0 \\ -1 & \underline{0} & 1 \\ 0 & 1 & 1 \end{bmatrix}, \ D_2' = \begin{bmatrix} 0 & -1 & -1 \\ 1 & \underline{0} & -1 \\ 1 & 1 & 0 \end{bmatrix}. \tag{6}$$

3. Sobel masks:

$$H_3 = \begin{bmatrix} -1 & 0 & 1 \\ -2 & \underline{0} & 2 \\ -1 & 0 & 1 \end{bmatrix}, \ V_3 = \begin{bmatrix} 1 & 2 & 1 \\ 0 & \underline{0} & 0 \\ -1 & -2 & -1 \end{bmatrix}, \ D_3 = \begin{bmatrix} -2 & -1 & 0 \\ -1 & \underline{0} & 1 \\ 0 & 1 & 2 \end{bmatrix}, \ D_3' = \begin{bmatrix} 0 & -1 & -2 \\ 1 & \underline{0} & -1 \\ 2 & 1 & 0 \end{bmatrix}. \tag{7}$$

4. Three-pixel wide neighborhood:

$$H_4 = \begin{bmatrix} -1 & 0 & 0 & 0 & 0 & 0 & 1 \\ -1 & 0 & 0 & \underline{0} & 0 & 0 & 1 \\ -1 & 0 & 0 & 0 & 0 & 0 & 1 \end{bmatrix}, \ V_4 = \begin{bmatrix} 1 & 0 & 0 & 0 & 0 & 0 & -1 \\ 1 & 0 & 0 & \underline{0} & 0 & 0 & -1 \\ 1 & 0 & 0 & 0 & 0 & 0 & -1 \end{bmatrix}^T,$$

$$D_4 = \begin{bmatrix} 0 & 0 & 0 & 0 & 1 & 0 & 0 \\ 0 & 0 & 0 & 0 & 0 & 1 & 0 \\ 0 & 0 & 0 & 0 & 0 & 0 & 1 \\ 0 & 0 & 0 & \underline{0} & 0 & 0 & 0 \\ -1 & 0 & 0 & 0 & 0 & 0 & 0 \\ 0 & -1 & 0 & 0 & 0 & 0 & 0 \\ 0 & 0 & -1 & 0 & 0 & 0 & 0 \end{bmatrix}, \ D_4' = \begin{bmatrix} 0 & 0 & 1 & 0 & 0 & 0 & 0 \\ 0 & 1 & 0 & 0 & 0 & 0 & 0 \\ 1 & 0 & 0 & 0 & 0 & 0 & 0 \\ 0 & 0 & 0 & \underline{0} & 0 & 0 & 0 \\ 0 & 0 & 0 & 0 & 0 & 0 & -1 \\ 0 & 0 & 0 & 0 & 0 & -1 & 0 \\ 0 & 0 & 0 & 0 & -1 & 0 & 0 \end{bmatrix}. \tag{8}$$

The four directional gradients are combined together in order to obtain single cornerness function. They are used to compute the second-order matrix as defined in Eq. 2. Since two pairs of perpendicular gradients are used here, two matrices are computed: M based on gradients with masks H and V, and M' based on D and D'. The elements of these matrices are filtered using the lineal low-pass filter with the mask W (Eq. 3). Type of such mask has influence on pixel neighbor context taken while processing cornerness response. We have examined the impact of the change of a filter mask on the quality of detected corners. The following masks ware investigated:

1. Gaussian

$$W_1 = \begin{bmatrix} 0.12 & 0.23 & 0.29 & 0.23 & 0.12 \end{bmatrix}^T \cdot \begin{bmatrix} 0.12 & 0.23 & 0.29 & 0.23 & 0.12 \end{bmatrix}. \tag{9}$$

2. Box filter

$$W_2 = \begin{bmatrix} 0.2 & 0.2 & 0.2 & 0.2 & 0.2 \end{bmatrix}^T \cdot \begin{bmatrix} 0.2 & 0.2 & 0.2 & 0.2 & 0.2 \end{bmatrix}. \tag{10}$$

3. Triangle-shaped filter

$$W_3 = \begin{bmatrix} 0.1 & 0.2 & 0.3 & 0.2 & 0.1 \end{bmatrix}^T \cdot \begin{bmatrix} 0.1 & 0.2 & 0.3 & 0.2 & 0.1 \end{bmatrix}. \tag{11}$$

Matrices M and M' was used to compute the cornerness function. We assumed here the following variants:

1. Smallest of eigenvalues of M (similar to KLT method, further referred to as **klt**):

$$r = \min\{\lambda_1, \lambda_2\}, \tag{12}$$

   where λ_1,λ_2 are the eigenvalues of the matrix M.
2. Smallest of eigenvalues of M and M' (normal and diagonal, **minsv**)

$$r = \min\{\lambda_1, \lambda_2, \lambda'_1, \lambda'_2\}, \tag{13}$$

   where λ'_1 and λ'_2 are eigenvalues of matrix M'. This estimation takes into account the smallest of all eigenvalues to improve corner response detection.
3. Mean of smallest eigenvalues of M and M' (**meansv**)

$$r = \frac{\min\{\lambda_1, \lambda_2\} + \min\{\lambda'_1, \lambda'_2\}}{2}. \tag{14}$$

   This estimation takes into account both responses (classical and diagonal gradient) to compute more accurate result.
4. Determinant of the matrix M (**determinant**)

$$r = det(M) = I_{HH} \cdot I_{VV} - I_{HV}{}^2. \tag{15}$$

   This is simplified, low computational version of cornerness function.

We did not modify in our research the final step of processing i.e. extraction of points from the cornerness function—this step is performed using the approach used in the Harris method (see previous chapter).

4 Tests

4.1 Methodology of Testing

Evaluation of the quality of corner detectors requires two principal factors to be taken into account: correctness and stability. The first factor—correctness is usually verified by comparing, for a particular image, automatically detected corners with the indication of a human observer.

The second factor used to determine the quality of the detector—stability—refers to insensitivity of the detector to typical image transforms that may happen in case of images presenting the same visual scene from various points of view. In our testing scheme we've chosen several image transforms that may simulate the real distortions that may occur in case of various images of the same scene. Among these transforms are: blurring, speckle noise, gamma correction, illumination change, zoom and rotation. Moreover, to validate the level of dependence the invariance from the strength of the given transform, we prepared series of transforms with increasing strength. For example, a series of blur transforms consists of Gaussian blurs computed for increasing standard deviation value. The possible image transforms are of two kinds—with and without the geometric transformation. The presence of the geometric transformation is important, because in this case, the position of the corner points changes.

In order to measure the invariance to particular image transform we compare the corner points before and after applying this transform on the input image. This comparison is made by computing the factor equal to the number of corner points matched divided by the total number of them.

In Fig. 1, the position of corner points before and after the geometric transform of rotation is shown. Cross mark indicates the corner found on original image (a) and the same corner transformed to coordinates of rotated image (b). Circle mark represents corners found on transformed image (b).

Let P be a set of corner points on image I: $P = \{p_1, p_2, \ldots, p_n\}$, where n is the number of points detected on image I. Let T_A is the operator of geometric transformation with parameters described by the matrix A, and P' is set of corner points transformed by T_A:

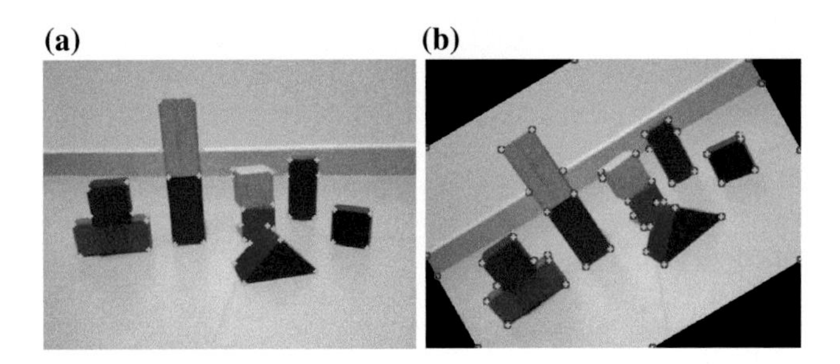

Fig. 1 Detected corners before **a** and after **b** image rotation

$$P' = T_A(P) = \left\{p_i' : \forall_{i\in(1,\ldots,n)}\ p_i' = T_A(p_i)\right\}. \tag{16}$$

We also define the I' as the image I transformed by T_A:

$$I' = T_A(I) \tag{17}$$

The size of image is same as transformed image, which means that after geometric transform the image is cut to original size (see Fig. 1).

Let Q be a set of corner points on image I': $Q = \{q_1, q_2, \ldots, q_m\}$, where m is the number of points detected on image I'.

Since the corner points of P' and Q are in the same coordinate system, it is possible to compute the Euclidean distance D between them. A pair $(p_i^{'} \in P^{'}, q_j \in Q)$ represents corresponding corner points (point $p_i^{'}$ has the counterpart of q_j) in the context of the transformation T_A if and only if the following stability conditions are met:

$$D(p_i', q_j) \leq D(p_i', q_k),\ \forall_k : k \in (1, \ldots, m) \wedge k \neq j\ ;\ D(p_i', q_j) \leq \epsilon. \tag{18}$$

Therefore, for each "original" point p_i we are looking for points being closest to "new" corner point q_i from transformed image such that its distance to p_i' is shorter than given ϵ value. Consequently it is possible that some of the "original" corner points do not have a corresponding point in the transformed image. It may happen that the "original" point has several equally distanced corresponding points in the transformed image.

To measure the overall stability rate of detectors, the function S is introduced:

$$S(P') = \frac{k}{n'}, \tag{19}$$

where $k \leq n$ is the number of points $p' \in P'$ which has corresponding points in Q and $n' \leq n$ is the number of points in P' with coordinates that are located in transformed image range.

To measure the quality of detectors we use $S(P^{'})$ functions for sequence of image transformations.

4.2 Results

Considering all 4 types of gradients (Eqs. 5–8), 4 formulas of cornerness function (Eqs. 12–15) and 3 low-pass filters (Eqs. 9–11), the total number of 48 detectors has been tested. Classic Harris detector, KLT and some proposed by us have been examined against minimal and mean values of $S(P^{'})$ of the same sequence of affine transformations. Sequence consists of 170 image transforms including 72 rotations (every 5°) of image, 20 different scale factors of image (from 95 to 105 %), 18

Table 1 Chosen results of tests (best methods are indicated in bold) and comparison with classic Harris and KLT methods

Gradient mask	Corner function	Blurring mask	Mean	Min
H_1, V_1, D_1, D_1'	Minsv	W2	94.53	88.18
H_2, V_2, D_2, D_2'	Minsv	W1	92.83	86.15
H_2, V_2, D_2, D_2'	Meansv	W2	93.70	87.94
H_3, V_3, D_3, D_3'	Minsv	W1	95.03	88.42
H_3, V_3, D_3, D_3'	Meansv	W1	95.13	88.70
H_4, V_4, D_4, D_4'	Minsv	W2	95.91	90.97
H_4, V_4, D_4, D_4'	Meansv	W2	89.58	81.02
D_3, D_4'	Determinant	W1	92.74	83.4
D_3, D_4'	Determinant	W2	95.82	90.13
Harris detector			87.77	73.13
Kanade-Lucas-tomasi (KLT) detector			93.25	85.11

blurring factors, 20 levels of noise, 40 brightness values (60–150 %). Some of the chosen results (taking into account mean value of $S(P^{'})$ results) are presented in Table 1. The measures included in the table refers to mean and minimum value of $S(P')$. Some of the results performs better than classic Harris and KLT method.

5 Conclusions

The aim of a study presented in this paper was to compare various combinations of directional gradient cornerness functions and linear filters within the classic three-step schema of corner detection. As the results of experiments have shown some of these combinations provide us with methods witch perform better than classic Harris and KLT method. Taking into consideration diagonal gradient responses gives more accurate information about possible edge/corner presence. Also, using determinant version of cornerness function allows improving computational efficiency.

Acknowledgments This work was co-financed by the European Union within the European Regional Development Fund.

References

1. Chen, J., hui Zou, L., Zhang, J., hua Dou, L.: The comparison and application of corner detection algorithms. J. Multimedia **4**(6) (2009)
2. Harris, C., Stephens, M.: A combined corner and edge detector. In: Proceedings of Fourth Alvey Vision Conference. pp. 147–151 (1988)

3. Lucas, B.D., Kanade, T.: An iterative image registration technique with an application to stereo vision. In: Proceedings of the 7th International Joint Conference on Artificial Intelligence, vol. 2. pp. 674–679. IJCAI'81, Morgan Kaufmann Publishers Inc., San Francisco, CA, USA (1981)
4. Mikolajczyk, K., Schmid, C.: An affine invariant interest point detector. In: Heyden, A., Sparr, G., Nielsen, M., Johansen, P. (eds.) Computer Vision—ECCV 2002. Lecture Notes in Computer Science, vol. 2350, pp. 128–142. Springer, Berlin Heidelberg (2002)
5. Mikolajczyk, K., Schmid, C.: Scale & affine invariant interest point detectors. Int. J. Comput. Vision **60**(1), 63–86 (2004)
6. Miksik, O., Mikolajczyk, K.: Evaluation of local detectors and descriptors for fast feature matching. Pattern Recognition (ICPR), 2012 21st (2012)
7. Moravetz, H.: Obstacle avoidance and navigation in the real world by a seeing robot rover. Technical Report CMU-RI-TR-3, Carnegie-Melon University, Robotics Institute (Sept. 1980)
8. Overgaard, N.: On a modification to the harris corner detector. In: proceedings of Symposium Svenska Sallskapet for Bildanalys, Stockholm. pp. 6–7 (2003)
9. Pei, S.C., Ding, J.J.: Improved harris' algorithm for corner and edge detections. In: Image Processing, 2007. ICIP 2007. IEEE International Conference on. vol. 3, pp. III-57–III-60 (Sept. 2007)

Sensor Fusion Enhancement for Mobile Positioning Systems

Karol Ciążyński and Artur Sierszeń

Abstract Recently, a great amount of mobile devices are equipped with various sensors such as accelerometers, GPS receivers, gyroscopes, microphones and cameras. Moreover, all these devices use embedded Bluetooth and wireless network cards for mobile communication. One of the most significant development of recent years is GPS. This system has become the standard for navigation for many people all over the world. A huge variety of applications employ GPS from car positioning to mobile phone navigation for emergency purposes. In spite of the fact that this system is commonly used in everyday's actions it has some limitations. GPS is unable to determine the position inside the buildings, because of the loss of signal from satellites. Recently, due to existence of ubiquities mobile devices the curiosity in location based services has considerably increased. The application of Bluetooth and wireless network was employed for portable positioning systems. However, those systems considerably increase the accuracy for indoor localization, still the outcome is not satisfactory. Thus, the combination of a variety of signals measured by mobile equipments could provide an enhancement for mobile system used in positioning.

1 Introduction

Mobile positioning techniques have been very popular recently. It is related to the existence of ubiquities computers and context awareness of mobile phones. The significance of the information about the current accurate position plays an crucial role in today's world. Position localization system is employed for indoor positioning, context-aware applications, augmented reality and mobile guides in public buildings such as hospitals, university campuses or shopping malls. The inspiration is

K. Ciążyński (✉) · A. Sierszeń
Institute of Applied Computer Science, Lodz University of Technology,
ul. Stefanowskiego 18/22, 90-924 Łódź, Poland
e-mail: karol.ciazynski@p.lodz.pl

A. Sierszeń
e-mail: artur.sierszen@p.lodz.pl

R.S. Choraś (ed.), *Image Processing and Communications Challenges 7*,
Advances in Intelligent Systems and Computing 389,
DOI 10.1007/978-3-319-23814-2_4

to design an enhancement for indoor positioning system that is able to obtain the instantaneous position more precisely. The categorization of positioning techniques relies on whether extra external infrastructure is required or not. The first group of approaches includes WLAN triangulation, Assisted-GPS and RFID tags. It is possible to use features of wireless network system and its existing infrastructure to determine the position indoors. Analyzing the propagated signal mobile device could estimate the distance between the terminal itself and the access point. Then, combining the signals from more access points with the utilization of triangulation the system could estimate the current position. The position estimated by WLAN location system could be accurate but requires prior installed infrastructure in the buildings and is very prone to any environmental changes even such as other people movement. The second category of indoor positioning systems includes inertial sensors. This kind of approaches are significantly cheaper and simpler solutions but often much less accurate and requires frequent recalibration in order to decrease errors of estimations. Both approaches separately are not perfect and could not guarantee the satisfactory accuracy. Thus, the idea arises to enhance the WLAN positioning system with inertial components. The combination of both approaches could significantly increase the precision of position determination.

2 Related Works

The application of positioning system for smartphone based on dead-reckoning method employing step counters is presented in [4]. This is navigation process based on calculating current position by application of previously determined position and advancing that position based upon identified or estimated speed over elapsed course and time. Adaptive Kalman filters with activity based map matching are applied to enhance the position estimation. Dual approach that is an advancement of previous solution is presented in [2]. This method combines step detection and step heading with matching the detected step onto expected path to the destination with the application of sequence alignment algorithms. Two complementary methods for position estimation using mobile phone are presented in [5]. Accelerometer and magnetometer are used in the first method. The reference direction for gravity with the component of the electromagnetic field that is perpendicular to the gravity provides the reference for magnetic north. Each measurement is collected externally so the orientation errors are not accumulated over the time. The second method for tracking the orientation uses the gyroscope sensor that is less noisy than two previous sensors (accelerometer and magnetometers) and is not prone to external factors. The movement pattern recognition in urban street behavior is described in [3]. The data receiving from the accelerometer of the pedestrians mobile device was collected. It involved dealing with complex events which inevitably lead to an increase number of errors. The methods employed for the recognition method considered various filters based on solely threshold settings that proved beneficial [1].

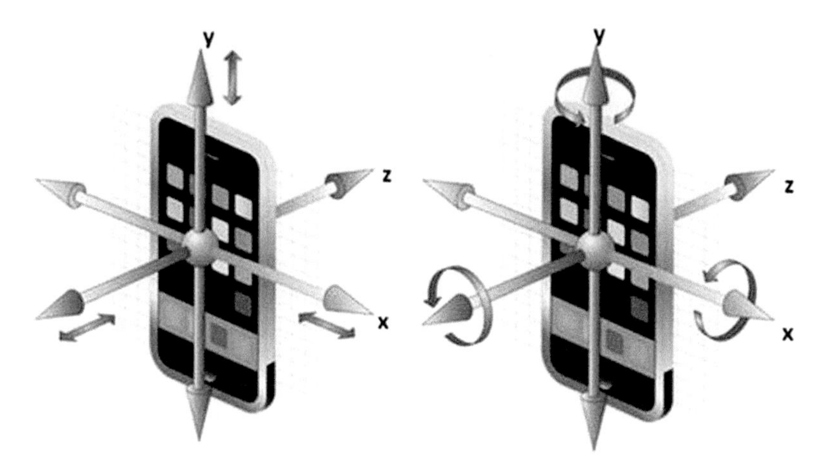

Fig. 1 Accelerometer measures the acceleration and Gyroscope measures the rotation

3 Background Theory

A gyroscope is a sensor that uses gravity of Earth to help the measurement of current orientation. Its design includes freely-rotating disk that is called a rotor. This piece of equipment is mounted onto a spinning axis in the center of much larger and more stable wheel. When the axis turns, the rotor remains inactive in order to indicate the central gravitational pull. An accelerometer is an equipment that is designed to determine non-gravitational acceleration. When the device starts moving in any direction, an accelerometer is responding to vibrations associated with this movement. The sensor employs microscopic crystals that stress when vibrations occur. Next, voltage from that stress is generated in order to create a data corresponding to the acceleration. Two schemes of sensors are presented in Fig. 1.

4 Measurement Results

Acceleration evaluation have been performed using iPhone 4 mobile phone with the operational system iOS 7.1.1. The number of experiments with various velocity, acceleration and the direction of the smartphone movement have been carried out. From the number of the results with low and high acceleration the one with high acceleration is presented. The sample frequency for data collection is stable and is equal to 75 Hz. The device was moved towards the y-axis for 100 cm. First experiment was performed with the law acceleration that was equal to about 20 m/s^2. In this step the velocity has reached the level of 0.8 m/s. The device with accelerometer was moved for a distance of 100 cm. This kind of motions have been performed 10 times. Figure 3 presents the data of the accelerations without any filtering. Figure 4 presents the data of acceleration with the usage of low pass filter. The low pass filter has been used in order to reduce the noise of the accelerometer sensor. The data was gathered separately in three dimensions.

Figures 2 and 3 present the current acceleration in three directions. However, the mobile phone was moved only in y-direction, the acceleration in x and z direction is not equal to 0. Nevertheless, the acceleration values in y-direction are much higher than two other ones. Obtaining the instant acceleration it is possible to determine the current velocity of the device. The Fig. 4 presents the instantaneous velocity in three directions. The results present that the signal is disturbed with random noises. Figure 4 presents that when the equipment is stopped at third second (when the acceleration is about 0) the velocity still has not reached 0. This results in the error that is increasing during the time. In order to limit this error an enhancement has been introduced in y-direction. Figures 4 and 5 present "y-direction (enhanced)" when the

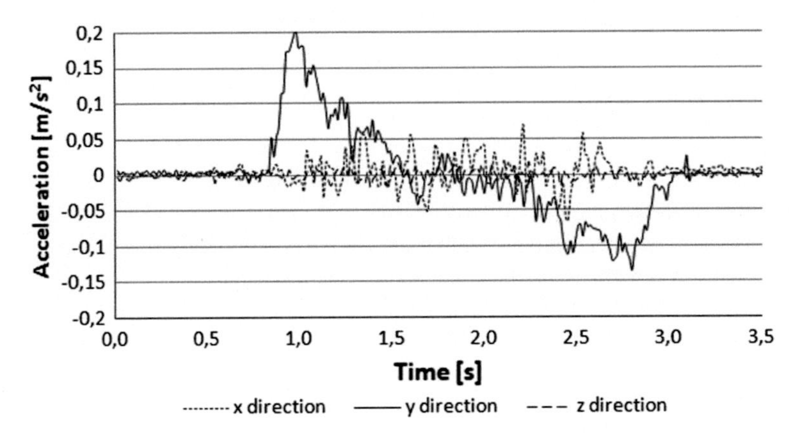

Fig. 2 Raw acceleration data within time

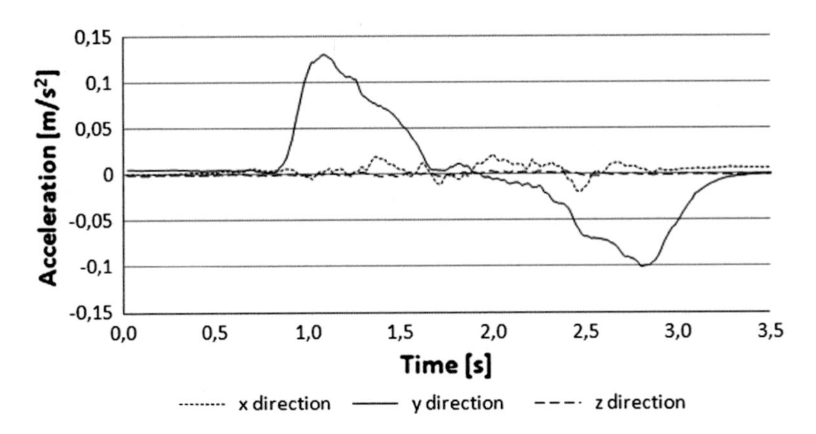

Fig. 3 Acceleration data with low pass filter within time

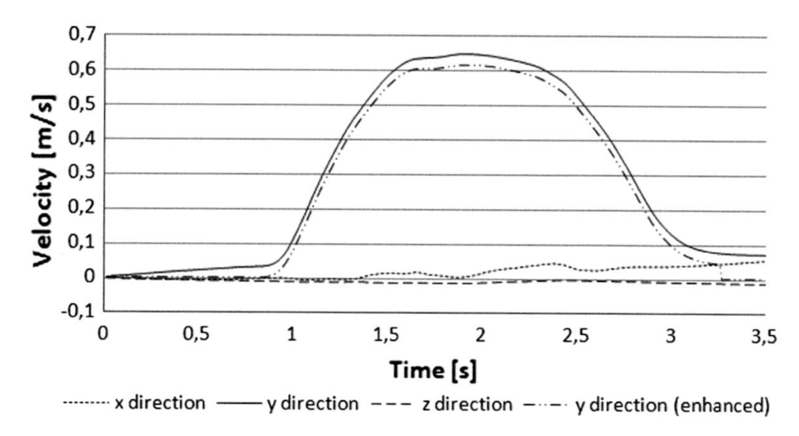

Fig. 4 Instantaneous velocity within time

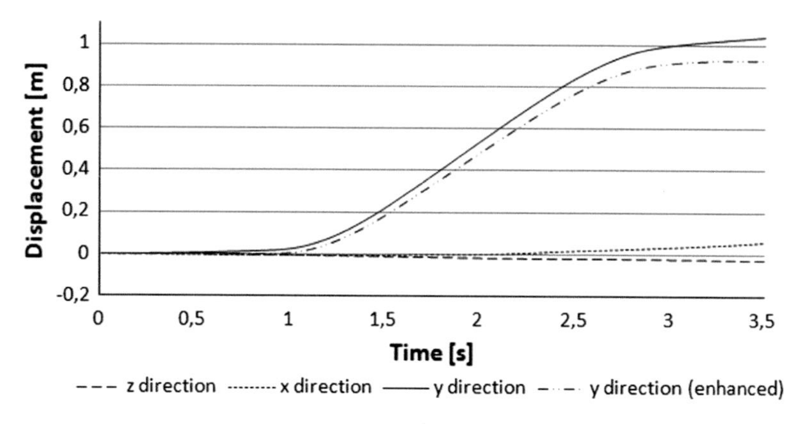

Fig. 5 Displacement within time

error is reduced by calibration of the non-accelerated moments in time. Figure 5 presents the calculated displacement in three directions.

The similar experiments have been performed for gyroscope evaluation. The sample frequency for gyroscope data collection is equal to 10 Hz. The device was rotated towards the y-axis once (360). The experiment was carried out with the angular velocity about 3 rad/s. The device in this step was rotating towards one axis. This kind of rotation was performed 10 times with different angular velocities. Figure 6 presents the instantaneous velocity in three directions. The experiment was carried out with the angular velocity about 3 rad/s. The device in this step was rotating towards one axis. This kind of rotation was performed 10 times with different angular velocities. Figure 6 presents the instantaneous velocity in three directions. The estimation of rotation could be determined with the usage of known angular velocity. Figure 7 presents the calculated rotation in three directions. The rotation in y-direction is obtained properly. However one could notice that the rotation in other two dimensions is not equal to 0. It's cause by the noise.

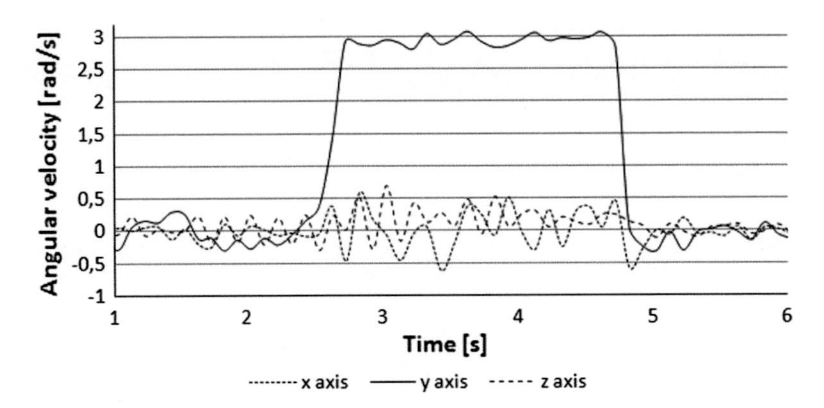

Fig. 6 Instantaneous angular velocity within time

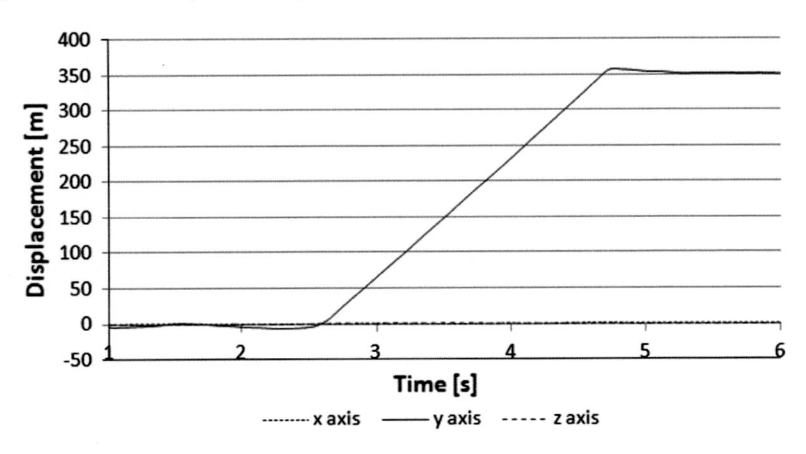

Fig. 7 Angular rotation within time

5 Conclusions

The paper presents the possibility of enhancements for a precise indoor positioning system for smartphones employing an accelerometer and gyroscope sensors that are embedded in mobile device. The indoor positioning system that are based on WLAN received signal strength comparison are very prone to any environmental distortions. This article presents the opportunities of embedded sensors in today's mobile phones like gyroscope and accelerometer. Those two sensors could provide an enhancement for WLAN positioning system in order to obtain more precise current position. The performance of the inertial sensors could be satisfactory as a short duration distance measurements for mobile platform. Nevertheless, in long term positioning the values obtained with those sensors is fulfilled with random noise. Thus, further research will be focused on the reduction of undesirable noise.

References

1. Bay, H., Tuytelaars, T., Gool, L.: SURF: speeded up robust features. In: European Conference on Computer Vision, pp. 404–417 (2006)
2. Bitsch, J., Smith, P., Wehrle, K.: FootPath: accurate map-based indoor navigation using smartphones. In: International Conference on Indoor Positioning and Indoor Navigation, Guimaraes, Portugal (2011)
3. Bujari, A.: Movement pattern recognition through smartphone's accelerometer, Univ. degli Studi di Padova (2011)
4. Gusenbauer, D., Isert, C., Krosche, J.: Self-contained indoor positioning on off-the-shelf mobile devices. In: International Conference on Indoor Positioning and Indoor Navigation, Guimaraes, Portugal (2011)
5. Kothari, N., Kannan, B., Dias, M.B.: Robust indoor localization on a commercial smart-phone. In: Technical Report of the Robotics Institute, Carnegie-Mellon University (2001)

Thermal Face Recognition

Ryszard S. Choraś

Abstract This paper presents a method for recognition of infrared images of faces. In this paper, we propose two steps to realize the recognition faces based on InfraRed images. Firstly, find out the signature in face infrared face thermograms using marphological operations. Secondly, measure the texture parameters of face regions.

Keywords Face recognition · InfraRed face images · Signature face images · Texture features

1 Introduction

Thermal Infrared (IR) imaging has been used extensively in a wide range of applications in particular biometric applications. Thermal infrared images represent the heat patterns emitted from an object. The camera absorbs the IR energy emitted by the object and converts it into electrical signal [8]. Sensors in the IR camera respond to thermal radiation in the infrared spectrum range at $0.7-14.0\,\mu\text{m}$. The infrared (IR) spectrum can be further divided into different spectral bands. The IR spectrum is divided into [5]:

- the active IR band:
  - Near-IR,
  - Short-Wave IR,
- the passive IR band:
  - the Mid-Wave (MWIR),
  - Long-Wave InfraRed (LWIR) band.

R.S. Choraś (✉)
Institute of Telecommunications and Computer Sciences, UTP University of Science and Technology, S. Kaliskiego 7, 85-796 Bydgoszcz, Poland
e-mail: choras@utp.edu.pl

R.S. Choraś (ed.), *Image Processing and Communications Challenges 7*,
Advances in Intelligent Systems and Computing 389,
DOI 10.1007/978-3-319-23814-2_5

The wavelength ranges of different infrared spectrums are shown in Fig. 1. The wavelength ranges of different infrared spectrums are shown in Table 1.

Biometrics refers to the automatic recognition of a person based on physiological and/or behavioral traits. Face recognition is widely used applications of biometric techniques. The vessels transport warm blood throughout the body; the thermal patterns of faces are derived primarily from the pattern of blood vessels under the skin. The vein and tissue structure of the face is unique for each person, and therefore the IR images are also unique. Thermal face recognition analyzes the thermal patterns in an infrared image band [3, 7]. Advantages of thermal IR imaging in face recognition include the invariance to illumination changes. Facial feature extraction has become an important issue in automatic recognition of human faces.

In these work, we used the frontal face IR images from 35 different persons.

The main stages in the feature extraction process are:

1. Image acquisition,
2. Image preprocessing
3. Extraction of thermal face signatures,
4. Extraction of texture features.

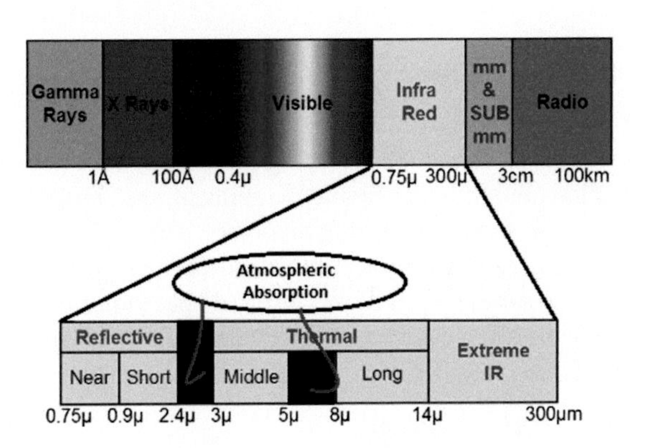

Fig. 1 Infrared spectrums

Table 1 Wavelength ranges for different infrared spectrums

Spectrum	Wavelength range (µm)
Visible spectrum	0.4–0.7
Near infrared (NIR)	0.7–1.0
Short-wave infrared (SWIR)	1–3
Mid-wave infrared (MWIR)	3–5
Thermal infrared (TIR)	8–14
Far-infrared (FIR)	15–1,000

This paper is organized as follow. In Sect. 2 very briefly the image acquisition techniques are discussed. Section 3 presents the geometrical face descriptor vector based on morphological operations and face signature. In Sect. 4 thermal face texture features are presented.

2 Image Acquisition

Process of acquire a thermal image is shown in Fig. 2. By using a thermal imaging camera, the system performs an infrared scan of a persons face. Infrared cameras detect the heat distribution in the image.

Thermal imaging cameras will have multiple color palettes, that are user selectable. In many applications (e.g. in medicine, biometrics), where the temperature range is more limited a "rainbow" palette is preferred, with red as hot and blue/black as cold (Fig. 3a and b). The rainbow palette has the best thermal sensitivity for displaying the differences in temperature.

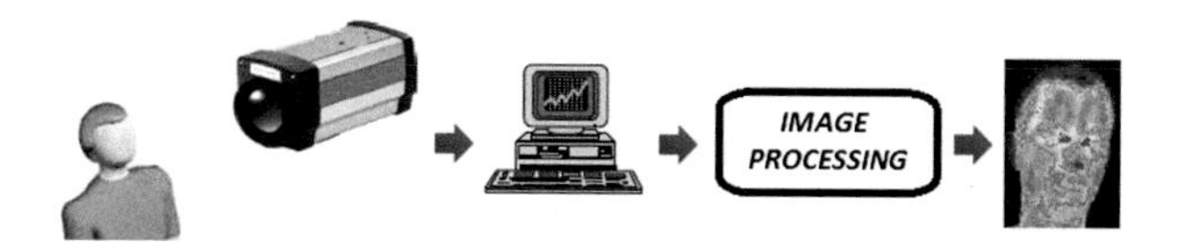

Fig. 2 Capture a thermal image

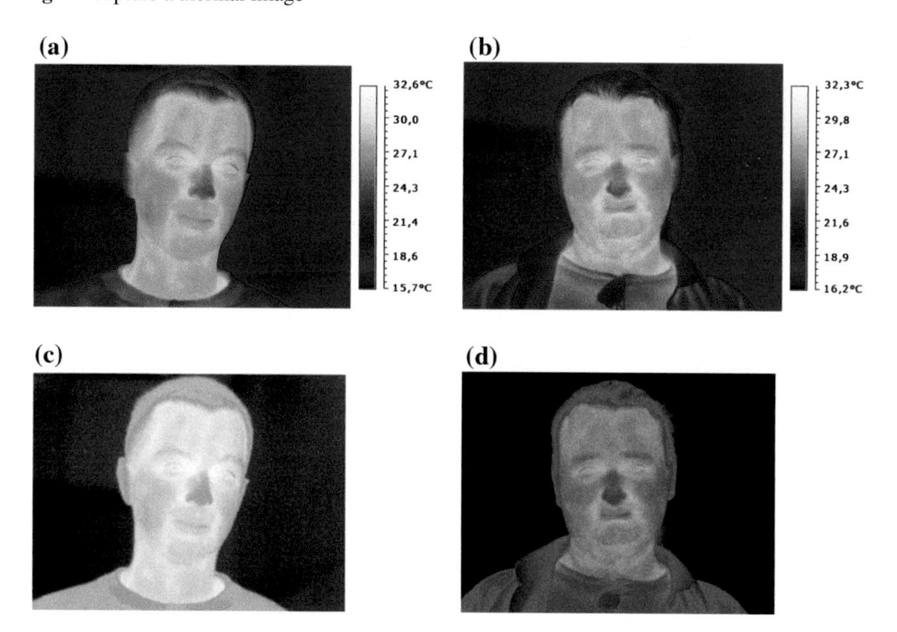

Fig. 3 Face images: "rainbow" *color* palette (**a** and **b**) and *gray* level palette (**c** and **d**)

Another palette is the black/white palette where the intensity of the gray level is proportional to the temperature on the object surface and where the brighter pixels have higher temperature values (Fig. 3c and d).

3 Image Preprocessing

Before performing feature extraction, the original thermal IR images are subjected to some image processing operations, such as:

- face segmentation (Fig. 4),
- morphological operation on face ROI image,
- face signature extraction.

Morphological operations work with two sets [9]. One set is the image to be processed. Another set is called structuring element. Mathematical Morphology is a tool for extracting image components, that are useful in the representation and description of region shape.

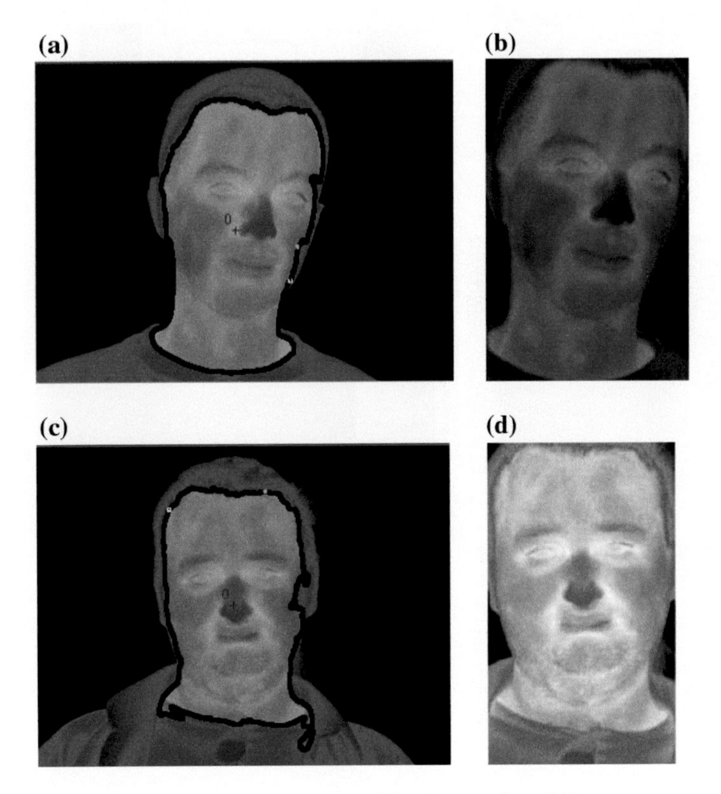

Fig. 4 Original face images (**a** and **c**) and cropped face images (**b** and **d**)

In gray scale images we deal with digital image functions of the form $f(x, y)$ as an input image and $b(x, y)$ as a structuring element. (x, y) are integers from $Z \times Z$ that represent a coordinates in the image. $f(x, y)$ and $b(x, y)$ are functions that assign gray level value to each distinct pair of coordinates.

Dilation is the maximum pixels set union when structuring element overrides image, while erosion is the minimum pixels set union when image is overlapped by structuring element.

Greyscale Dilation defined as

$$\delta_b(f(x, y)) = f \oplus_g b = \max\{f(x - z) + b(z)\}$$

This process consists stages: (1) reflect the structuring element, (2) position the structuring element at position x, (3) pointwise add the structuring element over the neighborhood, and (4) take the maximum of that result over the neighborhood.

Greyscale Erosion defined as

$$\epsilon_b(f(x, y)) = f \ominus_g b = \min\{f(x + z) - b(z)\}$$

This process consists stages: (1) position the structuring element at position x, (2) pointwise subtract the structuring element over the neighborhood, and (3) take the minimum of that result over the neighborhood.

Top-Hat transform (TH) is a well known morphological operation for locally extracting structures from given images:

$$TH = f(x, y) - \gamma_b(f(x, y)) = f(x, y) - \delta_b[\epsilon_b(f(x, y))].$$

After (TH) morphological operation the face ROI images is obtained and in next step we utilizing a standard thining algorithm.

The thinning algorithm modifies each pixel according to its initial value and to those its neighborhood.

A pixel is considered deletable if it satisfies the following conditions:

1. $1 < B(p) \leq 7$ where $B(p) = \sum_{i=1}^{8} p_i$
2. $A(p) = 1$ is number of 01 transitions in the eight neighbors of pixel p.

Additional conditions may permit the removal of element if it is on south or east edge or if is on a corner

3. $(p_1 = 0)$ or $(p_7 = 0)$ or $(p_3 = 0$ and $p_5 = 0)$

The next conditions may permit the removal of element if it is on north or west edge or if it is on a corner

4. $(p_3 = 0)$ or $(p_5 = 0)$ or $(p_1 = 0$ and $p_7 = 0)$ (Fig. 5).

Fig. 5 Pixel notation

p_4	p_3	p_2
p_6	p	p_1
p_6	p_7	p_8

Based on signature image the feature vector is determined using the following algorithm. The centroid of binary signature image (after the normalization) is placed in the point (0, 0) of the coordinates system. The centroid also becomes the centre of the concentric circles of the radius r_i. The algorithm uses the surrounding circle of signature image for partitioning it to $o = \lfloor \frac{r\prime_{max}}{\Delta r} \rfloor$ radial partitions, where $r\prime_{max}$ is the maximal possible radius, Δr is the distance between the consecutive circles and $\lfloor\rfloor$ is the maximal integer number less than o [2]. The intersection points of signature image with the created concentric circles of radius r_i are calculated and this extraction phase results is thus represented by vector of intersection point coordinates with each point representing the position of a intersection landmarks (Fig. 6).

The feature vector created on the basis of the presented feature extraction algorithm can be formulated as:

$$V = \{V_1, \ldots, V_l, \ldots, V_o, \} \tag{1}$$

where V_l is a n dimensional vector of point coordinates $V_l = [(x_1, y_1), \ldots, (x_n, y_n)$ with each point representing the position of a landmark (the intersection points of signature image with the concentric circles of radius r_i) and n is the predefined constant number of landmarks in the each r_i. In our work we arbitrary used $i = 1, 2, 3$ (three circles) and $n = 8$ for each circle (Fig. 7).

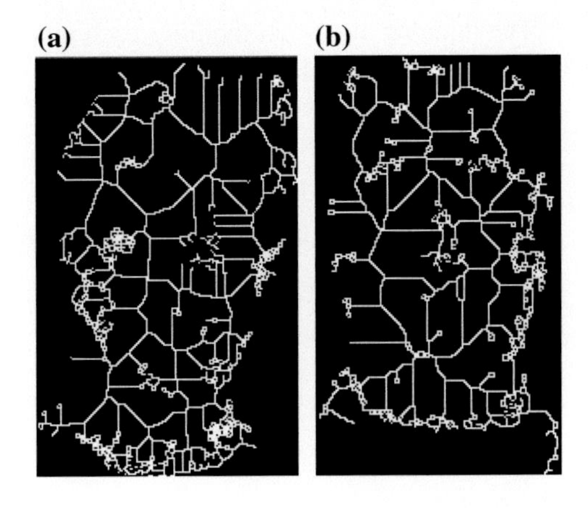

Fig. 6 Signatures of images with (**a**) Fig. 4(b) and (**b**) Fig. 4(d)

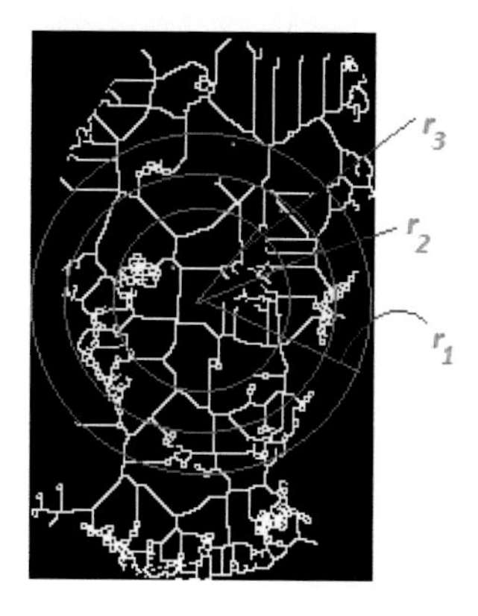

Fig. 7 The intersection points of signature image with the created concentric circles [2]

4 Texture Feature

Texture is a powerful descriptor that helps in the feature extraction for face analysis and recognition process [1, 4]. Texture can be characterized by the statistical distribution of the image intensity using Grey Level Co-occurrence Matrix (*GLCM*), proposed by Haralick [6].

The co-occurrence matrix $C(i, j)$ counts the co-occurrence of pixels with gray values i and j at a given distance d. The distance d is defined in polar coordinates (d, α), with discrete length and orientation. In practice, α takes the values 0°; 45°; 90°; 135°. The co-occurrence matrix $C(i, j)$ can now be defined as follows:

$$C(i, j) = Pr(P(p_1) = i \wedge P(p_2) = j \mid |p_1 - p_2| = d) \tag{2}$$

where Pr is probability, and p_1 and p_2 are positions in the gray-scale image P.

Let r be the number of gray-values in the image, then the dimension of the co-occurrence matrix $C(i, j)$ will be $r \times r$. So, the computational complexity of the co-occurrence matrix depends quadratically on the number of gray-scales used for quantization.

Features can be extracted from the co-occurrence matrix to reduce feature space dimensionality and the formal definitions of eight features from the co-occurrence matrix are done

Energy

$$\sum_{i,j} C(i; j)^2 \tag{3}$$

Entropy

$$-\sum_{i;j} C(i; j) log C(i; j) \tag{4}$$

Inverse Difference Moment

$$\sum_{i;j} \frac{1}{1+(i-j)^2} C(i;j) \tag{5}$$

Inertia (or contrast)

$$\sum_{i;j} (i-j)^2 C(i;j) \tag{6}$$

Correlation

$$\sum_{i;j} \frac{(i-\mu_i)(j-\mu_j)C(i;j)}{\sigma_i \sigma_j} \tag{7}$$

Angular Second Moments

$$\sum_i \sum_j C(i,j)^2 \tag{8}$$

Variance

$$\sum_i \sum_j (i-j)^2 C(i,j) \tag{9}$$

Inverse Difference Moment

$$\sum_i \sum_j \frac{1}{1+(i-j)^2} C(i,j) \tag{10}$$

Homogeneity

$$\sum_{i;j} \frac{1}{1-(i-j)^2} C(i;j) \tag{11}$$

where

$C(i;j)$ the $(i;j)$th entry in a co-occurrence matrix C;
$\sum_i$ defined as: $\sum_{i=1}^{i=M}$ where M is the number of rows;
$\sum_j$ defined as: $\sum_{j=1}^{j=N}$ where N is the number of columns;
$\sum_{i;j}$ means $\sum_i \sum_j$;
μ_i defined as: $\mu_i = \sum_i i \sum_j C(i;j)$;
μ_j defined as: $\mu_j = \sum_j j \sum_i C(i;j)$;
σ_i defined as: $\sigma_i = \sum_i (i-\mu_i)^2 \sum_j C(i;j)$;
σ_j defined as:$\sigma_j = \sum_j (j-\mu_j)^2 \sum_i C(i;j)$;
μ_x, μ_y the mean of row and column sums respectively;
σ_x, σ_y the standard deviation of row and column sums respectively.

Four directional *GLCM* are computed and for each *GLCM* above features are computed (Table 2).

Table 2 Texture features of faces

	Face image 4(b)				Face image 4(d)			
Features	0°	45°	90°	135°	0°	45°	90°	135°
ASM	0.007	0.006	0.008	0.006	$7.060E^{-4}$	$5.164E^{-4}$	$7.064E^{-4}$	$5.054E^{-4}$
Contrast	33.296	49.507	21.078	51.941	57.377	96.587	45.958	93.115
Correlation	$6.828E^{-4}$	$6.798E^{-4}$	$6.734E^{-4}$	$6.803E^{-4}$	$5.339E^{-4}$	$5.318E^{-4}$	$5.190E^{-4}$	$5.323E^{-4}$
IDM	0.407	0.346	0.442	0.328	0.271	0.204	0.280	0.202
Entropy	6.849	7.088	6.762	7.127	7.761	8.050	7.760	8.051
Energy	0.007	0.006	0.008	0.006	$7.060E^{-4}$	$5.164E^{-4}$	$7.064E^{-4}$	$5.054E^{-4}$
Inertia	33.296	49.507	21.078	51.941	57.377	96.587	45.958	93.115
Homogeneity	0.471	0.418	0.502	0.403	0.353	0.289	0.361	0.289
Variance	1450.251	1446.515	1472.352	1439.816	1846.191	1829.502	1899.377	1828.530

5 Conclusions

In this paper, a thermal face recognition method is proposed based on face signatures and texture features. Signatures and texture features of the thermal image are most successful approaches for face recognition. The proposed features vector can be used for identifying the individual in an image. Recognizing the faces eith visible images is difficulty in low illumination conditions. Thermal images are less sensitive to illumination variations. The experimental results indicate that, thermal face recognition improves better recognition accuracy than visible faces under wide range illumination changes.

References

1. Choras, R.S.: Image feature extraction techniques and their applications for cbir and biometrics systems. Int. J. Biol. Biomed. Eng. **1**, 6–16 (2007)
2. Choras, R.S., Choras, M.: Geometric algorithms of ear contour shape representation and feature extraction. In: Sixth International Conference on Intelligent System Design and Applications ISDA-2006, vol. II, 451–456. IEEE Computer Society Press (2006)
3. Etehadtavakol, M., Ng, E.Y.K., Chandran, V., Rabbani, H.: Separable and non-separable discrete wavelet transform based texture features and image classification of breast thermograms. Infrared Phys. Technol. **61**, 274–286 (2013)
4. Gabor, D.: Theory of communication. J. Inst. Elect. Eng. **93**, 429–459 (1946)
5. Hammoud, R.I. (Ed.).: Augmented Vision Perception in Infrared. Springer, London (2009)
6. Haralick, R.M.: Statistical and structural approaches to texture. IEEE Trans. Syst. Man Cybern. **67**, 786–804 (1979)
7. Pavlidis, I., Tsiamyrtzis, P., Manohar, C., Buddharaju, P.: Biometrics: Face recognition in thermal infrared. In: Biomedical Engineering Handbook, 3rd edn. CRC Press, Boca Raton (2006)
8. Rogalski, A.: Recent progress in infrared detector technologies. Infrared Phys. Technol. **54**, 136–154 (2011)
9. Serra, J.: Image Analysis and Mathematical Morphology. Academic Press, New York (1982)

Feature Reduction Using Similarity Measure in Object Detector Learning with Haar-Like Features

Jerzy Dembski

Abstract This paper presents two methods of training complexity reduction by additional reduction the number of features to check in object detector training task with AdaBoost training algorithm. In the first method, the features with weak performance at first weak classifier building process are reduced based on a list of features sorted by minimum weighted error. In the second method the feature similarity measures are used to reduce that features which is similar to earlier checked features with high minimum error rates in possible weak classifiers for current step. Experimental results with MIT-CMU 19 × 19 face images show that the error presented by ROC curves is near the same for the learning with and without additional feature reduction during the computational complexity is well reduced.

Keywords Face detection · Adaboost classifier · Feature selection

1 Introduction

The AdaBoost face detector using rectangular Haar-like wavelet features first described by Viola and Jones [7] is the most popular in recent years. The detector learning process using AdaBoost algorithm is computationally complex mainly due to the need to check all possible rectangular features, the decision threshold and polarity for each feature in each weak classifier building process. The most of works focus on detection speed optimization which is important to make any pictures or video stream scanning be possible in real time. The AdaBoost cascade variant with dynamically acquiring negative examples also proposed in [7] accelerates detection by quick rejection of negative parts (without objects) of an image in compare with original AdaBoost scheme [6]. Generally as the learning process is more computa-

J. Dembski (✉)
Faculty of Electronics, Telecommunications and Informatics,
Gdansk University of Technology, Ul. Narutowicza 11/12, 80-952 Gdansk, Poland
e-mail: dembski@ue.eti.pg.gda.pl

R.S. Choraś (ed.), *Image Processing and Communications Challenges 7*,
Advances in Intelligent Systems and Computing 389,
DOI 10.1007/978-3-319-23814-2_6

tionally complex, the detection process is quicker. Besides there are some reasons to accelerate also the training process [9]:

- we can use more examples and higher sized images with more details to train classifier in the same time,
- in context detection tasks the higher sized training images are need and apart of this more types of objects need to be detected,
- in autonomous agents learning tasks high learning speed allow agents to learn to detect earlier unknown objects in real time,
- higher training speed increase research possibilities for new algorithms and feature sets investigation.

2 The Object Detector Training

2.1 Features

In the simplest case Haar-like wavelets are used as features in such a way that each feature is calculated as a difference of mean pixel intensities between two neighbouring rectangular regions. Each pair of rectangles can vary in vertical and horizontal size, vertical or horizontal neighbouring and can be placed in any possible location in the image window. For instance, in the case of 19×19 windows there are 34200 edge features like the one shown in Fig. 1a. Two other features are shown in Fig. 1b, c. The neighbourhood of edge and line features can be positioned vertically or horizontally. Only few arithmetic operations are need to calculate each of this feature thanks to integral images technique. The number of all possible rectangular features increase very quickly with the size of example images as was shown in Table 1. Another feature types are gathered in [9], especially rectangular features with flexible sizes and distances, joint Haar-like features which compensate greediness of standard singular rectangular feature choice at each weak classifier selection step. In [1, 4] the usability of Haar-like rotated rectangular features is checked. Quite another features described in [3] are illumination invariant and are based on modified census transform similar to local binary patterns (LBP).

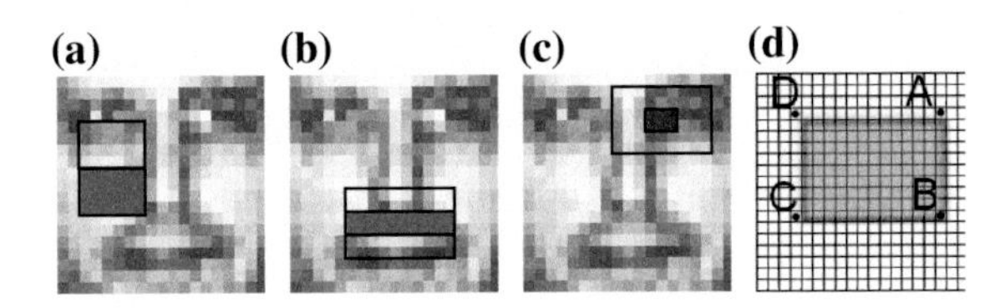

Fig. 1 Features based on intensity differences: **a** edge, **b** line, **c** center-surround, **d** calculation of the sum of pixel intensities using integral image

Table 1 Number of edge Haar-like features based on image sizes

Size (pixels)	19	24	28	32	48	64	128
Number of features	34200	86400	159152	270336	1354752	4259840	67633152

2.2 The AdaBoost Algorithm

The standard strong classifier AdaBoost learning algorithm presented in Table 2 can be treated as a cascade level building process. The weak classifier in general manner can be shown as a function with 4 arguments $h(\mathbf{x}, f, p, \theta)$, where $\mathbf{x}$ is an input image, f—selected feature, p—polarity e.g. the direction of class 1 and θ—threshold. The number of possible weak classifiers $L = 2N(K-1)$ depends on N—number of features and K—number of examples, which is equal to number of thresholds, because only thresholds between neighbouring examples due to current feature value are significant. The set of training examples must be sorted by values of each considered feature with complexity $O(K\log(K))$, so the computational complexity of all possible weak classifiers checking process is $O(NK\log(K))$ with memory complexity $O(NK)$. Practically this process is often placed inside the main loop of the algorithm with $O(TNK\log(K))$ computational complexity but without memory usage for example-feature value table. In both cases the complexity reduction can be done by feature reduction.

Table 2 Adaboost classifier (cascade layer) learning algorithm

for each training example $\{(\mathbf{x}_1, d_1), (\mathbf{x}_2, d_2) \ldots (\mathbf{x}_K, d_K)\}$

- initialize weights $w_i = 1/K_c$, where K_c is a number of examples which belong to the same class c as i-th example
- generate all possible weak classifiers $h_j(\mathbf{x}), j = 1 \ldots L$

while false positive error $f > f_{min}$ **or** true positive rate $d < d_{max}$

1. normalize weights: $w_i \leftarrow \frac{w_i}{\sum_{j=1}^{K} w_j}$
2. select the weak classifier $h_t(\mathbf{x})$, which minimizes the weighted classification error: $\epsilon_t = \min_j \sum_{i=1}^{K} w_i |h_j(\mathbf{x}_i) - d_i|$
3. decrease the weights of properly classified examples: $w_i \leftarrow w_i \frac{\epsilon_t}{1-\epsilon_t}$
4. find the best threshold value Θ for strong classifier with minimum false positive error if false negative error is under expected value for node (cascade layer) $d >= d_{max}$

end while

The strong classification function $H(\mathbf{x})$ consisted with T weak classifiers is a result of AdaBoost learning process and is described by the formula:

$$H(\mathbf{x}) = \begin{cases} 1 & \text{if } \sum_{t=1}^{T} \alpha_t h_t(\mathbf{x}) \geq \Theta \\ 0 & \text{otherwise,} \end{cases} \tag{1}$$

where $\alpha_t = \log(1 - \epsilon_t) - \log \epsilon_t$ is a weight of t-th weak classifier.

2.3 The Classifier Cascade Learning

The main idea of AdaBoost cascade version (Fig. 2) is to reduce time complexity during detection due to huge asymmetry between the number of picture regions which contain object e.g. face (positive examples) and without it. The asymmetric classification error rates are assumed: small for positive examples (false negative rate) and big for negative examples (false positive rate). It is desirable to reject non-objects (non-faces) regions as quick as possible but leave almost 100 % of true positive examples. The final false positive error rate can be estimated as a product of all layer's false positive error rates because the succeeding layers are independent and are trained without examples rejected in previous layers. For instance in my experiments the assuming false positive error rate $f_{max} = 0.4$, true positive rate $d_{min} = 0.995$. For 10-layers cascade classifier estimated final false positive error $F_{max} = f_{max}^{10} = 0.4^{10} = 1.0486e-004$ and the true positive rate $D_{max} = 0.99^{10} = 0.9044$. Another idea described in [7] is negative examples acquisition procedure using a set of images e.g. photographs which don't contain detected objects (faces).

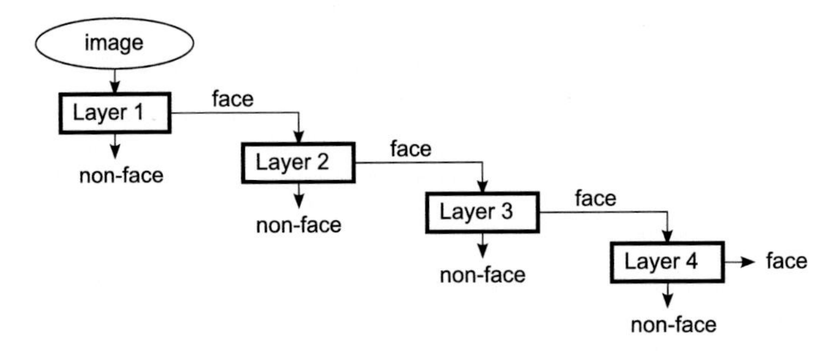

Fig. 2 Cascade of classifiers

3 Feature Reduction

The quick growth of the number of rectangular Haar-like features presented in Table 1 is the main reason for taking into account feature reduction methods due to weak classifier building process computational complexity which depends on the number of features which was described in Sect. 2.2.

3.1 Related Works

In [5] the Viola-Jones full exhaustive search was replaced by simple heuristic search. The next feature was chosen as the best from neighbouring features and so on it reduces the number of features about 300 times. Other fast learning presented in [8] is based on precomputing classification table for each training example provided that each example in each weak classifier evaluation process have the same weight during cascade node training process.

3.2 Methods Based on a Subset of the Best Features from First Weak Classifier Building Step in Cascade Layer

In this group of methods features are sorted using weighted classification error at the first weak classifier building step of cascade layer learning process. In following steps only a subset of the best features from first step is using for weak classifier building procedure. This method seems to be very effective in spite of its simplicity as was shown in Sect. 4.

3.3 Methods Based on Features Similarity Measures

Two similarity measures are taking into account in this work:

- the spatial similarity based on the area of overlapping rectangular fields,
- permutational similarity based on example permutations comparison.

The spatial similarity measure is specific for rectangular features only and is calculated as $R/(S_1 + S_2 - R)$, where S_1, S_2-areas of rectangle 1 and 2, R-area of overlapped part of rectangles as is shown in Fig. 3.

For the permutational similarity calculation the set of examples is sorted by first and second feature value to obtain two permutations of example indices and next mean distance between example positions on sorted lists is evaluated. It is due to the way of weak classifier optimal threshold finding described in Sect. 2.2: similar

Fig. 3 Overlapped rectangular features

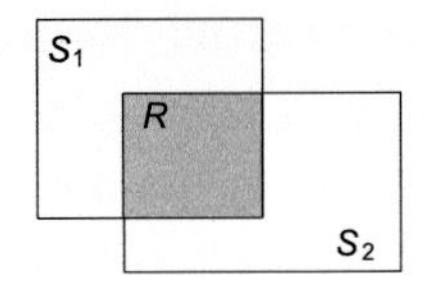

permutation is correlated with similar weighted error values for each feature. The standard algorithm (Table 2) is supplemented with a step of building a list of each feature L-nearest neighbours using one or both measures. Thanks to this list during weak classifier selection at step 2, the similar features to earlier checked features are leaved. Additionally the list of M best (from the least weighted error weak classifier) features is created during step 2. In the next step of the modified algorithm the weak classifier selection (step 2) is repeated with the unchecked nearest neighbours of the M best features. This step can be named condensation step. The weak classifier generation step in the new algorithm was replaced by on-line feature generation process which need to sort examples during each of T weak classifier selection procedures.

4 Experiments and Results

The object detection learning system [2] with modular parallel computing architecture was used in experiments with face detector training. The MIT-CMU 19×19 images from file svm.train.normgrey[1] was divided into training and testing subsets. Each subset contains 1200 positive and 1200 negative examples. Additionally beginning from cascade level 2 the 10,000 negative examples from non-face image repository were added to training subset. Each negative example has been rejected from training set only when it was classified as non-face, so the last cascade layers were trained with the subset of the most difficult negative examples. Only the edge rectangular features were used for classifier building. For each cascade level $f_{max} = 0.4, d_{min} = 0.995$. The theoretical false positive error for cascade $F_{max} = 10^{-6}$. The similarity measure method parameter $M = 10$. The ROC curves for each of reduction method are presented at Fig. 4. Figure 4a shows that using a subset of features 4 times smaller that the full set of features (8550/34200), the output classifier is the same. If the number of features is reduced 1024 times to only 33, the classification errors are near the same as in the case without any reduction but the number of weak classifiers in final cascade is 2 times bigger which increase the detection computing time. Figure 4b shows that regardless of spatial similarity degree the generalization results are similar as in Fig. 4a. This similarity can be explained by constant requirements (f_{max}, d_{min}) for each cascade level and algorithmic ability to complement small feature quality by quantity of weak classifiers in cascade level.

[1]The file svm.test.normgrey is not used because contains quite different examples which can disturb a comparison of reduction methods.

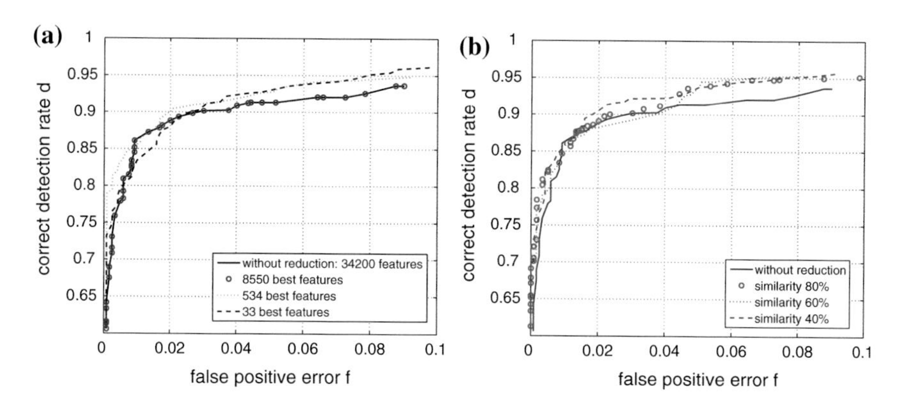

Fig. 4 ROC curves of the AdaBoost classifiers trained with two rectangular edge features reduction methods: **a** based on the feature performance at first weak classifier building step in a cascade layer (Sect. 3.2), **b** based on feature similarity measure (Sect. 3.3)

Table 3 Training time and classifier complexity comparison for two reduction methods

	No reduction	(a) Features performance			(b) Features similarity		
Reduction factor	1	1/4	1/64	1/1024	80 %	60 %	40 %
Number of features	34200	8550	534	33	14800	5500	3600
Number of weak classifiers	119	119	124	286	130	113	122
Number of levels	17	17	15	16	17	16	16
Training time [s]	13990	6071	3349	3679	7324	3481	2845

(a) based on the feature performance in the first weak classifier building step in a cascade layer building process (Sect. 3.2), (b) based on feature similarity measure (Sect. 3.3)

The training times, general numbers of weak classifiers and numbers of cascade levels are compared in Table 3. The training time is not proportional to reduction factor in the first method which is due to additional negative examples preparation time and other actions but the time reduction is noticeable in both methods and probably will be higher with training images size.

5 Conclusions

The experiments show that the generalization error curves with and without reduction is near the same. The speed of detection which depends on the distribution of weak classifiers in subsequent cascade layers (Table 3) is also similar except the most extremely case with 1/1024 feature reduction factor.

The exhaustive "brute-force" rectangular Haar-like features checking during each weak classifier building process due to experimental results seems to be unnecessary in practice and an amount of time which is saved in this way should be used in my opinion rather for higher sized images as learning examples, context based learning or learning with bigger sets of examples.

The heuristic Haar-like feature searching presented in this paper can be replaced by another more efficient method like simulated annealing or genetic algorithm.

Acknowledgments The research leading to these results has received partial funding from the Polish-Norwegian Research Programme operated by the National Centre for Research and Development under the Norwegian Financial Mechanism 2009–2014 in the frame of Project Contract No Pol-Nor/210629/51/2013.

References

1. Dembski, J.: Feature type and size selection in AdaBoost face detection. In: International Conference on Image Processing and Communication—IPC'2010 (2010)
2. Dembski, J., Smiatacz, M.: Modular machine learning system for training object detection algorithms on a supercomputer. In: Advances in System Science, pp. 353–361 (2010)
3. Küblbeck, C., Ernst, A.: Face detection and tracking in video sequences using the modified census transformation. Image Vis. Comput. **24**(6), 564–572 (2006)
4. Lienhart, R., Maydt, J.: An extended set of haar-like features for rapid object detection. IEEE ICIP **2002**, 900–903 (2002)
5. McCane, B., Novins, K.: On training cascade face detectors. In: Proceedings of Image and Vision Computing New Zealand, pp. 239–244 (2003)
6. Schapire, R.E., Freund, Y.: Boosting the margin: a new explanation for the effectiveness of voting methods. Ann. Stat. **26**(5), 1651–1686 (1998)
7. Viola, P., Jones, M.: Robust real-time face detection. Int. J. Comput. Vis. **57**(2), 137–154 (2004)
8. Wu, J., Brubaker, S., Mullin, M., Rehg, J.: Fast asymmetric learning for cascade face detection. IEEE Trans. Pattern Anal. Mach. Intell. **30**(3), 369–382 (2008)
9. Zhang, C., Zhang, Z.: A survey of recent advances in face detection, Technical Report MSR-TR-2010-66 (2010)

Assessment of the Brain Coverage Ratio in the Postoperative Craniosynostosis Based on 3D CT Scans

Anna Fabijańska, Jarosław Gocławski and Wanda Mikołajczyk-Wieczorek

Abstract This paper considers the problem of numerical quantification of the post-operative craniosynostosis. In particular, the method for assessment of the brain coverage ratio based on 3D CT brain scans is proposed. The method determines the percentage of brain surface which is not covered by skull and is intended to measure the rate of skull regeneration after the cranial vault remodelling. Results of applying the method to the postoperative cases of the craniosynostosis are presented and discussed.

Keywords Craniosynostosis · Brain segmentation · Brain coverage ratio

1 Introduction

Craniosynostosis is a birth defect which manifests itself as a serious malformation of the skull shape. It is caused by the premature closure of the cranial sutures which changes the growth pattern of a skull. In particular, when two or more sutures close, the growth of a skull in the direction of the suture fusion is stopped, but the compensatory growth in other directions occurs [1].

The only method of the craniosynostosis treatment is the surgical correction of a skull shape called the cranial vault remodelling. It restores the normalized skull shape via the excision of the fused suture, removal of the abnormal or fused bones

A. Fabijańska (✉) · J. Gocławski
Institute of Applied Computer Science, Lodz University of Technology,
18/22 Stefanowskiego Str., 90-924 Lodz, Poland
e-mail: anna.fabijanska@p.lodz.pl

J. Gocławski
e-mail: jaroslaw.goclawski@p.lodz.pl

W. Mikołajczyk-Wieczorek
Polish Mother's Memorial Hospital, Research Institute,
281/289 Rzgowska St., Lodz, Poland
e-mail: drwanda@tlen.pl

R.S. Choraś (ed.), *Image Processing and Communications Challenges 7*,
Advances in Intelligent Systems and Computing 389,
DOI 10.1007/978-3-319-23814-2_7

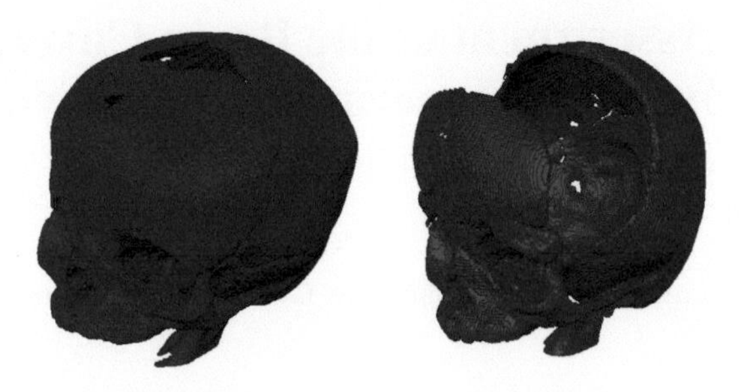

Fig. 1 A sample skull before and after the corrective surgery

and their replacement by the remodelled bones [4]. However, after the surgery some parts of the brain remain uncovered, until the final regeneration of the skull. This is presented in Fig. 1 which shows a sample skull before and after the surgery.

The postoperative assessment of the craniosynostosis is mainly based on the measurement of skull characteristic dimensions obtained from CT scans. These can only be used for a skull shape monitoring. However, there are no tools for measuring the rate of a skull regeneration. In particular, it would be beneficial to investigate how quickly the parts of brain uncovered during the surgery are afresh covered in the treatment process.

Having in mind the above mentioned needs, this paper proposes a method for assessment of the brain coverage ratio (BCR). In particular, the method determines the percentage of brain surface which is not covered by a skull. The method is dedicated to 3D CT brain images where image intensities represent Hounsfield units.

The introduced approach is described in details in Sect. 2. This is followed by the results provided by the method shown in Sect. 3. Finally, Sect. 4 concludes the paper.

2 The Proposed Approach

The main idea behind the introduced approach is to calculate the ratio of pixels on the surface of a brain which are not covered by a skull. Therefore, after the preprocessing (see Sect. 2.1), the brain segmentation is performed (see Sect. 2.2) [2, 6]. This is followed by the identification of pixels which belong to the brain surface but have no neighbours in the skull (see Sect. 2.3).

2.1 Preprocessing

In the preprocessing step, intensities of the input 3D image $I(x, y, z)$ are changed in accordance to Eq. 1.

$$I_2(x, y, z) = \begin{cases} \frac{1}{150} I(x, y, z) & \text{if } I(x, y, z) \in [0, 150] \\ 0 & \text{otherwise} \end{cases} \tag{1}$$

The preprocessing highlights intensities within the range of [0–150] HU which relate to the region of brain and skull. Remaining intensities are suppressed to the range limits. In particular, after the preprocessing, the background and the skull are represented by 0 and 1 respectively, while the region of brain obtains values from the range (0, 1).

Additionally, in the preprocessing step, linear interpolation of intensity is performed across Z direction, in order to equalize a slice thickness and a pixel spacing, and thus ensure uniform processing in all directions.

2.2 Brain Segmentation

The brain segmentation is performed via a series of morphological operations [5, 7].

Firstly, based on image I_2 binary masks of the skull s (see Figs. 2c and 3b) and the head h (see Figs. 2b and 3a) are created by thresholding described in Eqs. 2 and 3 respectively.

$$s(x, y, z) = \begin{cases} 1 & \text{if } I_2(x, y, z) \in [0.9, 1] \\ 0 & \text{otherwise} \end{cases} \tag{2}$$

$$h(x, y, z) = \begin{cases} 1 & \text{if } I_2(x, y, z) > 0 \\ 0 & \text{otherwise} \end{cases} \tag{3}$$

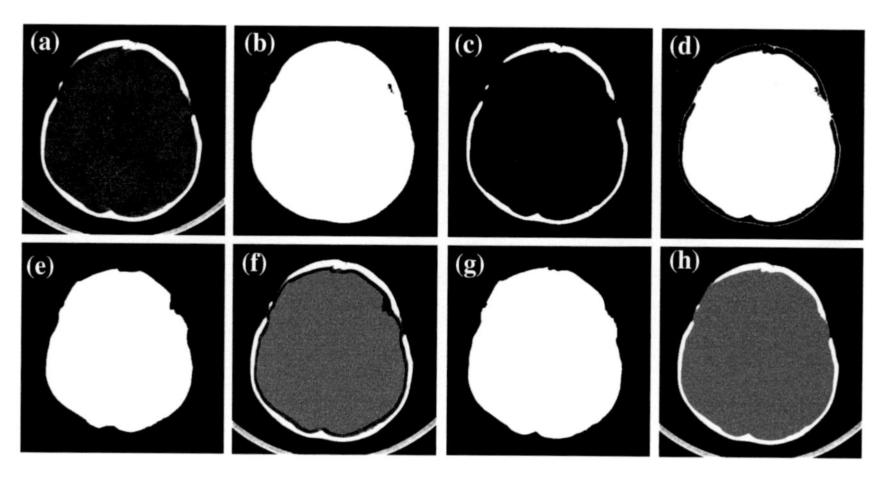

Fig. 2 The consecutive steps of the brain segmentation shown on a sample 2D slice; **a** original slice; **b** bead region; **c** skull region; **d** brain region before erosion; **e** brain region after erosion; **f** brain region after erosion compared to skull region; **g** brain region after dilation; **h** final brain region compared to skull

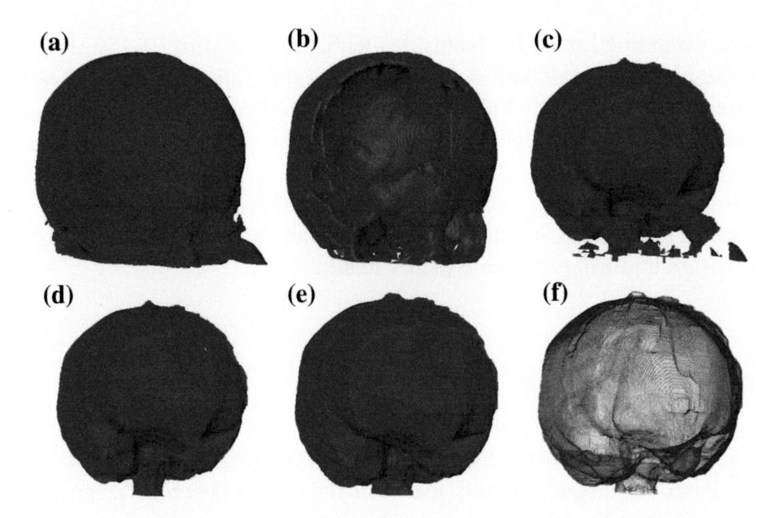

Fig. 3 The consecutive steps of the brain segmentation visualized in 3D; **a** a head; **b** a skull; **c** a head after skull removal; **d** the coarse brain; **e** the final brain; **f** the surface of brain

Then, the coarse image of brain b_1 (see Figs. 2d and 3c) is obtained by subtracting the binary image of skull from the binary image of head, as described in Eq. 4 [2, 3].

$$b_1 = h - s \tag{4}$$

However, the binary image b_1 covers not only the brain, but also includes its peripherials such as the nose, ears and skin as well (see Fig. 3c). Therefore, in the next step image erosion with a big cube structuring element $s_e = ones(15, 15, 15)$ is performed in accordance to Eq. 5 in order to remove skin and disjoint brain and eye regions.

$$b_2 = b_1 \ominus s_e \tag{5}$$

The largest connected component b_{max} in the image b_2 representing the brain space is subsequently selected (see Figs. 2e and 3d) and then back-dilated with the same structuring element s_e (see Eq. 6) in order to produce binary image b_3 (see Figs. 2h and 3e).

$$b_3 = b_{max} \oplus s_e \tag{6}$$

The final binary image of brain b is still defined as the largest connected component in image b_3. The size of the structuring element was determined empirically (for slices of the resolution 512×512 pixels and slice thickness equal to pixel spacing).

The partial results of consecutive steps of the proposed brain segmentation approach are illustrated in Figs. 2 and 3. In particular, Fig. 2 presents the results applied to a sample slice, while Fig. 3 visualises them in 3D.

2.3 Determination of Brain Coverage Ratio (BCR)

After the brain region is detected, the pixels which belong to the surface of a brain are identified. In particular, the one-pixel-width outline of the brain region is found in each slice as shown in Fig. 4a–c. The intensities of the outline and remaining pixels are set to the values 1 and 0 respectively. Two dimensional outlines obtained for each slice are next stacked into a volumetric space in order to build 3D outline of brain $o(x, y, z)$.

Next, the spherical neighbourhood of each outline pixel is investigated in order to determine if the pixel is covered by a skull or not. In particular, the outline pixel

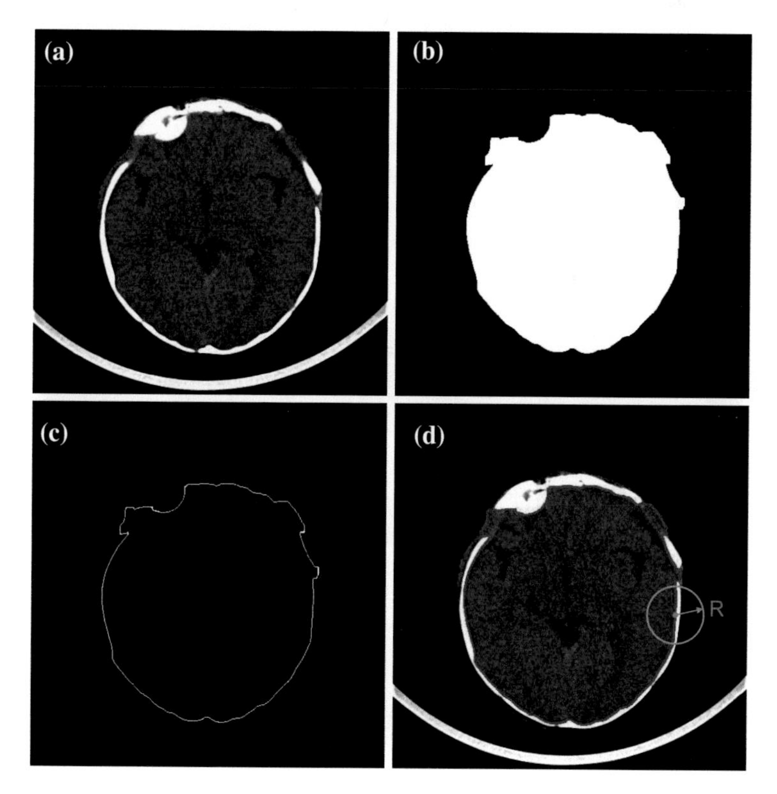

Fig. 4 The brain outline determination; **a** a sample CT brain slice; **b** the brain mask; **c** the brain outline; **d** a spherical neighbourhood of a sample outline pixel

is considered covered by skull if there is at least one pixel of the intensity 1 (which corresponds to the skull) in its spherical neighbourhood in the input image $I_2(x, y, z)$.

With the above assumptions the brain coverage ratio BCR is defined as a fraction of the covered outline pixels to the total number of pixels in the brain outline. This is described by Eq. 7.

$$BCR = \left(1 - \frac{\sum u(x, y, z)}{\sum o(x, y, z)}\right) \cdot 100\,\%, \tag{7}$$

where:

$$u(x, y, z) = \begin{cases} 1 & \text{if } o(x, y, z) \text{ is uncovered by skull,} \\ 0 & \text{otherwise.} \end{cases} \tag{8}$$

The radius R of the spherical neighbourhood of the outline pixel is equal to 7 pixels. The radius size was determined empirically (for slices of the resolution 512×512 pixels and the slice thickness equal to the pixel spacing).

3 Results

The results of applying the proposed approach to 5 sample 3D CT brain datasets are shown in Fig. 5. The cases of the decreasing brain cover ratio (BCR) were selected. For each case the skull s (in blue), the uncovered brain region u (in red), the combi-

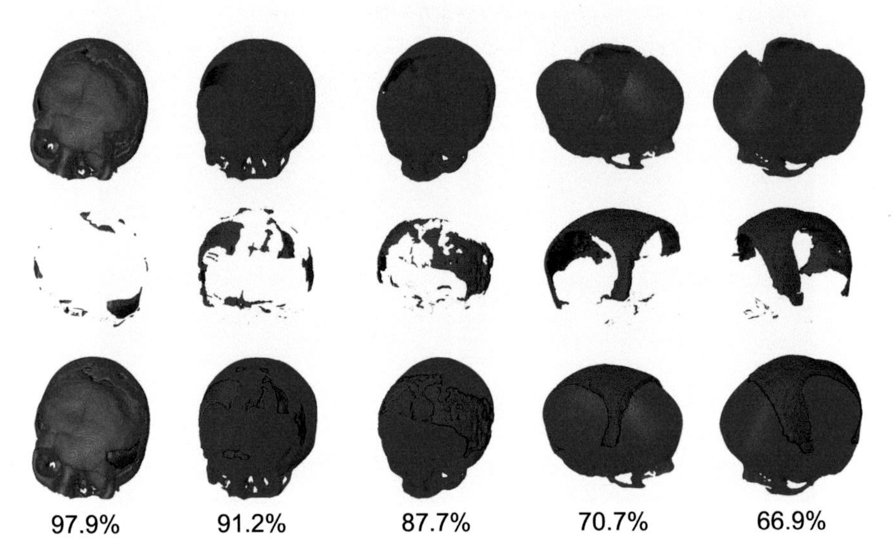

Fig. 5 The results of brain cover ratio determination; the consecutive rows show: the skull, the uncovered brain region, the comparison of these two regions and the resulting BCR

nation of these two regions and the resulting BCR value are shown in the row-wise order.

Based on the visual results shown in Fig. 5 it can be visually assessed that the uncovered regions of brain were mostly correctly identified by the proposed approach. This can be best seen when compared to skulls. The comparison shows that skull discontinuities were completely filled by the regions identified as uncovered.

Some erroneous pixels can be observed in the bottom part of a skull. They correspond to the regions where the brain is connected with the spinal cord and the brain segmentation is extremely difficult. These pixels however are less than 0.5 % of the outline pixels and thus may be neglected.

Unfortunately, the accuracy of the proposed approach cannot be assessed numerically due to the lack of the reference results.

4 Conclusions

The approach proposed is this paper is intended to measure the ratio of brain coverage by the skull in the postoperative craniosynostosis. The measurement is performed based on 3D CT brain scans. The introduced BCR index can be easily applied to investigate the rate of skull regeneration after the cranial vault remodelling. The most sensitive part of the method is the morphological erosion of prototype brain region cutting-off its peripherials. It requires careful, interactive selection of the structuring element size, which can vary for different cases. Although the approach is very simple it can be a very useful diagnostic tool in the case of the assessment at the craniosystosis treatment. Such functionality is not a part of basic CT software support typically including only linear, angle or aerial measurements. The computational complexity of the algorithm was not tested, because the method is still under development and will be accelerated by 3D GPU computing.

The introduced method is simple, fast and efficient, and thus can be easily introduced to everyday clinical routine.

Acknowledgments This research was funded by the Ministry of Science and Higher Education of Poland from founds for science in years 2013–2015 in a framework of Iuventus Plus Programme (project no. IP 2012 011272).

References

1. David, D.J., Poswillo, D., Simpson, D.: The Craniosynostoses: Causes, Natural History, and Management. Springer, Berlin (2013)
2. Gonzalez, E.R., Woods, R.E.: Digital Image Processing. Prentice Hall, Upper Saddle River (2008)
3. Gonzalez, E.R., Woods, R.E., Eddins, S.L.: Digital Image Processing Using MATLAB. Prentice Hall, Upper Saddle River (2004)

4. Hayward, R., Jones, B., Dunaway, D.: Clinical Management of Craniosynostosis. Mac Keith Press, London (2004)
5. Malina, W., Smiatacz, M.: Metody cyfrowego przetwarzania obrazow. Akademicka Oficyna Wydawnicza EXIT, Warszawa (in Polish) (2005)
6. Nikolaidis, N., Pitas, I.: 3-D Image Processing Algorithms. Wiley, New York (2001)
7. Serra, J.: Introduction to mathematical morphology. Comput. Vis. Graph. Image Process. **35**(3), 283–305 (1986)

Combined Imaging System for Taking Facial Portraits in Visible and Thermal Spectra

Piotr Jasiński and Paweł Forczmański

Abstract This work explores the subject of thermal imagery in the context of face recognition. Its aim is to create a database of facial images taken in both thermal and visual domains. To achieve this, a specialized photographic stand was designed, which allows simultaneous capture of images from IR thermal camera and SLR digital camera. To ensure precision, stability and fluency of photographic sessions, a Matlab application has been developed, through which it is possible to remotely control both devices, as well as automatically download captured images onto a hard drive and save them within an organized folder structure. Additionally, several image fusion techniques have been implemented in order to effectively combine visual and thermal data for use in face recognition algorithms.

Keywords Thermovision · Biometrics · Face · Database · Image fusion

1 Introduction

Biometric identification systems are becoming more and more accessible to average citizens and are no longer perceived only as a sophisticated way of guarding military secrets. Even low-cost laptops are equipped with webcams and pre-installed software capable of face recognition, which is still one of the most popular biometric features, easy to capture, describe and recognize [7]. It will not be long before biometric information replaces logins and passwords as means of securing our virtual accounts. However, there are situations when standard biometric solutions are not enough, e.g. when environmental conditions are not fully controlled [10] or we need a higher level of security. One of the possible solutions is thermal imaging. Data from

P. Jasiński (✉) · P. Forczmański
Faculty of Computer Science and Information Technology, West Pomeranian University of Technology, Żołnierska Str. 52, 71–210 Szczecin, Poland
e-mail: pjasinski@wi.zut.edu.pl

P. Forczmański
e-mail: pforczmanski@wi.zut.edu.pl

R.S. Choraś (ed.), *Image Processing and Communications Challenges 7*,
Advances in Intelligent Systems and Computing 389,
DOI 10.1007/978-3-319-23814-2_8

infrared sensors allows face recognition algorithms to disregard the necessity of visible light properly illuminating the subject [8]. Additionally, thermal imaging also provides great protection against spoofing attempts (using a photo instead of a real face [18]), because heat signatures of objects are completely different from real facial ones [16]. Such developments call for more ways of collecting data for researching combined thermo-visual personal identification systems, as well as means of testing and benchmarking said systems.

Hence, we propose a unified capture procedure based on a Matlab application and a photographic stand of our own design. Through its use it is possible to create a database of fused visual and thermal facial images, ready to serve any testing or research purposes. Similar idea has been presented in [5].

1.1 Existing Databases

Among the vast variety of existing face databases [9, 14] only a few include images captured using infrared technology. We selected the following as most prominent ones (see Fig. 1 for exemplary images):

- Equinox HID—one of the oldest, officially no longer available. It is divided in 3 parts, which differ by lighting angle (above, right, left). It includes 90 classes, each contains 10 shots (30 if we count all lighting cases) in visual and thermal domains. There is little variation in subject position (only expression) and some images suffer from minor pixel artifacts.
- Surveillance Cameras face database (SCface) [15]—contains 130 classes, 32 shots each. A lot of different positions are meticulously captured, in both controlled and uncontrolled environments, however apart from visual ones, the images are only near-infrared (non-thermal).
- IRIS Thermal/Visible Face Database [19]—30 classes of simultaneously acquired unregistered thermal and visible face images under variable illuminations, expressions, and poses. Unfortunately, affected by poor quality of old equipment.

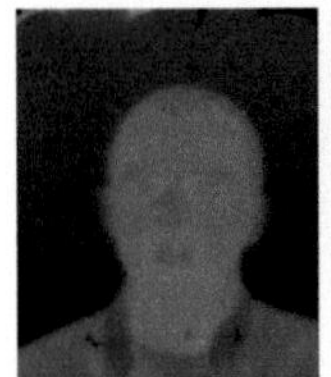

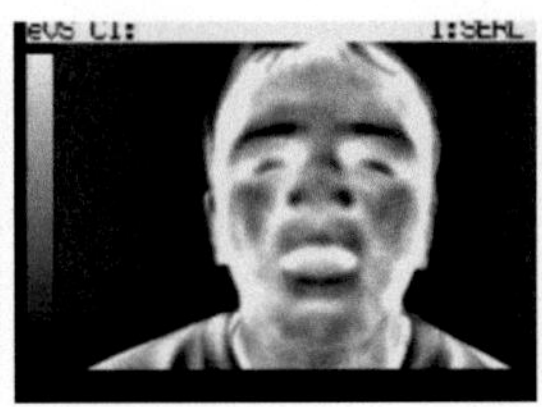
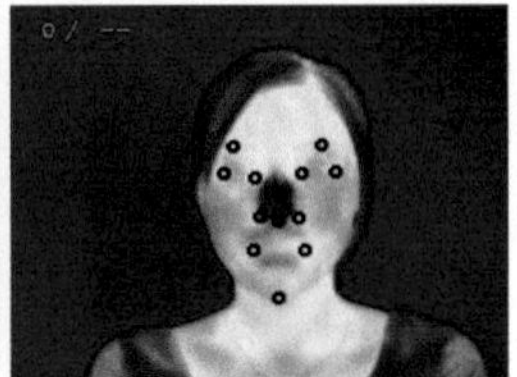

Fig. 1 Example infrared images from existing databases: Equinox HID, SCface, IRIS, and USTC-NVIE, respectively

- Natural Visible and Infrared facial Expression (USTC-NVIE) [20]—perhaps the closest to our concept, this database is comprised of more than 100 classes recorded simultaneously by a visible and an infrared thermal camera, with illumination provided from three different directions.

Out of all the databases presented above none mention the problem of capturing visual and thermal images in a way to ensure their proper alignment for an eventual fusion. They either use a complicated device with one lens and two types of sensors (e.g. [6]), or ignore said dilemma whatsoever. Thus our unified capture system (being cheap, accessible and precise at the same time) will make a valuable addition to the existing pool of combined visual-infrared face database solutions.

2 Capture System

Developed system is composed of two separate layers. The hardware layer is a photographic stand, which consists of (the numbers are referenced in Figs. 2 and 3):

1. Infra-Red (IR) camera mounted on a 108 cm tall tripod with tubular and bull's eye spirit levels
2. Digital Single Lens Reflex Camera (DSLR) mounted on a 116 cm tall W803 Light Stand
3. CY-25WT fluorescent lamp mounted on a 116 cm tall W806 Light Stand (2 total)
4. fluorescent tube reflector lamp mounted on a 118 cm tall W806 Light Stand (2 total)
5. focus marker (5 total)
6. approx. 2 m long piece of string
7. a 50 cm tall chair with back support
8. calibration board mounted on a 152 cm tall W803 Light Stand
9. a 275 cm × 200 cm greenscreen.

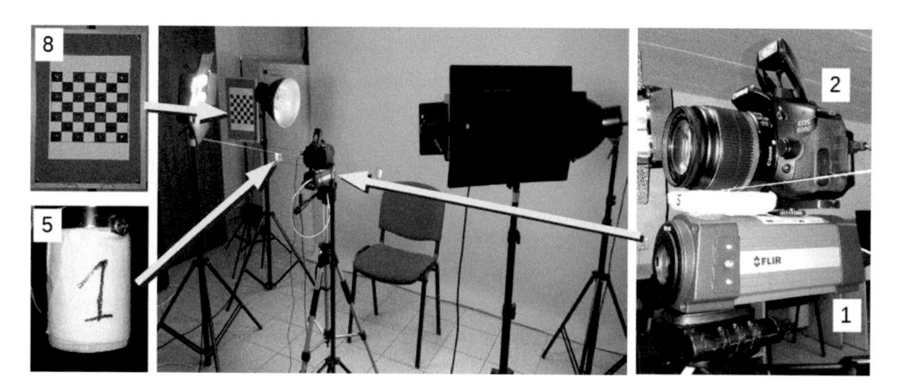

Fig. 2 The photographic stand

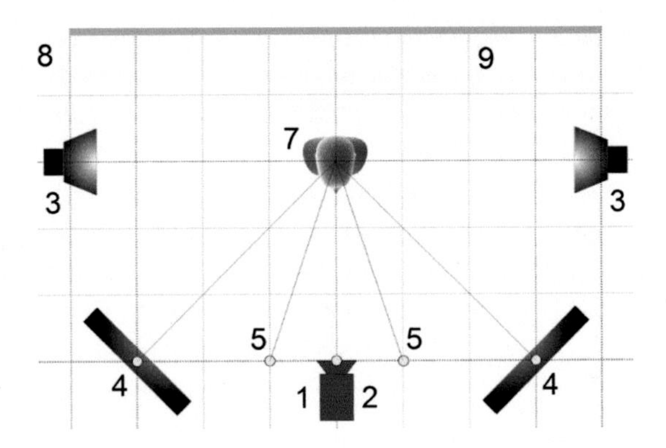

Fig. 3 Photographic stand layout. *Top view. One dotted square* is the equivalent of a floor tile measuring 30 cm × 30 cm

In our case, the DSLR is Canon EOS 600D, with a 14-bit CMOS sensor registering 5184 × 3456 pixels with 74.2° × 27.5° field of view (FOV), interfaced by USB 2.0 (see Fig. 2), and the IR camera is FLIR SC325, with 16-bit sensor of 320 × 240 pixels working at 60 Hz, having 25° × 18.8° FOV, interfaced by Ethernet.

The construction of the stand is referenced to the tile pattern on the room's floor, because of the need for periodic disassembly due to other activities in the room (see Fig. 3). With only the central point marked it is not a problem to quickly and accurately reconstruct the stand, as all elements are positioned directly on, or referenced to tileset fugue crossings. One such tile is a 30 cm × 30 cm square.

The central point is the one which marks the position of both cameras, but only the IR camera's stand. While the IR camera is mounted directly on top of its stand, the DSLR camera is placed above using a sideways extension from its stand. It is worth noting that DSLR's stand exact position is not relevant, because it acts as a support, with DSLR camera's placement being relative only to IR camera. When setting cameras, the most important factor is proper alignment of the lenses. We try to make them as close as possible, while preserving their coplanarity and maintaining the littlest rotation. Of course achieving a perfect setup is not possible with our equipment, hence the requirement for calibrating the system.

To conduct calibration, we must register four points of reference, captured by both cameras. We propose using a pinboard with four metal tacks and a paper sheet (shown as element no. 8 in Fig. 2). Due to low emissivity of metal-like materials their temperature in standard conditions (20 °C) will always be considerably lower when compared to the surrounding paper (emissivity: 0.7–0.9 [4]). As a result the tacks can be clearly distinguishable in visual and thermal images alike. For the purpose of calibration, the stand with pinboard mounted on it takes the place of subject's chair in a way that the pinboard's surface is perpendicular to the floor and its middle is at 115 cm height (intended position of subject's face). After the images for calibration have been captured we switch the stand with a chair. It is of utmost importance to avoid any change in both camera's position or calibration will have to be redone.

The whole station is lit by four lamps, with three being minimum as per [17]. The two large tube reflector lamps illuminate the subject widely from 45° angles and the CY-25WT lamps provide lighting from left and right sides (90°). Additionally, the DSLR camera has its own built-in flash light (approx. 13/43 ISO 100, in meters/feet [1]) which is used as direct frontal illumination. No extra light sources are required for the IR camera nor do the existing lamps have any effect on thermal measurement, apart from their heat generation. That however, considering their distance from subject and from camera, is low enough to be omitted in our case.

An important element of the photographic stand are five focus markers (shown in Fig. 2), which serve as stimulation for the subject when instructed to turn his/her head for achieving different shooting angles. The optimal subject's linesight height for our stand is regulated at 115 cm and such is the required height of the markers. The markers themselves are improvised from pieces of yellow sticky tape. Their positioning is coherent with other photographic stand elements. Two outer markers are wrapped around tube reflector lamp stands and two inner markers are wrapped around paper clips suspended on a piece of string, rigidly tethered between the same lamp stands. The central marker is essentially a piece of tape carefully placed in a small space between the camera lenses. With this setup, all markers are collinear and placed on a line perpendicular to the subject's line of sight. The following angles of head rotation come as a result of consecutively focusing on all markers: $-45°$, $-20°$, $0°$, $20°$, $45°$.

The software layer comprises of several interacting modules. It is built with a Matlab GUI application at its core. Using only a single window (see Fig. 5) it is possible to control both cameras, manipulate the database and fuse captured images after easily pre-processing them (Fig. 4).

The remote control part is possible thanks to SDK's provided by imaging hardware manufacturers. FLIR ThermoVision SDK [3] allows for a simple ActiveX formant to be embedded into our GUI, and through it commands can be sent using Matlab's invoke routine. The case is more complicated with Canon's EDSDK [2], because its functionality is oriented around C++ libraries. The most common way to integrate C++ and Matlab is to use a MEX function [11], so we implemented one which includes image capture and download procedures.

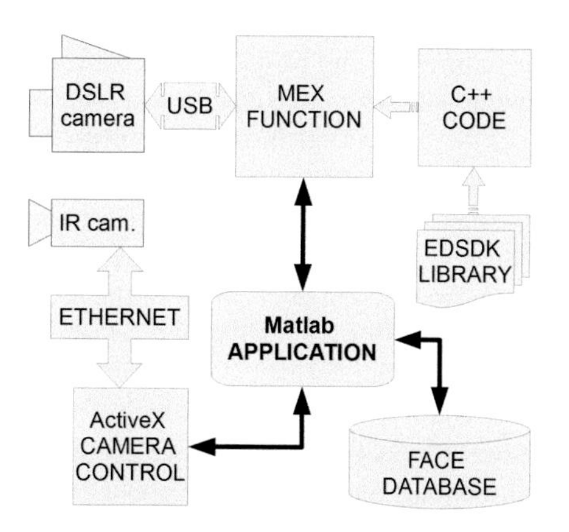

Fig. 4 System elements overview

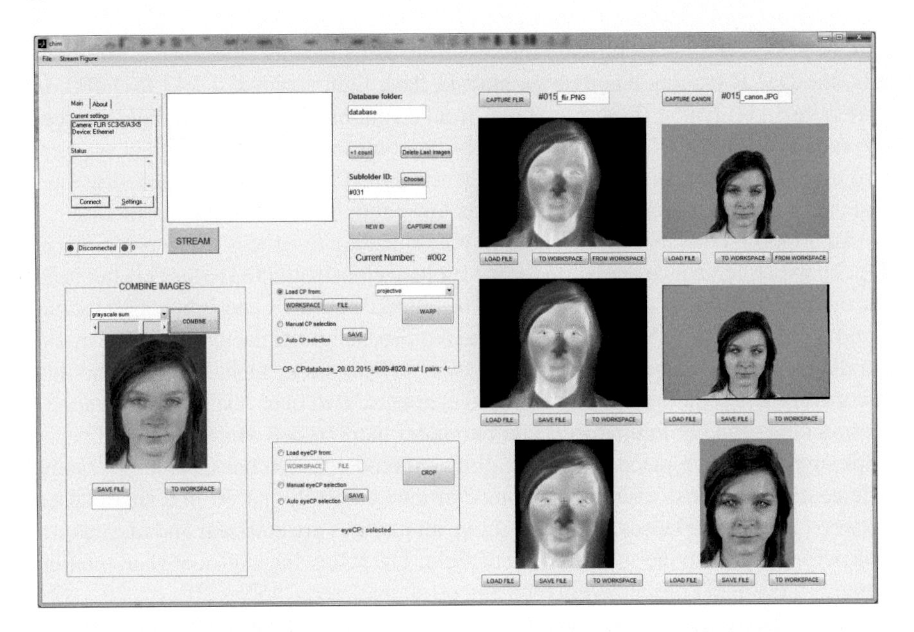

Fig. 5 Matlab application GUI

3 Data Acquisition

3.1 Operating Procedure

The first step when using our application to create a database would be obtaining four calibration points. It can be done by clicking the 'Manual CP selection' option (having first captured images of the calibration board) and consecutively selecting the tacks on both thermal and visual pictures. Four pairs of their coordinates will be stored for the purpose of inferring a chosen spatial transformation [12, 13]. We recommend using projective transform (default), because it fits best to our case of images shot from two different perspectives.

After calibration the system is ready to take a sequence of mugshots from multiple subjects. The procedure for remote shooting from both cameras was coded into 'CAPTURE CHIM' button. Each new subject has to be signified by the 'NEW ID' button which prepares a separate folder.

Having collected all images, the pre-processing step takes place. With a pair of chosen pictures and calibration points loaded we need to press the 'WARP' button, which will transform the visual image so that it matches the thermal's control points. Afterwards comes cropping which is relative to subject's eye position (indicated using 'Manual eyeCP selection'). Eye distance (d) determines the size of crop rectangle ($3d$ horizontally and $4d$ vertically), and the point in between eyes marks its centre. Those measures are adjusted automatically by the application when clicking 'CROP'. The warping and cropping actions can also be done straight away after each capture, but we recommend postponing them to keep the photo sessions fluent.

The images which have been processed and are coplanar, centered and similar in size can be fused ('COMBINE' button). The default ratio of visual/thermal pixel value is 0.7, but can be freely adjusted with a slider.

3.2 Data Structure

As the preferred format for face databases is just a simple folder structure, we did not use any database management engines, but rather made our application automatically handle catalogue and file creation/placement/naming. The main folder's default name is 'database' and within it, numerated subfolders are created, each representing a different subject (class). Every class folder will store the raw images captured from IR and DSLR cameras, naming them accordingly: *#imageNumber_deviceID.fileExtension* (e.g. *#002_flir*). Subfolders within the class folder contain absolute temperature values (abs_temp, saved in Matlab array), images after warping (warp), eye coordinates on both images (eyeCP, saved in Matlab structure), images after cropping (crop), and images fused together from cropped faces, thermal and visual (fusion). Exemplary aligned images taken in visible and thermal spectra are presented in Fig. 6.

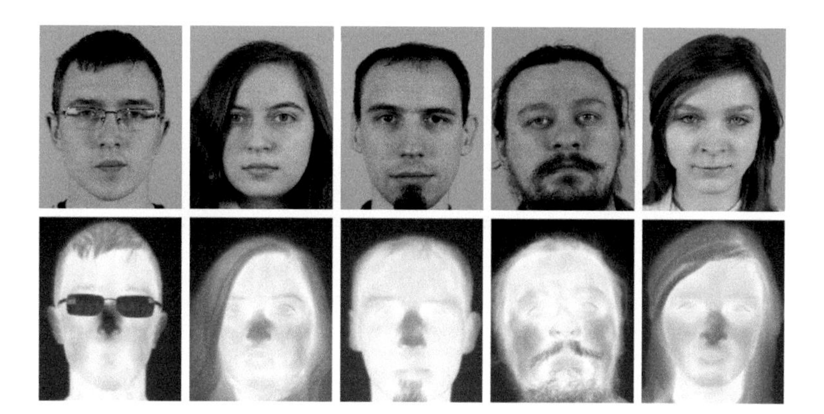

Fig. 6 Example images taken during experimental run

4 Summary

In the paper a dedicated, low-cost photographic stand has been presented, which allows simultaneous capture of images from IR thermal and DSLR cameras. It has a modular architecture consisting of interacting hardware and software layers and may be employed to create a database of facial portraits in visible and thermal spectra. As a control unit an application in Matlab has been developed, which remotely controls both devices, and automatically saves captured images within an organized folder structure. In order to provide a data fusion, several geometrical transforms have been tested in order to effectively combine visual and thermal data for use in face recognition algorithms.

References

1. Canon Inc. CANON EOS 600D Instruction Manual (2012)
2. Canon Inc. EDSDK 2.13 API Programming Reference, 2.13 edition (2013)
3. FLIR Instruments. Thermovision SDK User's manual, 2.6 sp2 edition (2010)
4. Bramson, M.A.: Infrared Radiation. A Handbook for Applications, 2nd edn. Plenum Press, New York (1971)
5. Bhattacharjee, D., Bhowmik, M.K., Nasipuri, M., Basu, D.K., Kundu, M.: Classification of Fused Face Images Using Multilayer Perceptron Neural Network. Published online at Cornell University Library. arXiv:1007.0633 (2010)
6. Chang, H., Koschan, A., Abidi, M., Kong, S.G., Won, C.-H.: Multispectral visible and infrared imaging for face recognition. In: 2008 IEEE Computer Society Conference on Computer Vision and Pattern Recognition Workshops, pp. 1–6 (2008)
7. Forczmański, P., Kukharev, G.: Comparative analysis of simple facial features extractors. J. Real-Time Image Proc. **1**(4), 239–255 (2007)
8. Forczmański, P., Kukharev, G., Kamenskaya, E.: Application of cascading two-dimensional canonical correlation analysis to image matching. Control Cybern. **40**(3), 833–848 (2011)
9. Forczmański, P., Furman, M.: Comparative analysis of benchmark datasets for face recognition algorithms verification. In: International Conference on Computer Vision and Graphics (ICCVG), LNCS, vol. 7594, pp. 354–362 (2012)
10. Forczmański, P., Kukharev, G., Shchegoleva, N.: Simple and robust facial portraits recognition under variable lighting conditions based on two-dimensional orthogonal transformations. In: 7th International Conference on Image Analysis and Processing (ICIAP), LNCS, vol. 8156, pp. 602–611 (2013)
11. Getreuer, P.: Writing MATLAB C/MEX Code (2010)
12. Goshtasby, A.: Piecewise linear mapping functions for image registration. Pattern Recogn. **19**, 459–466 (1986)
13. Goshtasby, A.: Image registration by local approximation methods. Image Vis. Comput. **6**, 255–261 (1988)
14. Grgic, M., Delac, K.: Face recognition homepage, database section. http://www.face-rec.org/databases/
15. Grgic, M., Delac, K., Grgic, S.: Surveillance cameras face database. Multimedia Tools Appl. J. **51**(3), 863–879 (2011)
16. Kukharev, G., Tujaka, A., Forczmański, P.: Face recognition using two-dimensional CCA and PLS. Int. J. Biometrics **3**(4), 300–321 (2011)
17. McCabe, R.M.: Best practice recommendation for the capture of mugshots. In: Best Practice and Face Pose Value Documents. 2.0 edition (1997)

18. Smiatacz, M.: Liveness measurements using optical flow for biometric person authentication. Metrol. Meas. Syst. **19**(2), 257–268 (2012)
19. OTCBVS benchmark dataset collection, iris thermal/visible face database. http://www.vcipl.okstate.edu/otcbvs/bench/
20. Wang, S., Liu, A., Lv, S., Lv, Y., Wu, G., Peng, P., Chen, F., Wang, X.: A natural visible and infrared facial expression database for expression recognition and emotion inference. IEEE Trans. Multimedia **12**(7), 682–691 (2010)

The PUT Surveillance Database

Michał Fularz, Marek Kraft, Adam Schmidt and Jakub Niechciał

Abstract In this paper we present a new, publicly available database of color, high resolution images useful in evaluation of various algorithms in the field of video surveillance. The additional data provided with the images facilitates the evaluation of tracking, recognition and reidentification across sequences of images.

Keywords Video surveillance · Image dataset · Object recognition · Object tracking · Reidentification

1 Introduction

The demand for vision-based surveillance has seen a significant increase over the last years, as it has a wide range of possible applications. These include public and private spaces monitoring for crime prevention, traffic monitoring and control, overseeing airports, train stations, stores, parking lots, offices and many more.

Image data is a rich source of useful information on the observed objects and places. Moreover, the costs of cameras and other necessary hardware decreases, while the capabilities and performance of the deployed systems constantly improve, so the existing video surveillance systems are scaled up, and the number of newly implemented systems grows. This however is not without its own challenges. An ever-increasing amount of data produced by these systems is overwhelming. The human operators are no longer capable of keeping up with the incoming data. Tiredness and loss of vigilance becomes a real problem, as maintaining focus is hard or even impossible with such a monotonous task [7, 10]. These shortcomings of human-operated large scale systems draw increased attention to automated surveillance.

The need for dedicated machine vision algorithms in the field of video surveillance has sparked a lot of interest in the research community and many solutions to common

M. Fularz (✉) · M. Kraft · A. Schmidt · J. Niechciał
Institute of Control and Information Engineering,
Poznań, University of Technology, Piotrowo 3A, 60-965 Poznań, Poland
e-mail: michal.fularz@put.poznan.pl

R.S. Choraś (ed.), *Image Processing and Communications Challenges 7*,
Advances in Intelligent Systems and Computing 389,
DOI 10.1007/978-3-319-23814-2_9

problems e.g. moving object detection, tracking, object recognition and identification have been proposed [1, 3, 17]. The multitude of available algorithms and solutions and the large scale of systems make the testing of performance of developed methods a non-trivial task. High quality evaluation data is therefore in high demand. Motivated by this need, this paper describes a publicly available database of high quality, high resolution images with rich annotation, prepared with the goal of autonomous video surveillance algorithm evaluation.

2 Review of Existing Solutions

Nowadays, multiple databases are made publicly available by the research community, but the vast majority of these databases focus on a single aspect of video surveillance. Examples of databases focusing on providing a rich set of objects of a single class for evaluation of detection and recognition algorithms are pedestrian databases [8, 9, 13, 15]. A large portion of pedestrian datasets are acquired with automotive applications in mind and focus on detection alone, without labeling the individual instances of an object across the whole sequence. Such information is usually available in datasets designed for the testing of object reidentification algorithms, but the detailed information on object's location and size in the form of a bounding box is often unavailable, and the registered images contain only one type of objects [4, 12].

An interesting approach to database generation is presented in [16]. The authors generated the test images using a video game engine and have shown, that using synthetic data for detector/classifier training enables the detection of real-world objects of the same class. The database is constrained to pedestrian images only, but the approach shows great promise, as the acquisition and annotation procedure can be automated to a great extent, and the number of observed object classes can be easily increased by virtual world object introduction and manipulation.

3 Acquisition Procedure

A high resolution Basler ace acA1600-60gc GigE Vision camera was used for image acquisition. The camera is capable of capturing color images of 1600×1200 pixel resolution with the speed of 60 frames per second. A zoom lens with the focal length range of 4.4–11 mm was used to adjust the field of view to match the observed scene. The images were saved using a lossless image compression format to ensure that no additional artifacts or noise are introduced and will not influence the results of tests performed using the database images [11]. The images were acquired with a frame rate of 10 frames per second. The observed scenes reflect setups of typical surveillance systems—urban areas with sidewalks, pedestrian crossings and intersections. Thus, the presence of both pedestrians as well as vehicles in the registered sequences was assured (Figs. 1 and 2).

Fig. 1 Example images from database with objects of interest marked

Fig. 2 Images of one person in different poses retrieved from the database

4 Additional Data

Each registered frame in the database is supplemented with an XML data file, containing information useful for evaluation of a wide range of automated surveillance algorithms.

To maintain consistent file format for future extensions of the database containing data registered collaboratively using multiple cameras, the viewpoint/camera ID is encoded into the file.

The objects of interest on all the registered frames were manually marked with a minimum bounding box. Although time consuming, manual annotation was preferred over automated methods for its accuracy and reliability. Information on the object's class (human, car, cyclist) and object's unique ID is also stored. The IDs remain consistent throughout all frames in a single sequence, enabling reliable evaluation of algorithms in the presence of occlusions or other events causing loss of sight of an object. Moreover, the description of objects enables the evaluation of object recognition [6] and object re-identification [2] algorithms.

The center of gravity and the dimensions of the bounding box are stored in the corresponding frame description XML file. This enables tracking performance evaluation by computing the average Euclidean distance between the center locations of the tracked targets and the manually labeled ground truths. The bounding box overlap is another frequently used performance metric for tracking evaluation. Given the tracked object bounding box r_t and the manually annotated ground truth bounding box r_{GT}, the score is computed as $S = \frac{r_t \cap r_{GT}}{r_t \cup r_{GT}}$ [14, 18]. Minimum bounding box can also be useful in evaluation of background subtraction algorithms [5].

To facilitate the testing of algorithms that require a fixed aspect ratio of detected regions, another set of bounding boxes is created by creating a minimum bounding box of proportions assigned to the particular class and centered at the same point. This is especially useful for testing the classification algorithms, as they typically use a fixed-size window for feature extraction [8, 19]. With this information, image samples serving as both positive and negative examples can be cropped and used in classifier training.

5 Potential Applications of the Database

The presented database contains images and additional data useful in research and evaluation of image and video processing algorithms for surveillance, in particular:

- evaluation of algorithms performing the task of background subtraction,
- evaluation of single target and multi-target tracking, also in the presence of occlusions,
- evaluation of object detection and classification/recognition algorithms based on various learning methods,
- evaluation of methods for object reidentification,
- work on the fusion of two or more of the above.

Other potential uses of the database include applications in object recognition outside the field of surveillance, e.g. object (pedestrian, vehicle) recognition in automotive driver assistance.

6 Conclusions and Future Work

An annotated image database aimed at facilitating the testing of algorithms in the field of video surveillance is presented in this paper. The database contains high quality, high resolution images with rich annotation. As the database is intended as a work in progress, it will be gradually extended with further data. Particular emphasis will be placed on registering sequences of heterogeneous camera setups observing a single scene, as such a system is the most general reflection of real-life surveillance networks.

7 Obtaining the Database

The database is freely available for research purposes. To access the database, please contact one of the authors or visit the website at:
www.vision.put.poznan.pl/surv_database

Acknowledgments This research was financed by the Polish National Science Centre grant funded according to the decision DEC-2011/03/N/ST6/03022, which is gratefully acknowledged.

References

1. Aghajan, H., Cavallaro, A.: Multi-Camera Networks: Principles and Applications. Academic press, London (2009)
2. Bedagkar-Gala, A., Shah, S.K.: A survey of approaches and trends in person re-identification. Image Vis. Comput. **32**(4), 270–286 (2014)
3. Bhanu, B., Ravishankar, C., Roy-Chowdhury, A., Aghajan, H., Terzopoulos, D.: Distributed Video Sensor Networks. Springer (2011)
4. Bialkowski, A., Denman, S., Lucey, P., Sridharan, S., Fookes, C.B.: A database for person re-identification in multi-camera surveillance networks. In: Proceedings of the 2012 International Conference on Digital Image Computing Techniques and Applications (DICTA 12), pp. 1–8. IEEE (2012)
5. Brown, L.M., Senior, A.W., Tian, Y.l., Connell, J., Hampapur, A., Shu, C.F., Merkl, H., Lu, M.: Performance evaluation of surveillance systems under varying conditions. In: IEEE International Workshop on Performance Evaluation of Tracking and Surveillance, Colorado, USA. Citeseer (2005)
6. Clapés, A., Reyes, M., Escalera, S.: Multi-modal user identification and object recognition surveillance system. Pattern Recognit. Lett. **34**(7), 799–808 (2013)
7. Dadashi, N., Stedmon, A., Pridmore, T.: Semi-automated CCTV surveillance: the effects of system confidence, system accuracy and task complexity on operator vigilance, reliance and workload. Appl. Ergon. **44**(5), 730–738 (2013)
8. Dalal, N., Triggs, B.: Histograms of oriented gradients for human detection. In: IEEE Computer Society Conference on Computer Vision and Pattern Recognition. CVPR 2005, vol. 1, pp. 886–893 (2005)
9. Dollar, P., Wojek, C., Schiele, B., Perona, P.: Pedestrian detection: a benchmark. In: IEEE Conference on Computer Vision and Pattern Recognition. CVPR 2009, pp. 304–311 (2009)
10. Donald, F.M., Donald, C.H.M.: Task disengagement and implications for vigilance performance in CCTV surveillance. Cogn. Technol. Work **17**(1) (2015)
11. Fauquet, J.: Algorithm for removing dct noise patterns in block encoded images. Electron. Lett. **34**, 2322–2323(1) (1998)
12. Gray, D., Tao, H.: Viewpoint invariant pedestrian recognition with an ensemble of localized features. In: Forsyth, D., Torr, P., Zisserman, A. (eds.) Computer Vision—ECCV 2008. Lecture Notes in Computer Science, vol. 5302, pp. 262–275. Springer, Berlin (2008)
13. Li, J., Huang, L., Liu, C.: Robust people counting in video surveillance: dataset and system. In: 2011 8th IEEE International Conference on Advanced Video and Signal-Based Surveillance (AVSS), pp. 54–59 (2011)
14. Milan, A., Schindler, K., Roth, S.: Challenges of ground truth evaluation of multi-target tracking. In: 2013 IEEE Conference on Computer Vision and Pattern Recognition Workshops (CVPRW), pp. 735–742 (2013)

15. Overett, G., Petersson, L., Brewer, N., Andersson, L., Pettersson, N.: A new pedestrian dataset for supervised learning. In: Intelligent Vehicles Symposium, 2008 IEEE, pp. 373–378 (2008)
16. Vazquez, D., Lopez, A., Marin, J., Ponsa, D., Geroimo, D.: Virtual and real world adaptation for pedestrian detection. IEEE Trans. Pattern Anal. Mach. Intell. **36**(4), 797–809 (2014)
17. Wang, X.: Intelligent multi-camera video surveillance: a review. Pattern Recognit. Lett. **34**(1), 3–19 (2013)
18. Wu, Y., Lim, J., Yang, M.H.: Online object tracking: a benchmark. In: 2013 IEEE Conference on Computer Vision and Pattern Recognition (CVPR), pp. 2411–2418 (2013)
19. Zhu, C., Bichot, C.E., Chen, L.: Multi-scale color local binary patterns for visual object classes recognition. In: 2010 20th International Conference on Pattern Recognition (ICPR), pp. 3065–3068 (2010)

Applying Image Features and AdaBoost Classification for Vehicle Detection in the 'SM4Public' System

Dariusz Frejlichowski, Katarzyna Gościewska, Paweł Forczmański, Adam Nowosielski and Radosław Hofman

Abstract The main goal of works described in the paper is to test and select algorithms to be implemented in the 'SM4Public' security system for public spaces. The paper describes the use of cascading approaches in the scenario concerning the detection of vehicles in static images. Three feature extractors were used along with benchmark datasets in order to prepare eight various cascades of classifiers. The algorithms selected for feature extraction are Histogram of Oriented Gradients, Local Binary Patterns and Haar-like features. AdaBoost was used as a classifier. The paper briefly introduces the 'SM4Public' system characteristics, characterizes the employed algorithms and presents sample experimental results.

Keywords Video surveillance · Vehicle detection · Cascades of classifiers

1 Introduction

In the paper, the problem of selecting appropriate algorithms for the 'SM4Public' system is considered. The system is being developed within the framework of EU cofounded project which is now at the research stage . The main goal of the project is

D. Frejlichowski (✉) · K. Gościewska · P. Forczmański · A. Nowosielski
Faculty of Computer Science, West Pomeranian University of Technology,
Szczecin, Żołnierska 52, 71-210 Szczecin, Poland
e-mail: dfrejlichowski@wi.zut.edu.pl

K. Gościewska
e-mail: katarzyna.gosciewska@smartmonitor.pl

P. Forczmański
e-mail: pforczmanski@wi.zut.edu.pl

A. Nowosielski
e-mail: anowosielski@wi.zut.edu.pl

K. Gościewska · R. Hofman
Smart Monitor sp. z o.o., Niemierzyńska 17a, 71-441 Szczecin, Poland
e-mail: radekh@smartmonitor.pl

R.S. Choraś (ed.), *Image Processing and Communications Challenges 7*,
Advances in Intelligent Systems and Computing 389,
DOI 10.1007/978-3-319-23814-2_10

the construction and implementation of innovative video content analysis-based system prototype that will ensure the safety of various public spaces. The 'SM4Public' system will use real-time solutions and typical computer components. It will be a highly customizable solution and offer features that enable its adaptation to the actual situation. The idea of the system is based on the previously developed and successfully finalized creation of the 'SmartMonitor' system which is an intelligent security system based on image analysis, designed for individual customers and home use [5, 6].

Because 'SM4Public' system is being created for public space video surveillance, it should effectively detect events threatening public safety, especially in places characterized by simultaneous movement of large number of people and vehicles. Therefore, specific system scenarios will be implemented, such as scenarios associated with vehicle traffic, such as accident detection, infrastructure protection e.g. for devastation or theft detection, breaking the law or detecting treats to life or health, such as a fight or a fall.

Various solutions for vehicle detection in static images have been reported in the literature so far. For instance in [13] a general approach to vehicle detection on highways, road intersections and parking lots has been presented. This approach uses only static images and employs a cascaded multichannel classifier based on corner features, edge features and wavelet coefficients. Another solution to vehicle detection in static images was presented in [1]. The proposed approach projects all input pixels colours to a new feature space using colour transformation. The Bayesian classifier is applied to distinguish between vehicle and non-vehicle pixels, and the Harris corner detector is used to detect corners. The detected extreme corners enable to identify and mark a vehicle with the rectangular shape.

In this paper, vehicle detection in static scenes is performed using AdaBoost-based classification and three different feature extractors, namely Haar-like features, Histogram of Oriented Gradients (HOG) and Local Binary Patterns (LBP). The employed algorithms are described in the second section. In the third section experimental conditions and results are given. The experiments included the study on eight cascades of classifiers which were built using features obtained by employing above-mentioned feature extractors and various learning datasets known from the literature. The fourth section concludes the paper.

2 The Description of the Applied Approaches

Visual Content Analysis algorithms enable automatic detection of vehicles without human intervention. The vehicle detection from static images scenario do not require complex calculations and high computational power, and is characterized by a high detection rate and a small error probability. Therefore, it is possible to perform static image analysis with a time interval of a few or more seconds. In the paper the attention is focused on cascades of classifiers that were selected based on the computational complexity, simplicity of implementation and the declared effectiveness. The selected feature extractors are, namely, Haar-like features, Histogram of Oriented Gradients and Local Binary Patterns, and a cascading approach for object classification is based on AdaBoost.

AdaBoost ('Adaptive Boosting') was proposed in [7] and is a method of training boosted classifier using positive and negative samples. During the training process only those weak classifiers that would be able to improve the prediction are selected. For each weak classifier AdaBoost determines an acceptance threshold, but a single weak classifier is not able to classify objects with low error rate. The initial weights are equal to $\frac{1}{2}$. In each iteration the weights are normalized and the best weak classifier is selected based on the weighted error value. In the next step weight values are updated and it is determined if an example was classified correctly or not. After all iterations a set of weak classifiers characterized by a specified error rate is selected and a resulted strong learned classifier is obtained. The classification is performed iteratively and its effectiveness depends on the number of learning examples. During classification, an image is analysed using a sliding window approach. Features are calculated in all possible window locations. The window is slid with a varying step, which depends on the required accuracy and speed.

HOG was proposed in [3] as an approach for human detection. It is based on gradient information and therefore a greater detector window is required in order to provide a sufficient amount of context information. The derivation of HOG representation starts from gamma and colour normalization. The second step includes the calculation of oriented gradients using directional filters (the detection of vertical and horizontal edges). Then, the length and orientation of gradients are calculated. In the third step, an image is divided into cells and frequencies of oriented gradients of each cell are obtained. The fourth step includes cell grouping into larger overlapping spatial blocks which is essential here. Then cells are separately normalized and each cell is normalized several times, each time with respect to a different block. Such procedure constitutes for local variations in illumination and contrast. Furthermore, it significantly improves the performance [3]. The final HOG representation is obtained in the fifth step by concatenating oriented histograms of all blocks. The representation is invariant to translation within an image plane. An exemplary application of joint usage of HOG features and AdaBoost for human detection was presented in [16].

LBP is a texture classification feature [10] and a particular case of the Texture Spectrum model proposed in [9]. LBP labels pixels by thresholding the neighbourhood of each pixel and producing a binary number. Considering the eight- neighbourhood case, the derivation of LBP representation consists of several steps. In the first step, the analysed window is divided into equal cells, and each pixel in a cell is compared with all its neighbours in a clockwise or counter-clockwise manner. It is important to keep the direction and starting point of the comparison process consistent for all cells. In the second step, the pixels value is checked and a neighbourhood values are obtained as follows: if the central pixels value is equal or greater than the neighbouring pixels values, then number '1' is written; otherwise, '0' is put. This process produces a binary number, usually converted to decimal. The third step involves the computation of histogram over the cell, based on the occurrence frequency of each number. In the final step, the concatenation of histograms of all cells is performed and a feature vector for the window is obtained. The original LBP is robust to monotonic intensity changes, but is not invariant to object rotation within an image plane. In [12] the extended LBP-based features are used in the framework of AdaBoost learning for the detection of objects, e.g. frontal faces and side car views.

Haar-like features were proposed in [14] as a solution for face detection. A simple rectangular Haar-like feature is defined as the difference of the sum of pixels of areas inside the rectangle, which can be characterized by any position and scale in an input image. For instance, using two-rectangle feature, the borders can be indicated based on the pixel intensities under the white and black rectangles of the Haar-like feature. Such features could be calculated fast using the integral image approach. Individual Haar-like features are weak classifiers, therefore they are combined in a cascade using AdaBoost algorithm [14]. During training, the most important image features are selected in a way enabling to reject negative regions at early stage of recognition, what reduces the number of calculations. During classification subsequent features of an unknown object are calculated only if the answer of the previous feature is equal to the learned value. Otherwise, the examined region is rejected. Haar-like features were applied together with AdaBoost for vehicle detection based on static images e.g. in [11, 15]. They were also successfully applied to human silhouettes detection in [4].

3 Experimental Conditions and Results

Extraction of objects from static scenes is based on the determination of the image part containing an object under detection. The process of detecting objects in static images assumes the absence of information about the objects size and location. Several issues may arise during this process, namely an appropriate selection of characteristic object features, the mechanism for feature matching and the method for scanning the source image. Object detection using sliding window approach is relatively frequently utilized in the literature. An input image is appropriately scanned and the selected image parts are matched with the templates from training set. If information about an object are unknown, then the search process has to be performed in all possible locations and for all probable window or image scales. In each location of the search window the features of a particular image part are calculated. The obtained features are used during the classification process in order to indicate the most probable object location. This strategy is called a pyramid.

During the experiments, the detectors based on Haar-like, HOG and LBP features were used. For classification a standard boosted cascade algorithm was applied, namely AdaBoost. The main goal of the experiments was to verify the effectiveness of the selected detectors in the task of the vehicle detection from static scenes. Eight different cascades of classifiers were built using various learning databases presented in the literature [2, 8]. The cascades varied also under following parameters: the size of search windows, the scaling factor in the pyramid and candidates rejection based

Table 1 Results of performance tests—detection accuracy and number of false detection for particular cascades are provided

No.	Cascade	Feature type	Window size	View	Scaling factor	Minimum no. of neigh-bours	Detection accuracy	No. of false detections
1	UIUC 1 [2]	LBP	20 × 20	Front	1.05	15	0.43	Low
2	UIUC 2 [2]	HOG	24 × 48	Side	1.01	1	0.23	High
3	UIUC 3 [2]	LBP	25 × 50	Side	1.03	9	0.31	High
4	UIUC 4 [2]	LBP	24 × 48	Side	1.03	13	0.58	High
5	UIUC 5 [2]	LBP	24 × 48	Side	1.03	13	0.27	High
6	CALTECH SURF [8]	Haar-like	20 × 20	Front	1.01	2	0.92	Medium
7	OpenCV 1	Haar-like	40 × 40	Front	1.01	1	0.62	Low
8	OpenCV 2	Haar-like	40 × 40	Front	1.01	7	0.16	Low

on the number of neighbours. The characteristics of cascades are provided in Table 1. The experiments were performed using video frames taken from the project video sequence database. The results of vehicle detection for sample images containing various scenes including frontal and side car views are illustrated in Figs. 1, 2, 3 and 4. White rectangles correspond to the detected regions. The complete results including detection accuracy and number of false detections are also provided in Table 1 .

As can be concluded from the presented figures, there is a small number of false detections and missed objects, however the vehicle detection in static scenes gives acceptable results. The in-depth analysis of the results presented in Fig. 1 and a comparison with test data in Figs. 1, 2, 3 and 4 unveils, that the efficient vehicle

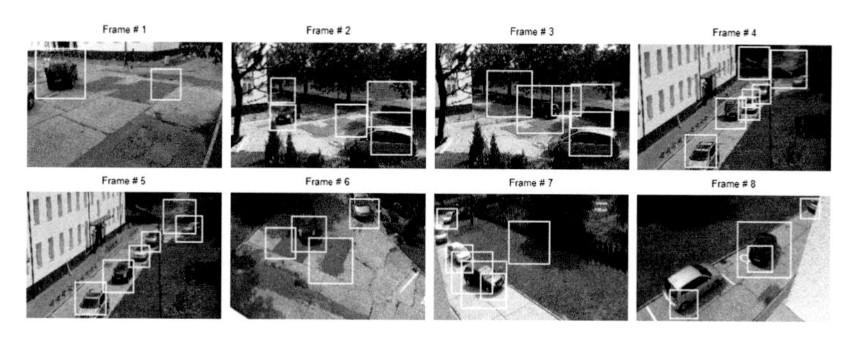

Fig. 1 Exemplary results of vehicle detection using cascade no. 6: Haar-like features and CALTECH SURF dataset

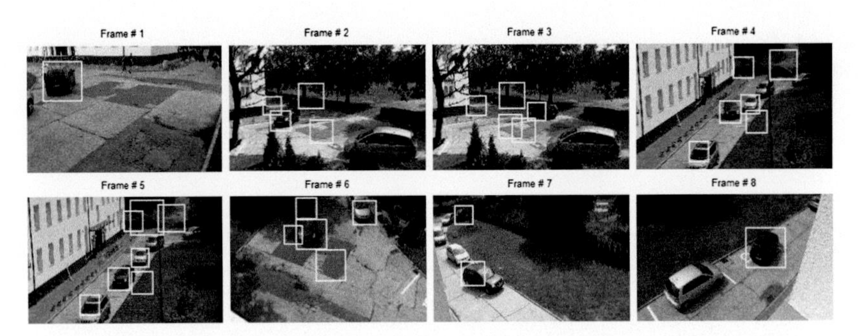

Fig. 2 Exemplary results of the vehicle detection using cascade no. 7: Haar-like features and OpenCV dataset

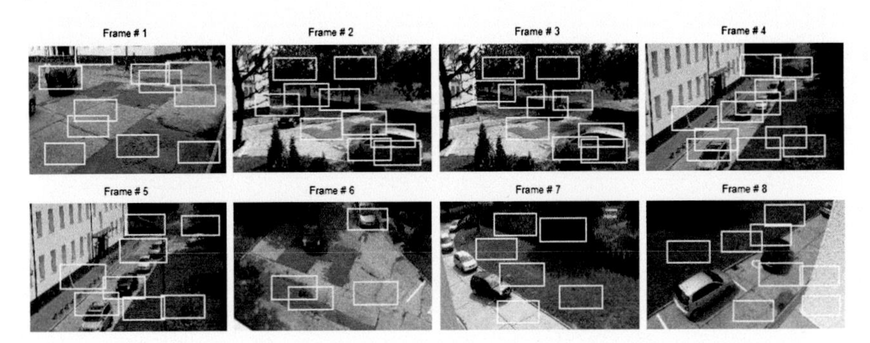

Fig. 3 Exemplary results of the vehicle detection using cascade no. 4: LBP features and UIUC dataset

Fig. 4 Exemplary results of the vehicle detection using cascade no. 2: HOG features and UIUC dataset

detection is not a trivial task. Vehicles (e.g. cars) as three-dimensional objects are hard to detect using presented appearance-based methods. It comes from the fact that the shape of such object varies with the changes of camera orientation. Extending the learning database with all possible projections of three-dimensional vehicles is complex, hence a construction of one common cascade may be impossible.

The proposed solution for the vehicle detection scenario is based on the coarse-to-fine strategy and involves a cascade of classifiers employing different features. The application of the classifier should be focused on one vehicle view only and as many potential vehicles as possible should be firstly detected, regardless of the occurrence of false detections. This task could be performed using Haar-like features at the initial stage, since this is the most satisfactory approach. Then LBP features to resulting objects should be applied. At the final stage HOG features should be included in order to adjust the results.

4 Summary and Conclusions

The paper covered the topic concerning the problem of vehicle detection from static scenes in the 'SM4Public' system. The main goal was to verify the possibility of implementing various cascading approaches based on three different feature extractors and AdaBoost classification. During the experiments eight cascades of classifiers were prepared using various benchmark learning datasets and tested using images extracted from the video sequences recorded for the project test database. Three feature extractors were experimentally tested, namely Histogram of Oriented Gradients, Local Binary Patterns and Haar-like features. The experimental results have shown that the vehicle detection is possible, however the classifier should be applied depending on the specific car view—the position and orientation of the camera strongly influence the detection accuracy. Moreover, the initial stage should detect the greatest possible amount of probable vehicle objects and more than one classifier can be used to improve the results. In addition it is advised to prepare an image sample set depicting vehicles from other orientations and simultaneously representing expected parameters of the surveillance camera.

Acknowledgments The project "*Security system for public spaces—'SM4Public' prototype construction and implementation*" (original title: *Budowa i wdrożenie prototypu systemu bezpieczeństwa przestrzeni publicznej 'SM4Public'*) is a project co-founded by European Union (EU) (project number PL: POIG.01.04.00-32-244/13, value: 12.936.684,77 PLN, EU contribution: 6.528.823,81 PLN, realization period: 01.06.2014–31.10.2015). *European Funds-for the development of innovative economy* (*Fundusze Europejskie-dla rozwoju innowacyjnej gospodarki*).

References

1. Aarthi, R., Padmavathi, S., Amudha, J.: Vehicle detection in static images using color and corner map. In: 2010 International Conference on Recent Trends in Information, Telecommunication and Computing, pp. 244–246 (2010)
2. Agarwal, S., Awan, A., Roth, D.: Learning to detect objects in images via a sparse, part-based representation. IEEE Trans. Pattern Anal. **26**(11), 1475–1490 (2004)
3. Dalal, N., Triggs, B.: Histograms of oriented gradients for human detection. IEEE Comput. Soc. Conf. Comput. Vis. Pattern Recognit. **1**, 886–893 (2005)

4. Forczmański, P., Seweryn, M.: Surveillance video stream analysis using adaptive background model and object recognition. In: Bolc, L. et al. (eds.) ICCVG 2010. LNCS, vol. 6374, pp. 114–121. Springer, Berlin (2010)
5. Frejlichowski, D., Gościewska, K., Forczmański, P., Hofman, R.: Application of foreground object patterns analysis for event detection in an innovative video surveillance system. Pattern Anal. Appl. 1–12 (2014)
6. Frejlichowski, D., Gościewska, K., Forczmański, P., Hofman, R.: 'SmartMonitor'—an intelligent security system for the protection of individuals and small properties with the possibility of home automation. Sensors **14**, 9922–9948 (2014)
7. Freund, Y., Schapire, R.E.: A decision-theoretic generalization of on-line learning and an application to boosting. In: Proceedings of the 2nd European Conference on Computational Learning Theory, pp. 23–37 (1995)
8. Google Project: OpenCV implementation of lane and vehicle tracking. https://code.google.com/p/opencv-lane-vehicle-track/source/browse/trunk/bin/haar/cars3.xml?r=2
9. He, D.C., Wang, L.: Texture unit, texture spectrum, and texture analysis. IEEE Trans. Geosci. Remote **28**, 509–512 (1990)
10. Ojala, T., Pietikinen, M., Harwood, D.: Performance evaluation of texture measures with classification based on Kullback discrimination of distributions. In: Proceedings of the 12th International Conference on Pattern Recognition, vol. 1, pp. 582–585 (1994)
11. Tang, Y., Zhang, C., Gu, R., Li, P., Yang, B.: Vehicle detection and recognition for intelligent traffic surveillance system. In: Multimedia Tools and Applications (online). Springer (2015)
12. Trefný, J., Matas, J.: Extended set of local binary patterns for rapid object detection. In: Špaček, L., Franc, V. (eds.) Computer Vision Winter Workshop 2010, Nové Hrady, Czech Republic, Czech Society for Cybernetics and Informatics (2010)
13. Tsai, L.-W., Hsieh, J.-W., Fan, K.-C.: Vehicle detection using normalized color and edge map. IEEE Trans. Image Process. **16**(3), 850–864 (2007)
14. Viola, P., Jones, M.J.: Robust real-time face detection. Int. J. Comput. Vis. **57**(2), 137–154 (2004)
15. Wu, C., Duan, L., Miao, J., Fang, F., Wang, X.: Detection of front-view vehicle with occlusions using AdaBoost. In: International Conference on Information Engineering and Computer Science, pp. 1–4, Wuhan (2009)
16. Zhu, Q., Avidan, S., Yeh, M.-C., Cheng, K.-T.: Fast human detection using a cascade of histograms of oriented gradients. IEEE Comput. Soc. Conf. Comput. Vis. Pattern Recognit. **2**, 1491–1498 (2006)

Automatic Analysis of Vehicle Trajectory Applied to Visual Surveillance

Adam Nowosielski, Dariusz Frejlichowski, Paweł Forczmański, Katarzyna Gościewska and Radosław Hofman

Abstract In this paper we discuss a problem of automatic analysis of vehicles trajectories in the context of illegal movements. It is crucial to detect restricted or security critical behaviour on roads, especially for safety protection and fluent traffic. Here, we propose an vision-based algorithm for vehicle detection and tracking, which is later employed to recognize patterns in resultant trajectories. Experiments were performed on real video streams. They gave encouraging results.

Keywords Vehicle detection · Vehicle tracking · Vehicle behaviour analysis · Trajectory analysis · Intelligent transportation systems (ITS)

1 Introduction

Road congestion as the result of poor planning of transport routes, existence of bottlenecks, and lack of existing infrastructure adaptation to the current traffic load is one of the biggest problems in modern cities [14]. The effective management of traffic emerges as one of the most important tasks. It requires proper approach and

---

A. Nowosielski (✉) · D. Frejlichowski · P. Forczmański · K. Gościewska
Faculty of Computer Science, West Pomeranian University of Technology,
Zołnierska 52, 71-210 Szczecin, Poland
e-mail: anowosielski@wi.zut.edu.pl

D. Frejlichowski
e-mail: dfrejlichowski@wi.zut.edu.pl

P. Forczmański
e-mail: pforczmanski@wi.zut.edu.pl

K. Gościewska
e-mail: katarzyna.gosciewska@smartmonitor.pl

K. Gościewska · R. Hofman
Smart Monitor sp. z o.o., Niemierzyńska 17a, 71-441 Szczecin, Poland
e-mail: radekh@smartmonitor.pl

R.S. Choraś (ed.), *Image Processing and Communications Challenges 7*,
Advances in Intelligent Systems and Computing 389,
DOI 10.1007/978-3-319-23814-2_11

good analysis. Of great importance here is traffic surveillance using monitoring cameras and automatic recognition of abnormal events with computer vision and image processing methods. These solutions allow an immediate (real-time) response to occurring events. In brief, the intelligent traffic video surveillance systems perform localization, tracking and recognition of vehicles in video sequences captured by road cameras [10]. By further analysis of the vehicle activities some more sophisticated schemes for traffic management are provided, for example for public transport planning purposes. All these contribute to the foundations concepts of Intelligent Transportation Systems (ITS) which goal is [19]: "protect the safety, increase efficiency, improve the environment and conservate energy".

There are intrusive and non-intrusive technologies used in Intelligent Transportation Systems (ITS) [13]. Intrusive techniques (such as inductive loop or piezoelectric cable) require the installation of the sensor directly onto or into the road surface [13]. With non-intrusive techniques there is no need for such interference since the components are mounted overhead or on the side of the roadway [13]. Among many sensing techniques (infrared-, ultrasonic-, acoustic-, GPS-, RFID- based) computer vision techniques seem to have the greatest potential and applicability. Image-based techniques can provide complete traffic flow information for [19]: traffic management system, public transportation systems, information service systems, surveillance systems, security systems and logistics management system.

In the paper the problem of trajectory of moving vehicle analysis is considered in the context of illegal vehicle movement. First, some overview of vision-based ITS solution is presented in Sect. 2. Then, in Sect. 3, the problem of possible use of trajectory analysis is outlined. The algorithm for trajectory of moving vehicles analysis is proposed in Sect. 4. Section 5 describes the evaluation process and application of the algorithm in the 'SM4Public' project. The article ends with a summary.

2 Vision-Based ITS Solutions

The best example of the use of computer vision methods in ITS is the ALPR (Automatic License Plate Recognition) also called ANPR (Automatic Number Plate Recognition). It uses optical character recognition to read vehicle registration plates. Performed in distinctive localizations can provide the users average time of travel information. It also forms the basis for the functioning of so called Limited Traffic Zones (LTZ) [12]. Other ALPR applications are usually connected with the law enforcement and include: border crossings, protection, speed enforcement, HIM (Height-in-Motion), WIM (Weigh-in-Motion).

Image processing techniques help monitor the movement and flow of vehicles around the road network providing: vehicle counting, congestion calculation, traffic jam detection, lane occupancy readings [20], road accident detection, traffic light control [15], comprehensive statistics. Some supplementary information is provided for safer and more efficient driving. Displayed on the variable message boards placed by the road can provide the driver with already mentioned time of travel notifications or parking occupancy [4]. Cameras, however, are not only used as part of the road infrastructure. Many vehicles are manufactured with on board driver assistant systems that base heavily on image processing and allow for example [11]: collision avoidance (thanks to detection and tracking of pedestrians or surrounding vehicles), adaptive cruise control, driver fatigue detection or recognition of road signalization.

In all the above mentioned examples the functioning of the system is largely based on automatic computer vision techniques. There is an abundance of work devoted to ITS problems solved with vision-based techniques and the reader can find some further references in literature surveys: [11, 19].

3 Trajectory Analysis and Its Potential Use

The use of moving vehicle trajectory analysis is extensive in Intelligent Transportation Systems. Motion trajectories obtained by vehicle tracking may be used off-line in generation of comprehensive statistics. In real-time systems the trajectory analysis is used for anomaly detection or even anomaly prediction. The task of analysis and identification the vehicles motion pattern is also known in the literature as Vehicle Behaviour Analysis [19]. The algorithms proposed so far in the literature can be divided into the following categories [19]: pattern recognition based, statistic based, traffic flow model based and artificial intelligence based.

The actual process of behaviour inference based on pattern matching and state estimating is, in fact, preceded by two steps [10]: vehicle detection and vehicle description using static and dynamic parameters.

The most commonly used segmentation methods for moving objects detection include [10]: optical flow method, frame difference method, background subtraction method. After the detection of candidates a classification step is required to differentiate between vehicles, pedestrians, bicycles or other moving objects. The classification usually bases on shape, motion and cooccurrence matrix [10]. It must be also noticed, that various solutions for vehicle detection have been proposed for static images (e.g. [1, 17] or Viola and Jones detector [18]). The Viola and Jones detector is interesting since its real-time implementation is possible. It is a sliding window algorithm which considers pixel intensity arranged in the form of integral image for fast summarizations. Adaboost procedure is used here to generate the strong classifier on the base of arranged best weak classifiers (which base on simple comparisons of summed regions of images). Features are calculated for all possible window scales and in all possible image locations and examined for the presence of the learned pattern (e.g. vehicle).

After the successive vehicle detection the succeeding tracking algorithm is commonly [10]: region-based, snake-based, feature-based, model-based or finally multi-thread. With region-based tracking (used in our solution introduced in Sect. 4) static features of the moving vehicle are extracted and matched between subsequent frames. As the result of the above steps the moving vehicle trajectory is obtained.

Vehicle trajectory analysis has a wide range of applications. Vehicle behaviour which may be directly connected with trajectory analysis includes:

- illegal parking,
- illegal left and right turns,
- illegal lane change and violation of traffic line,
- overtaking in prohibited places,
- illegal retrograde,
- traffic congestion.

There have been varied approaches to handle trajectory of moving objects analysis based on video and some solutions have already been proposed. In [10] fuzzy self-organizing neural network algorithm is applied to learn activity patterns from the sample trajectories. Then, vehicle activities are predicted based on observed partial trajectory and utilized for accidents predictions. The idea presented in [10] is to predict accidents accurately in advance for a real-time system and generate appropriate warning. In [16] trajectory analysis is conducted in order to detect the following traffic events: illegal changing lane, stopping, retrogradation, sudden speeding up or slowing down. The trajectories are obtained here by tracking the feature points through the image sequences with a specially designed template [16]. Trajectories seem to be modeled by straight lines here. Comparisons are based on the angle and variance between the benchmark lane line and the trajectory.

4 The Proposed Algorithm

As stated in Sect. 3 the process of vehicle behaviour analysis consists of three steps: vehicle detection, parametrization of features and behaviour recognition. We have already evaluated some contemporary approaches for the task of vehicle detection in context of static images. Various cascading approaches based on three different feature extractors and the AdaBoost classification have been examined. Altogether, eight cascades of classifiers were prepared using various benchmark learning databases presented in the literature and three feature extractors have been used: Haar-like features, Histogram of Oriented Gradients (HOG) and Local Binary Patterns (LBP). The choice was encouraged by literary reported computational complexity, simplicity of implementation and the declared effectiveness.

The main goal of our preliminary experiments was to verify the effectiveness of the selected detectors in the task of the vehicle detection from static scenes. The proposed solution based on the coarse-to-fine strategy and involved a cascade of classifiers employing different features. It occurred that Haar-like features are best at the initial stage, since they offer most satisfactory results. Then LBP features to resulting objects should be applied. At the final stage HOG features should be included in order to adjust the results. In video sequences, however, we deal with dynamic images and background modeling approaches and change detection can considerably enhance the process of vehicle detection reducing the false detection rate parameter.

For the task of vehicle behaviour analysis we combined GMM background modeling with AdaBoost classifier based on Haar-like features. The background model employs a pixel-based approach. Every pixel is modeled by a set of five mixtures of Gaussians in R, G and B channels. Similar approach has been successfully employed for human motion tracking [5, 7].

Background subtraction operation results in an image mask of possible foreground pixels which are grouped using connected components as Region of Interests (ROIs). This significantly reduces the search space and only on those locations (enhanced with dilation procedure) the classifier of vehicle detection is employed. Each ROI is handled independently. If classified as new vehicle object the tracking procedure is initialized and an initial value is assigned. Detected ROI that contain moving vehicle is further tracked using the Camshift algorithm. The Camshift algorithm was first proposed in [2] as a solution for face tracking and is actually widely used for tracking objects of miscellaneous type. It is a mean-shift algorithm and models the object using the colour distribution which is backprojected on new frames. Camshift originally bases on features calculated in HSV colour space. In our experiments we discovered that relatively small amount of cars in benchmark datasets are characterized by the clear and unambiguous colour (red or yellow for example). Most vehicles are characterized by shade of gray, black or white which are poorly described in HSV colour space. What works well for objects of different type (e.g. faces [2]) does not necessarily give satisfactory results with vehicles. For low and high intensity values the hue calculation is subjected to significant error. That is the reason of color space change to HSL model.

With new frames processed the new coordinates are added to the trajectory of moving vehicle. The recognition of trajectory requires a reference model. Many literature approaches are limited to straight lines or left and right turns (e.g. [16, 19]). We opted for the opportunity to draw the trajectory of any shape by the user. This approach will allow a wide practical applications since the drawn forbidden trajectory can model huge range of restricted or security critical behaviour. The forbidden trajectory and obtained real trajectory consists of different number of points coordinates and for the comparison purposes we adopted the Modified Hausdorff Distance (MHD), first introduced in [3].

The example depicted in Fig. 1 demonstrates diversification of distance between the forbidden trajectory and the one created on the fly from vehicle movement. In presented situation a left turn has been assumed as forbidden trajectory. Figure presents two examples. The upper one shows the case where trajectories are similar. The distance measure (shown above each individual image) decreases. At the bottom, trajectories at the beginning are similar and the distance measure decreases also. However, in the middle of the movement, the trajectories diverge, and the MHD value increases gradually. After the movement the measured value of MHD is compared with the threshold and if necessary triggers the alarm. Accompanied with ALPR technology the system can be a good deterrent from dangerous and illegal driving behaviour.

Fig. 1 Two examples of trajectories and characteristics of the distance measure

The main problem with the MHD measure is its separable treatment of trajectory points. Mutual relationships of coordinate points are important here, not their order in the sequence. Those points (of a given trajectory) are treated as a set. For that reason we introduced the start and stop areas (defined as patches in the calibration step). These areas are also used as appropriate markers for restricting the appropriate part of the actual vehicle trajectory included in the calculation of MHD.

5 Evaluation and Application in the 'SM4Public' Project

The proposed solution for trajectory of moving vehicles analysis was examined under the auspices of 'SM4Public' project. The project concerns the scientific research on the algorithms to be implemented in the prototype system for public spaces. The system is now being developed within the framework of EU cofounded project and is aimed at construction and implementation of innovative video content analysis-based system prototype that will ensure the safety of various public spaces using real-time solutions and typical computer components. The idea of the project was risen during the development of the previous system entitled SmartMonitor [6–9]—an intelligent security system based on image analysis, created for individual customers and home use. The analysis of alternative system applications has shown that there is a need to build other solution for public space video surveillance to effectively detect events threatening public safety and transportation environment constitutes a special case here. The problem addressed in the article concerns the specific system working scenario: illegal vehicle movement detection.

To evaluate the proposed solution some experiments have been conducted using the public dataset of the i-LIDS bag and vehicle detection challenge (resolution of 720 × 576 and 25 fps). Sequence of moving vehicles have been used under the assumption that forbidden trajectory is the left turn (as depicted in Fig. 1), we deal with left hand drive and consider vehicles driving the left lane upward. 63 individual trajectories from *AVSS_PV_Easy_Divx.avi* and *AVSS_PV_Hard_Divx.avi* sequences have been analyzed. From those 63 trajectories 5 were illegal and consistent with forbidden trajectory. For those 5 trajectories the value of MHD distance ranged from 8.61 to 15.06. For the trajectories different from the prohibited one the value of MHD systematically decreased to the breaking point (the smallest value observed equaled to 27.16) and then grew steadily. It is clear that with the appropriate threshold the perfect detection rate can be obtained. The main problem observed concerns tracking. Caused by occlusions three trajectories have been lost. Fortunately, partial trajectories are far from the given pattern.

6 Conclusions

In this paper we presented a novel algorithm for recognizing patterns in vehicle trajectory. The emphasis was put on the problem of moving vehicle detection, tracking and trajectory analysis in the context of illegal or dangerous movements. Detecting restricted or security critical behaviour on roads is needed for safety protection and traffic flow guarantee. As it was shown, the presented solution meets those requirements.

Acknowledgments The project "*Security system for public spaces—'SM4Public' prototype construction and implementation*" (original title: *Budowa i wdrożenie prototypu systemu bezpieczeństwa przestrzeni publicznej 'SM4Public'*) is a project co-founded by European Union (EU) (project number PL: POIG.01.04.00-32-244/13, value: 12.936.684,77 PLN, EU contribution: 6.528.823,81 PLN, realization period: 01.06.2014–31.10.2015). *European Funds-for the development of innovative economy* (*Fundusze Europejskie-dla rozwoju innowacyjnej gospodarki*).

References

1. Aarthi, R., Padmavathi, S., Amudha, J.: Vehicle detection in static images using color and corner map. In: 2010 International Conference on Recent Trends in Information, Telecommunication and Computing, pp. 244–246 (2010)
2. Bradski, G.: Computer vision face tracking for use in a perceptual user interface. Intel Technol. J. Microcomputer Research Lab, Intel Corporation (1998)
3. Dubuisson, M.P., Jain, A.K.: A modified Hausdorff distance for object matching. In: Proceedings of the 12th IAPR International Conference on Pattern Recognition ICPR94, pp. 566–568, Jerusalem, Israel (1994)
4. Fabian, T.: Parking lot occupancy detection using computational fluid dynamics. In: Burduk, R., et al. (eds.) CORES 2013. AISC 226, pp. 733–742 (2013)

5. Forczmański, P., Seweryn, M.: Surveillance video stream analysis using adaptive background model and object recognition. Comput. Vis. Graph., LNCS **6374**, 114–121 (2010)
6. Frejlichowski, D., Forczmański, P., Nowosielski, A., Gościewska, K., Hofman, R.: SmartMonitor: an approach to simple, intelligent and affordable visual surveillance system. In: Bolc, L., et al. (eds.) ICCVG 2012. LNCS, vol. 7594, pp. 726–734. Springer, Heidelberg (2012)
7. Frejlichowski, D., Gościewska, K., Forczmański, P., Nowosielski, A., Hofman, R.: Extraction of the foreground regions by means of the adaptive background modelling based on various colour components for a visual surveillance system. In: Burduk, R., et al. (eds.) CORES 2013. AISC 226, pp. 351–360. Springer, Heidelberg (2013)
8. Frejlichowski, D., Gościewska, K., Forczmański, P., Hofman, R.: SmartMonitor—an intelligent security system for the protection of individuals and small properties with the possibility of home automation. Sensors **14**, 9922–9948 (2014)
9. Frejlichowski, D., Gościewska, K., Forczmański, P., Hofman, R.: Application of foreground object patterns analysis for event detection in an innovative video surveillance system. Pattern Anal. Appl. 1–12 (2014). doi:10.1007/s10044-014-0405-7
10. Hu, W., Xiao, X., Xie, D., Tan, T.: Traffic accident prediction using vehicle tracking and trajectory analysis. In: Intelligent Transportation Systems. Proceedings. 2003 IEEE, vol. 1, pp. 220–225 (2003)
11. Kovacic, K., Ivanjko, E., Gold, H.: Computer vision systems in road vehicles: a review. In: Proceedings of the Croatian Computer Vision Workshop, Year 1, pp. 25–30. Zagreb, Croatia (2013)
12. Miklasz, M., Nowosielski, A., Kawka, G.: Automated supervision systems for limited traffic zones. Arch. Transp. Syst. Telemat. **6**(2), 41–45 (2013)
13. Mimbela, L.E.Y., Klein, L.A.: Summary of vehicle detection and surveillance technologies used in intelligent transportation systems. Federal Highway Administration's (FHWA) Intelligent Transportation Systems Joint Program Office (2003)
14. Munuzuri, J., Corts, P., Guadix, J., Onieva, L.: City logistics in Spain: why it might never work. Cities **29**(2), 133–141 (2012)
15. Pandit, V., Doshi, J., Mehta, D., Mhatre, A., Janardhan, A.: Smart traffic control system using image processing. Int. J. Emerg. Trends Technol. Comput. Sci. (IJETTCS) **3**(1), 280–283 (2014)
16. Song, H.-S., Lu, S.-N., Ma, X., Yang, Y., Liu, X.-Q., Zhang, P.: Vehicle behavior analysis using target motion trajectories. IEEE Trans. Veh. Technol. **63**(8), 3580–3591 (2014)
17. Tang, Y., Zhang, C., Gu, R., Li, P., Yang, B.: Vehicle detection and recognition for intelligent traffic surveillance system. Multimedia Tools and Applications (online). Springer (2015)
18. Viola, P., Jones, M.: Robust real-time object detection. In: Second International Workshop on Statistical and Computational Theories of Vision—Modeling, Learning, Computing, and Sampling, Vancouver, Canada (2001)
19. Wu, J., Cui, Z., Chen, J., Zhang, G.: A survey on video-based vehicle behavior analysis algorithms. J. Multimed. **7**(3), 223–230 (2012)
20. Zajac, W., Kołopieńczyk, M., Andrzejewski, G.: Traffic load detection system. Arch. Transp. Syst. Telemat. **6**(3), 46–48 (2013)

Algorithmically Optimized AVC Video Encoder with Parallel Processing of Data

Tomasz Grajek, Damian Karwowski and Jakub Stankowski

Abstract Algorithmically optimized AVC (MPEG-4 part 10/H.264) video encoder with parallel processing of data is presented in the paper. The paper reveals the architecture of the proposed encoder together with the description of the applied software optimization techniques. Conducted experiments show exactly the degree of parallelization of computations in the encoder and the compression performance of video encoding.

Keywords Optimized encoder · AVC video compression · MPEG-4 part 10 · H.264 · Encoding efficiency

1 Introduction

Hybrid video encoders are of a great importance in communication systems due to their ability to represent a video on relatively small number of bits. The most popular hybrid encoder that is currently used in areas such as IPTV and high definition (HD) television is AVC (MPEG-4 part 10/H.264) video compression [4]. This technique is well known and has already been described in many papers and books [6, 7]. As a matter of fact the newer technique of video compression has been worked out recently (called High Efficiency Video Coding—HEVC) [5], but the application of the new technique is an issue of further future. Therefore, the AVC technique is also the subject of ongoing studies.

T. Grajek (✉) · D. Karwowski · J. Stankowski
Chair of Multimedia Telecommunications and Microelectronics,
Poznan University of Technology, Poznań, Poland
e-mail: tgrajek@multimedia.edu.pl

D. Karwowski
e-mail: dkarwow@multimedia.edu.pl

J. Stankowski
e-mail: jstankowski@multimedia.edu.pl

R.S. Choraś (ed.), *Image Processing and Communications Challenges 7*,
Advances in Intelligent Systems and Computing 389,
DOI 10.1007/978-3-319-23814-2_12

The AVC allows 100 fold reduction of the encoded data stream for high quality of the encoded video. Such a strong compression of a video is possible by the use of the advanced intra- and inter-frame predictive coding of image blocks together with the sophisticated methods of transform and entropy coding of residual data [6, 7]. Nevertheless, applied in the AVC algorithms make both the encoder and the decoder computationally very complex. In the case of the HD videos and AVC, the real-time video encoding is a big challenge even for todays' high performance multimedia processors. It is particularly true for the video encoder side, whose complexity may be dozens of times greater than the complexity of the decoder. From that reasons, an important research problem is finding of such an architecture of highly optimized AVC compliant video encoder that will be able to exploit the potentials of todays' multimedia processors in an efficient manner. This makes the topic of this paper.

2 Research Problem

The architectures of optimized AVC encoders have already been the topic of the authors' previous research. As a result of this works the optimized structure of AVC video encoder was proposed, that was dedicated to x86-based platforms [3]. High computational performance of the encoder was achieved performing algorithmic optimization of encoder functional blocks, taking into account both the specificity of the AVC algorithms (application the context-based coding mechanisms) and features of x86-platforms (ability of using the vector operations and small size of fast cache memory). Although the proposed architecture increased the throughput of video encoder 77 times on average [3] (relative to the JM 13.2 reference software of AVC and at the virtually the same compression performance of the encoders) the encoder proposed earlier was intended for a single processor platforms with sequential processing of video data in general. This paper makes the continuation of the previous works and concerns the tuned version of the optimized AVC software encoder with additional possibility of performing computations in parallel by the use of multi-core/multi-threading technologies. It must be emphasized that parallelization of computations in the AVC encoder is a very difficult technical problem, especially in the context of the software realization of the encoder. It is commonly known that the using of the context-based coding paradigm in AVC (the use of the previously encoded data for encoding of the data of the currently processed image block) together with the raster scanning order of image blocks force in a large extent sequential processing of image data. The goal of the paper is to explore the possibilities of concurrent application of multiple processors in the encoder, analysis of such an encoder throughput and compression performance of video encoding. It must be stressed, that the assumed constraint is to preserve the full compatibility of the developed encoder with the AVC standard. The obtained results will be referenced to those obtained for the sequential version of the optimized AVC encoder described in [3].

3 Architecture of Optimized AVC Encoder with Parallel Processing of Data

3.1 Introduction

As stated before, the architecture of the optimized, sequential version of the AVC encoder (that was proposed by the authors' earlier [3]) was the starting point for works on the version of the encoder that is capable to perform computations in parallel. At the course of the study the authors found out that a possibility of dividing the frame into independent fragments called slices is the only real way for doing computations in parallel in the software version of the encoder. Since each slice is a self-contained unit which content can be fully encoded and decoded without referencing to data of other slices [6, 7], the individual slices of the image may be processed in parallel with the use of multiple processors (or processor threads) at the same time.

Therefore, division of the images into multiple slices is the main solution applied in the proposed video encoder. In this solution, the degree of computations parallelization in encoder depends directly from the number of slices within an image and the number of processor cores (or threads) available in the system. Nevertheless, not all of the functional blocks of AVC encoder can be parallelized in this way. Notable exception is deblocking filter and image interpolation procedure which, from the reason of data dependencies, operates on the entire image.

3.2 Parallel Processing of Slices

In order to get the possibility of parallel processing of data in encoder, two parts were extracted in the structure of the encoder. These are: (1) the management part, and (2) the execution part. It was realized in such a way that two groups of encoder program threads are created: master thread and a slave thread(s). There is only one master thread, while there may be a higher number of slave threads. The master thread is responsible for controlling the work of video encoder, that is to say allocation of memory buffers, assignment of individual structures to objects, controlling the global bitrate and the bitrate for individual images, dividing the image into slices, and finally running the slave threads. Whereas, the individual slave thread deals with encoding of assigned image slice, or the entire image (depending on configuration of the video encoder). In case of one slave thread and a single slice within an image (see Fig. 1), the master thread prepares all the data for encoding and runs the slave thread. After that, the master thread waits until the slave thread will encode the image and finish the operation. Then, the master thread performs deblocking filtering of decoded image, and does the image interpolation to 1/2 sampling period (only for the reference images), and manages the encoded image (i.e. inserting the image into the reference image list, writing the bitstream to the output data stream, etc.).

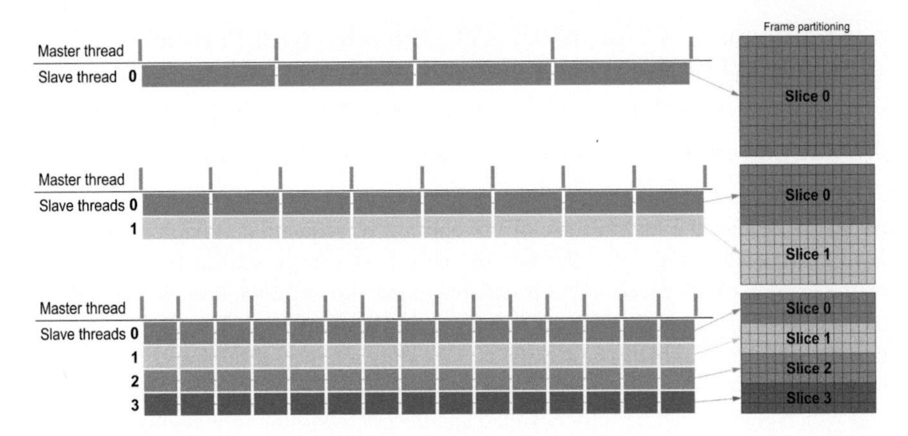

Fig. 1 Parallel processing of data in optimized AVC encoder

In case of two slave threads and two slices within a frame, the master thread prepares the relevant data for the first and the second slave threads and runs them. Each of the two slave threads encodes the image slices that were assigned to them. It must be emphasized that the slave threads work in parallel, due to the fact that they operate on a separate data sets. The master thread waits until the slave threads will finish the encoding and proceed with deblocking filtering and image interpolation (if necessary).

In the scenario of fewer number of image slices in comparison to the number of slave threads, some of the slave threads will not be utilized. In the opposite situation (the number of image slices exceeds the number of slave threads) when a given slave thread will finish the encoding, it will get another task of encoding the next image slice from the master thread. This process will continue until all image slices will be encoded.

In the situation when at least two image slices are concurrently encoded, typically there is a problem of simultaneous access of individual slave threads to a common memory. It is especially critical when at least one of threads tries write data to memory area that is used by another thread. In order to avoid such memory access conflicts, encoder uses separate memory cache for each thread and avoid race conditions by allowing slave threads only to read from common memory.

4 Methodology of Experiments

Coding efficiency of the optimized AVC encoder with parallel processing of data was thoroughly investigated with set of test video sequences. The goal was to explore the influence of allowed number of slices and allowed number of processor threads on encoding speed and efficiency. Experiments were done according to the following encoding scenario:

- Full HD test video sequences were used: *BasketballDrive*, *BQTerrace*, *Cactus*, *Kimono1*, *ParkScene*. The sequences were recommended by groups of experts ISO/IEC MPEG and ITU VCEG as a test material for research on new video compression technologies [2].
- Structure of group of pictures (GOP) was set to IBBPBBPBBPBBPBBP.
- Experiments were done for a wide range of bitrates (controlled by $QP = 22, 27, 32, 37$). This results in the quality of a reconstructed video from excellent ($QP = 22$) to very poor ($QP = 37$).
- Allowed number of slices and allowed number of processor threads was set to 1, 2, 4, or 8.
- CABAC entropy encoder was used.
- Testing platform: Intel(R) Xeon(TM) CPU 3.06 GHz (6 cores, 12 processor threads, Nehalem microarchitecture), 24 GB RAM, Windows 7 64-bit.

During experiments all combinations of number of slices and number of processor threads was tested.

Dividing frame into slices results in slightly different bitstreams (from the view-point of their size and quality of reconstructed videos). In order to compare results achieved for different number of slices and number of processor threads settings the Bjøntegaard metric was calculated [1]. The metric allows to compare the RD curves of two encoders in terms of bitrate reduction and PSNR gain based on four RD points (for $QP = 22, 27, 32, 37$ in experiments). Such tests were done for luma (Y) component. It should be noted, that the setting one slice per frame and one processor thread results in pure sequential encoder. The pure sequential encoder provides a benchmark for the performance of the parallel encoder.

5 Results

First of all we have evaluated influence of dividing frames into slices on encoding efficiency. Achieved results have been gathered in Table 1. Dividing frame into slices results in increase of the bitrate on average on 0.60, 1.68 and 3.50 % for 2, 4 and

Table 1 The average bitrate change (Bjøntegaard metric) resulted from dividing frame into 2, 4 and 8 slices per frame against 1 slice per frame

Sequence	2 slices per frame (%)	4 slices per frame (%)	8 slices per frame (%)
BasketballDrive	0.98	2.21	4.76
BQTerrace	0.69	1.81	3.30
Cactus	0.31	1.01	2.18
Kimono1	0.86	2.47	5.25
ParkScene	0.22	0.89	2.03
Average	**0.60**	**1.68**	**3.50**

Positive numbers correspond to bitrate increases

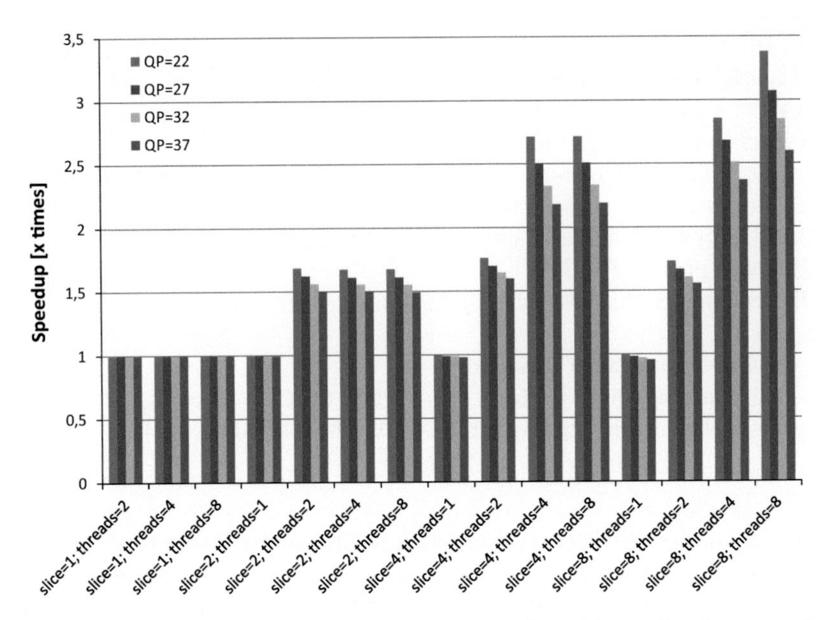

Fig. 2 Speedup (execution time of analyzed case to one slice with one thread case ratio) for *BasketballDrive* sequence for different QP values

8 slices respectively in comparison to case with only one slice per frame. Obtained results are consistent with those reported in [8].

Figure 2 presents achieved encoding speedup for all combinations of investigated number of slices and processor threads for exemplary video sequence. Achieved speedup depend on target bitrate (controlled by QP) due to different ratio between parallelized slave thread(s) tasks and sequential master thread processing.

Table 2 present results supplemented with achieved speedup resulted from using more processor threads (i.e. how many times encoder using N processor threads is faster that encoder with only one processor thread allowed). The first observation is that number of allowed processor threads dose not influence total encoding time in case of one slice per frame. The negligibly small increase of total encoding time for four and eight processor threads results from thread management overhead. For two slices and two and more processor threads we achieve 1.64 times faster encoder that for one slice case. For four slices and four and more threads the achieved speedup is about 2.5 and for the last case (eight slices and eight processor threads)—3.02.

Achieved results clearly confirm that some part of the encoder has not been parallelized. Otherwise, dividing frame into two slices and using two processor threads would result in approximately two times faster encoder.

Table 2 The average bitrate change (Bjøntegaard metric) and average speedup according to one thread with one slice per frame case

DB-rate/speedup	1 slice per frame	2 slices per frame	4 slices per frame	8 slices per frame
1 thread	**0.00 %/1.00x**	0.60 %/1.00x	1.68 %/1.00x	3.50 %/1.00x
2 threads	0.00 %/1.00x	**0.60 %/1.64x**	1.68 %/1.63x	3.50 %/1.63x
4 threads	0.00 %/0.99x	0.60 %/1.64x	**1.68 %/2.51x**	3.50 %/2.51x
8 threads	0.00 %/0.98x	0.60 %/1.66x	1.68 %/2.54x	**3.50 %/3.02x**

Positive numbers of DB rate correspond to bitrate increases. Speedup means how many times the execution time of analyzed case is lower than for one thread with one slice case

6 Conclusions and Final Remarks

The presented results prove that the use of the idea of frame slices can be the basis for substantial parallelism of computations in the video encoder. The exact impact of the number of processor threads and the number of slices in the image on the degree of computations parallelisation in the encoder was numerically presented in experimental section. In an exemplary scenario of four slices and four processor threads the encoder works 2.5-times faster relative to the sequential encoder. Taking into consideration the algorithmic optimizations that had been previously carried out in the sequential version of the encoder (see [3] for more details) gives almost 200 times faster encoding with respect to the complexity of the JM 13.2 reference software.

Since the use of N image slices (with the same number of processor threads) does not give in general the N-times faster encoder (see experimental results) there exists parts of the encoder that were not parallelized. Detailed analysis of the operations that are carried out in the encoder allowed for identification of such a fragments. These are mainly: image interpolation and deblocking of reconstructed images. Further optimization of these functional blocks can be a source of additional acceleration of the encoder. The results revealed also that the compression performance loss (expressed as BD-Rate increase) will not exceed 3.5 % in the scenario of 3-times encoder acceleration.

Acknowledgments The presented work has been funded by the Polish Ministry of Science and Higher Education for the status activity consisting of research and development and associated tasks supporting development of young scientists and doctoral students.

References

1. Bjøntegaard, G.: Calculation of average PSNR differences between RD curves. In: ITU-T SG16/Q6, 13th VCEG Meeting, Austin, USA, April 2001, Doc. VCEG-M33 (2001)
2. Bossen, F.: Common test conditions and software reference configurations, Joint Collaborative Team on Video Coding (JCT-VC) of ITU-T SG16 WP3 and ISO/IEC JTC1/SC29/WG11, Doc. JCTVC-J1100, Stockholm, Sweeden, July (2012)

3. Grajek, T., Karwowski, D., Łuczak, A., Maćkowiak, S., Domański, M.: Architecture of algorithmically optimized MPEG-4 AVC/H.264 video encoder. Lect. Notes Comput. Sci. **7594**, 79–86 (2012)
4. ISO/IEC 14496–10: Generic coding of audio-visual objects, Part 10: advanced video coding, March (2010)
5. ISO/IEC 23008–2 (MPEG-H Part 2)/ITU-T Rec. H.265: High efficiency video coding (HEVC), Apr (2013)
6. Richardson, I.E.: The H.264 advanced video compression standard, 2nd edn. Wiley (2010)
7. Special issue on H.264/AVC video coding standard. IEEE Trans. Circuits Syst. Video Technol. **13** (2003)
8. Sze, V., Chandrakasan, A.P.: A high throughput CABAC algorithm using syntax element partitioning. In: IEEE International Conference on Image Processing (ICIP) (2009)

Neural Video Compression Based on SURF Scene Change Detection Algorithm

Rafał Grycuk and Michał Knop

Abstract In this paper we present a new method for video compression. Our approach is based on a well known neural network image compression algorithm: predictive vector quantization (PVQ). In this method of image compression two different neural network structures are exploited in the following elements of the proposed system: a competitive neural networks quantizer and a neuronal predictor. It is important for the image compression based on this approach to correctly detect key frame in order to improve performance of the algorithm. For key frame detection our method uses a SKFD method based on the SURF algorithm.

Keywords SURF · Key frame detection · Video compression · Image key point

1 Introduction

Data compression is becoming an increasingly important issue in all areas of computing and communications. There are various techniques for coding the data to reduce the redundancy of the video. Most of algorithms combine spatial compensation of images as well as movement compensation in time. Currently, there are many compression standards for audiovisual data. They differ in the level of compression as well as application. The most popular of these are H.261, MPEG and JPEG. H.261 is the first of entire family H.26x video compression standards. It has been designed for handling video transmission in real time. More information about the family H.26x can be found in [2]. There is a whole family of international compression standards of audiovisual data combined in the MPEG standard, which is described in more details

R. Grycuk (✉) · M. Knop
Institute of Computational Intelligence, Czestochowa University of Technology,
Armii Krajowej 36, 42-200 Czestochowa, Poland
e-mail: rafal.grycuk@iisi.pcz.pl

M. Knop
e-mail: michal.knop@iisi.pcz.pl

R.S. Choraś (ed.), *Image Processing and Communications Challenges 7*,
Advances in Intelligent Systems and Computing 389,
DOI 10.1007/978-3-319-23814-2_13

in [5]. The best known standards are MPEG-1, MPEG-2, and MPEG-4. JPEG and JPEG2000 standards are used for image compression with an adjustable compression rate. They are also used for video compression. The method involved in the fact that each movie frames is compressed individually. In the proposed approach a PVQ video compression algorithm is used, which combines techniques of VQ (vector quantization) [7, 8], and DPCM (differential pulse code modulation). More information on the techniques can be found in [3, 4]. For detection of key frames was used SURF algorithm based on the key points of the image [12]. Correct detection of key frames allows to change the necessary compression parameters such as the predictor and the codebook [13, 14].

2 Related Works

2.1 Predictive Vector Quantization Method

The PVQ algorithm is a special approach to image compression. It combines vector quantization technique [7, 8] with the scalar differential pulse code modulation scheme [3, 4], and combined with Huffman coding. The block diagram of the PVQ algorithm consists of the encoder and decoder, each containing: an identical neural-predictor, a codebook, a neural vector quantizer and the Huffman coder.

The successive input vectors $\mathbf{V}(t)$ are introduced to the encoder. The differences $\mathbf{E}(t) = [e_1(t), e_2(t), \ldots, e_L(t)]^T$ given by the equation

$$\mathbf{E}(t) = \mathbf{V}(t) - \overline{\mathbf{V}}(t) \tag{1}$$

are formed, where: $\overline{\mathbf{V}}(t) = [\overline{v}_1(t), \overline{v}_2(t), \ldots, \overline{v}_L(t)]^T$ is the predictor of $\mathbf{V}(t)$. Statistically, the differences $\mathbf{E}(t)$ require fewer quantization bits than the original subimages $\mathbf{V}(t)$. The next step is vector quantization of $\mathbf{E}(t)$ using the set of reproduction vectors $\mathbf{G} = [\mathbf{g}_0, \mathbf{g}_1, \ldots, \mathbf{g}_J]$ (codebook), where $\mathbf{g}_j = [g_{1j}, g_{2j}, \ldots, g_{qj}]^T$ (codewords). For every L-dimensional difference vector $\mathbf{E}(t)$, the distortion (usually the mean square error) between $\mathbf{E}(t)$ and every codeword $\mathbf{g}_j$, $j = 0, 1, \ldots, J-1$ is determined. The codeword $\mathbf{g}_{j^0}(t)$ is selected as the representation vector for $\mathbf{E}(t)$ if

$$d_{j^0} = \min_{0 \le j \le J} d_j, \tag{2}$$

where we can take a measure d in expression (2) as e.g. the Euclidean distance. When adding the prediction vector $\overline{\mathbf{V}}(t)$ to the quantized difference vector $\mathbf{g}_{j^0}(t)$ we get the reconstructed approximation $\widetilde{\mathbf{V}}(t)$ of the original input vector $\mathbf{V}(t)$, i.e.

$$\widetilde{\mathbf{V}}(t) = \overline{\mathbf{V}}(t) + \mathbf{g}_{j^0}(t). \tag{3}$$

The predicted vector $\overline{\mathbf{V}}(t)$ of the input vector $\mathbf{V}(t)$ is made from past observation of reconstructed vector $\widetilde{\mathbf{V}}(t-1)$. In our approach, the predictor is a nonlinear neural network specifically designed for this purpose. In future research we plan to employ orthogonal series nonparametric estimates for the predictor design [9, 17], neuro-fuzzy predictor [15], and decision trees for mining data streams [18, 19].

The appropriate codewords $j^0(t)$ are broadcasted via the transmission channel to the decoder. In the decoder, first the codewords $j^0(t)$ transmitted by the channel are decoded using codebook and then inverse vector-quantized. Next, the reconstructed vector $\widetilde{\mathbf{V}}(t)$ is formed in the same manner as in the encoder (see relation (2)).

2.2 SURF Key Frame Detection Method

The SKFD method was firstly proposed by Grycuk et al. [12]. The algorithm is based on SURF [1, 16], which detects the frame keypoints and allow to compare two frames an match the detected keypoints. The SURF method is widely used in various systems: image recognition image databases, content-based image retrieval [10], image description [11], object tracking [6] an many others. The SKDF consists of several stages. In the first step the method divides input video on individual and labeled frames. The next step initiates two local variables $current$ and $next$ and loads the first frame to $current$ and second one to $next$ respectively. The following stage detects the keypoints on both images and match the corresponding keypoints. On output of this step we obtain the $factor$, which is the percentage relation of the matched of corresponding and all keypoints. The $factor$ is in range $0 \leq factor \leq 1$, where 0—zero similarity and 1—complete similarity. If the $factor$ is compared with t value (input parameter). If $factor$ is lower then t the $next$ frame is tagged as key frame. Then the $next$ frame is set as $current$ and to the $next$ frame the following frame is load. In the last step the previous steps are repeated until the $next$ frame is null. The correctness of SURF method is crucial, because founding the corresponding keypoints has much influence on the *factor*.

3 Proposed Method

The proposed method for video compression is based on the existing compression algorithm PVQ, and has been extended to key frame detection method based on SURF algorithm. During the first steps, the algorithm divides video input into individual frames. Next it checks whether the current frame is a key frame or not. In this stage SURF detector searches corresponding key points on both image (current, key frame). This step is crucial. When the result of this analysis is true, the algorithm create a new parameters of compression. These parameters are used by neural coder based

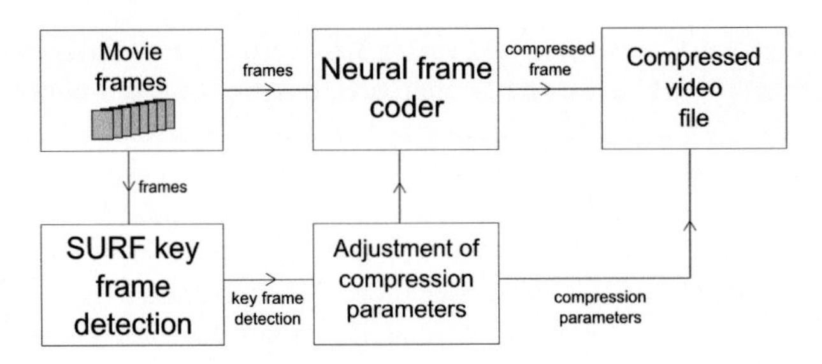

Fig. 1 Video compression algorithm

on PVQ algorithm, to compress all frames compatible with key frame. The next step of the algorithm is saving compression parameters and compressed frames to the output file. The diagram below Fig. 1 shows the proposed algorithm.

4 Experimental Result

Efficiency of our algorithm was tested using a set of frames extracted from uncompressed video file. We conducted several experiments. The purpose of the first two tests was to examine the behavior of algorithm when compressed frames were compatible with key frame (Fig. 2). During the first one, we checked the image quality of individual frames (Fig. 3). Separate codebook and predictor was created for each frame. In the second experiment, the same codebook and predictor was used for all frames compatible with the key frame (Fig. 4).

For the third and fourth experiments, we examined algorithm behaviour when the scene changed (Fig. 5). SKFD algorithm was used in order to detect scene transitions

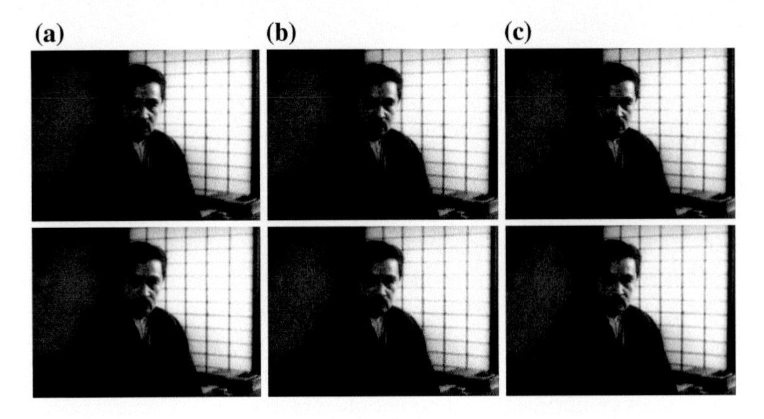

Fig. 2 **a** Original sequence. **b** Compressed sequence test 1. **c** Compressed sequence test 2

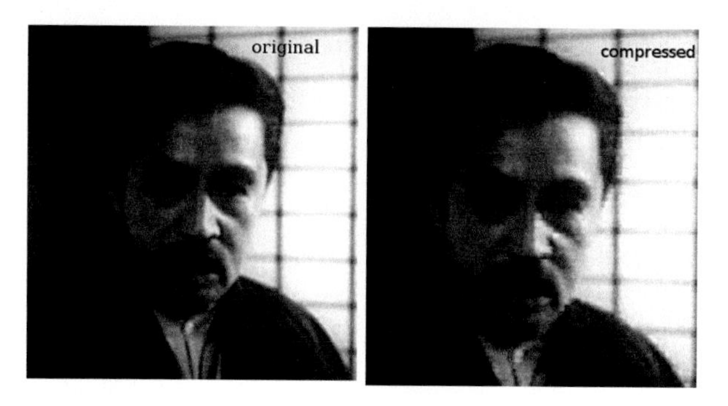

Fig. 3 Difference between frames in test 1

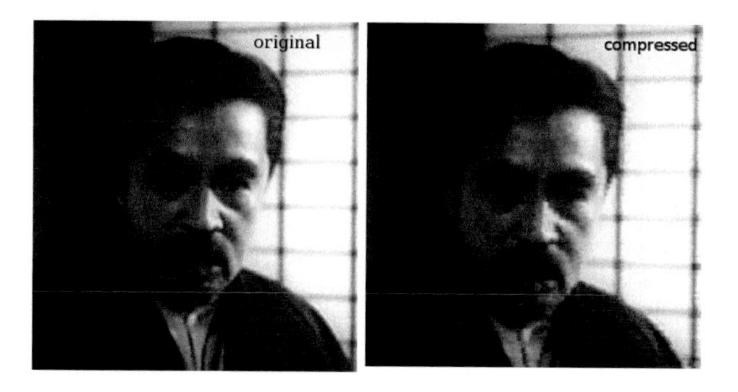

Fig. 4 Difference between frames in test 2

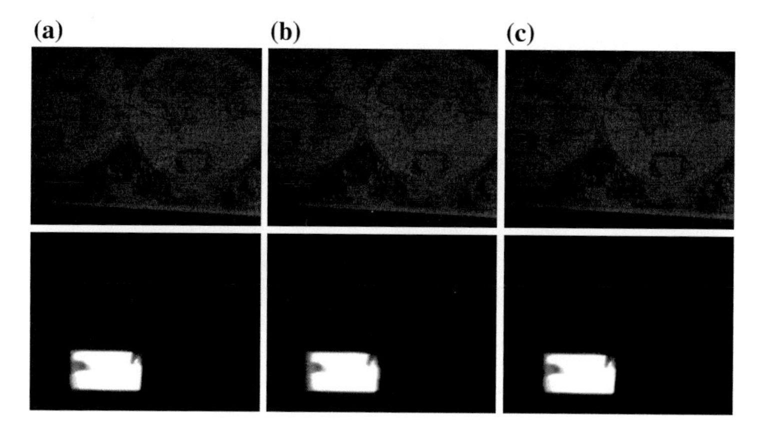

Fig. 5 **a** Original sequence. **b** Compressed sequence test 3. **c** Compressed sequence test 4

and label new key frame (Fig. 6). *minHessian* parameter for this algorithm was set to 500 similarly to our previous research [12]. Thanks to training of codebook and predictor for keyframe, algorithm adapts better compression parameters to the set of frames and improves the image quality after decompression (Fig. 7).

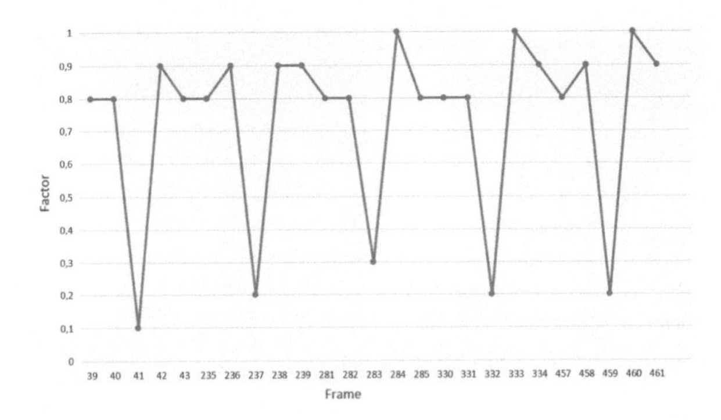

Fig. 6 Key frame detection

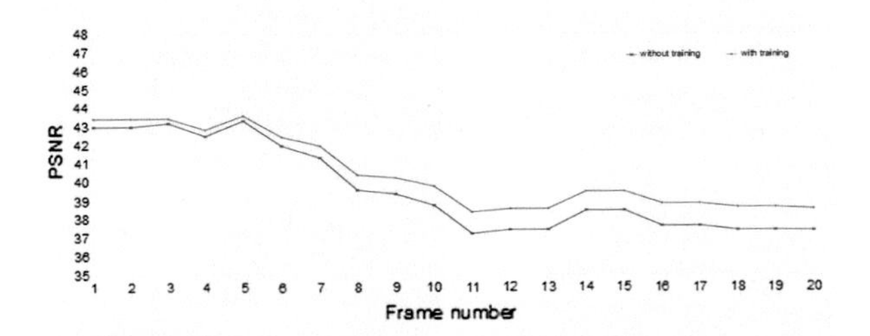

Fig. 7 PSNR change

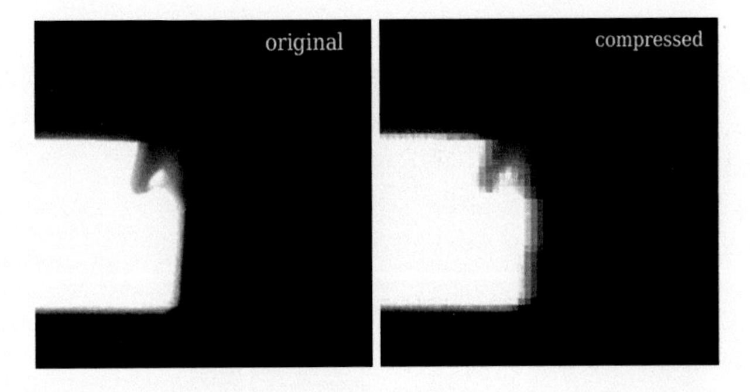

Fig. 8 Difference between frames in test 3

In third experiment we used the same codebook and predictor when the scene was changing. The results show that this approach is insufficient in case of a major scene changes (Fig. 8). In the fourth experiment, when detecting new key frame the algorithm creates new predictor and codebook. Image quality after compression is better than in the third experiment as shown in Fig. 9.

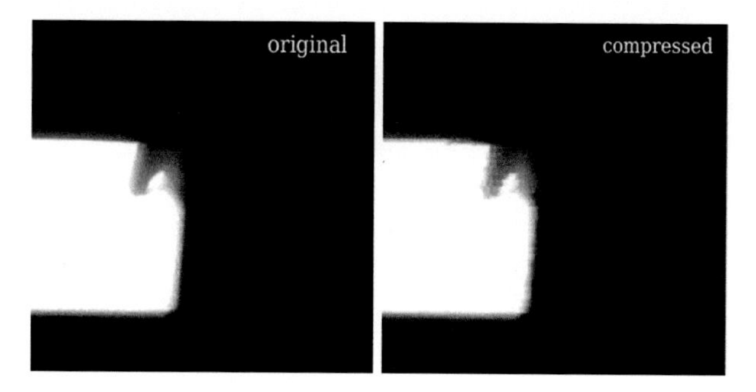

Fig. 9 Difference between frames in test 4

5 Conclusions

The presented method is a novel approach for video compression. Simulations verified the correctness of the algorithm. The presented approach shows that the key frame detection algorithm is especially useful for the presented compression algorithm. During the experiments we discovered, that without the key frame detection a video sequence compressed by our algorithm would exhibit a poor quality of frames after the scene transition. On the other hand, the number of data resulting from including the compression parameters for every frame would greatly impact on the output files size. In the future work we will try to compare our method with other compression algorithm such as MPEG4 or H264. We will perform various simulations that allows us to evaluate our method.

Acknowledgments The work presented in this paper was supported by a grant BS/MN-1-109-301/14/P "Clustering algorithms for data stream—in reference to the Content-Based Image Retrieval methods (CBIR)".
The work presented in this paper was supported by a grant BS/MN 1-109-302/14/P "New video compression method using neural image compression algorithm".

References

1. Bay, H., Ess, A., Tuytelaars, T., Van Gool, L.: Speeded-up robust features (surf). Comput. Vis. Image Underst. **110**(3), 346–359 (2008)
2. CCITT: Video codec for audio visual services at px 64 kbits/s (1993)
3. Cierniak, R.: An image compression algorithm based on neural networks. In: Artificial Intelligence and Soft Computing. Lecture Notes in Computer Science, vol. 3070, pp. 706–711. Springer, Berlin (2004)
4. Cierniak, R., Rutkowski, L.: On image compression by competitive neural networks and optimal linear predictors. Sign. Proces.: Image Commun. **15**(6), 559–565 (2000)
5. Clarke, R.J.: Digital Compression of Still Images and Video. Academic Press, Inc. (1995)

6. Gabryel, M., Korytkowski, M., Scherer, R., Rutkowski, L.: Object detection by simple fuzzy classifiers generated by boosting. In: Artificial Intelligence and Soft Computing. Lecture Notes in Computer Science, vol. 7894, pp. 540–547. Springer, Berlin (2013)
7. Gersho, A., Gray, R.M.: Vector Quantization and Signal Compression. Kluwer Academic Publishers, Boston (1991)
8. Gray, R.: Vector quantization. ASSP Mag., IEEE **1**(2), 4–29 (1984)
9. Greblicki, W., Rutkowska, D., Rutkowski, L.: An orthogonal series estimate of time-varying regression. Ann. Inst. Stat. Math. **35**(1), 215–228 (1983)
10. Grycuk, R., Gabryel, M., Korytkowski, M., Scherer, R.: Content-based image indexing by data clustering and inverse document frequency. In: Beyond Databases, Architectures, and Structures, Communications in Computer and Information Science, vol. 424, pp. 374–383. Springer International Publishing (2014)
11. Grycuk, R., Gabryel, M., Korytkowski, M., Scherer, R., Voloshynovskiy, S.: From single image to list of objects based on edge and blob detection. In: Artificial Intelligence and Soft Computing, Lecture Notes in Computer Science, vol. 8468, pp. 605–615. Springer International Publishing (2014)
12. Grycuk, R., Knop, M., Mandal, S.: Video key frame detection based on surf algorithm. In: Artificial Intelligence and Soft Computing. Lecture Notes in Computer Science, vol. 9119, pp. 572–583. Springer, Berlin (2015)
13. Knop, M., Dobosz, P.: Neural video compression algorithm. In: Image Processing and Communications Challenges 6, Advances in Intelligent Systems and Computing, vol. 313, pp. 59–66. Springer International Publishing (2015)
14. Knop, M., Cierniak, R., Shah, N.: Video compression algorithm based on neural network structures. In: Artificial Intelligence and Soft Computing, Lecture Notes in Computer Science, vol. 8467, pp. 715–724. Springer International Publishing (2014)
15. Korytkowski, M., Rutkowski, L., Scherer, R.: From ensemble of fuzzy classifiers to single fuzzy rule base classifier. In: Artificial Intelligence and Soft Computing—ICAISC 2008. Lecture Notes in Computer Science, vol. 5097, pp. 265–272. Springer, Berlin (2008)
16. Lowe, D.G.: Distinctive image features from scale-invariant keypoints. Int. J. Comput. Vis. **60**(2), 91–110 (2004)
17. Rutkowski, L.: A general approach for nonparametric fitting of functions and their derivatives with applications to linear circuits identification. IEEE Trans. Circuits Syst. **33**(8), 812–818 (1986)
18. Rutkowski, L., Jaworski, M., Pietruczuk, L., Duda, P.: Decision trees for mining data streams based on the gaussian approximation. IEEE Trans. Knowl. Data Eng. **26**(1), 108–119 (2014)
19. Rutkowski, L., Jaworski, M., Pietruczuk, L., Duda, P.: The cart decision tree for mining data streams. Inf. Sci. **266**, 1–15 (2014)

Cell Detection in Corneal Endothelial Images Using Directional Filters

Krzysztof Habrat, Magdalena Habrat, Jolanta Gronkowska-Serafin and Adam Piórkowski

Abstract The article presents an algorithm for the detection of corneal endothelium cells in images obtained with confocal microscopy (KH algorithm). Firstly, preprocessing issues are presented. The proposed methodology is based on image processing algorithms, especially filters. The method outputs images that are prepared for further analysis, e.g. stereological measurements. Each step of the algorithm is discussed in detail and other methods of digital images processing are compared to the research results.

Keywords Preprocessing · Binarization · Segmentation · Cell counting · Corneal endothelial

1 Methods for Cell Detection in Digital Images of Corneal Endothelium

Image processing methods are successfully implemented in various branches of science, for example medicine [15]. In computer image analysis (regardless of the analyzed discipline) the main purpose of transformation is to obtain a properly segmented binary image. In order to obtain such images a series of transformations is

---

K. Habrat · M. Habrat · A. Piórkowski (✉)
Department of Geoinfomatics and Applied Computer Science,
AGH University of Science and Technology, A. Mickiewicza 30 Av.,
30–059 Cracow, Poland
e-mail: pioro@agh.edu.pl

J. Gronkowska-Serafin
Department of Ophthalmology, Pomeranian Medical University,
Powstańców Wielkopolskich 72 Av., 70–111 Szczecin, Poland

J. Gronkowska-Serafin
Oejenafdelingen, Regionshospitalet Holstebro, Laegaardvej 12a,
7500 Holstebro, Denmark

R.S. Choraś (ed.), *Image Processing and Communications Challenges 7*,
Advances in Intelligent Systems and Computing 389,
DOI 10.1007/978-3-319-23814-2_14

carried out on the input image. This results in a contrast image, in which objects (e.g. specific objects such as cells) are clearly distinguishable from the background.

1.1 Input Data

A watershed algorithm is used frequently to separate individual cells in images. Unfortunately, this algorithm is sensitive to artifacts in the input images. These artifacts might exist due to image acquisition issues and may be manifested, for example, by the occurrence of stripes or other noise which confound essential information in the image (Fig. 1). A detailed description of the structure and performance issues of the proposed system is available in the literature [7]. The other issue is the cell bindings which are visible in pictures, or which cause artifacts (e.g. nuclei are shown as a "dimples" which are visible during analysis as a cell division which makes one cell looks like two). In order to eliminate such issues it is necessary to perform some transformations that reduce noise and normalize image illumination. This has a crucial significance when analyzing corneal endothelium images as they are characterized by a number of disturbances in the form of numerous horizontal stripes caused by interlacing (Fig. 1b), especially in the upper area of the image.

Figure 2 shows disturbances of illumination. The presented graphs (Fig. 2) show an average level of pixel brightness in the original image in rows (Fig. 2a) and columns (Fig. 2b).

2 Methods for Data Preprocessing

Uneven illumination leads to a situation in which the level of brightness tallies in one place with the cell surface, but in another denotes intercellular space. An exceptional irregularity is visible along the columns. Significantly, these issues make it

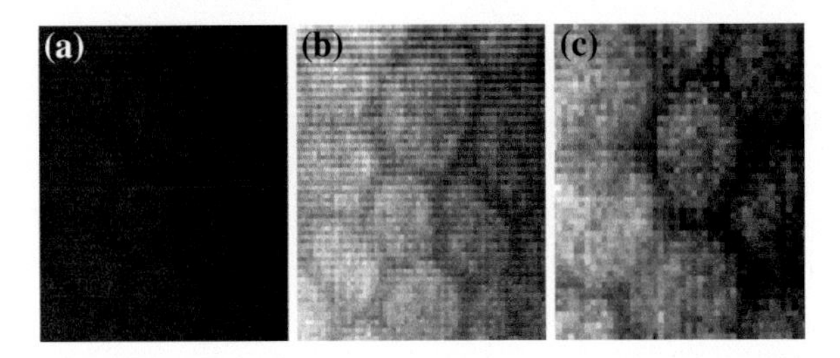

Fig. 1 Fragments of the original (**a**) and normalized (**b**) image of corneal endothelium with visible disturbances in the form of *horizontal* stripes, normalized (**c**) image of corneal endothelium where noise is visible

difficult to bring the images to a binary form which is sufficient for further analysis. Consequently, a noise removal methodology has been introduced and compared to commonly used methods (median filtering). These factors make the use of the Watershed algorithms inapplicable [10].

Method of adjusting the brightness levels for images of corneal endothelium cells The last stage of the initial transformation is adjusting the level of image brightness in columns and rows. It is possible to use one of the two proposed methods according to the following formulae. The first formula estimates the average value in each row (2) or column (3) (depending on whether the image is adjusted in rows or columns), compares the calculated value with the average brightness of the image (1). The next step, depending on the obtained result, is to add or subtract a certain value so the average brightness in a row or a column is equal to the average brightness of the image (4), (5) [9]. The second formula differs from the first in the following way: when in a given row or column the average brightness of a pixel is less than the average brightness of the image, a certain number is multiplied instead of added to get a value which is equal to the average brightness of the image (6), (7). This method produces higher contrast (Fig. 4) when the average value in a row or column is lower than the average image brightness The last method performs illumination adjustment

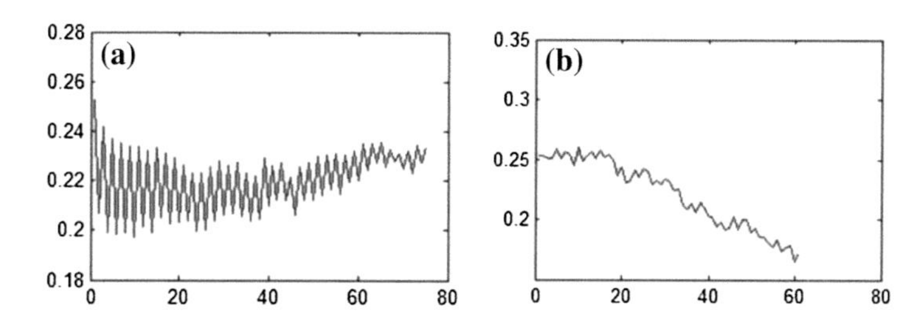

Fig. 2 Values of average levels of brightness in rows (**a**) and columns (**b**). The x–axis shows the location of a pixel against the *upper left corner* of the image—profile, the y–axis shows the illumination value of a given pixel

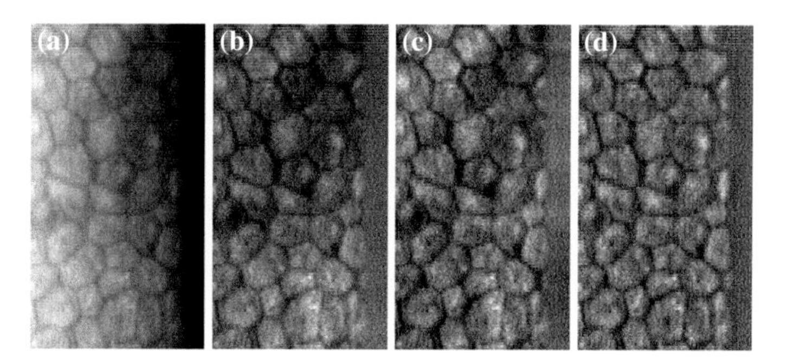

Fig. 3 Normalized (**a**) image of corneal endothelium with adjusted illumination along the columns (**b**) and in both directions (**c**) and in the round neighborhood using $r = 10$ (**d**)

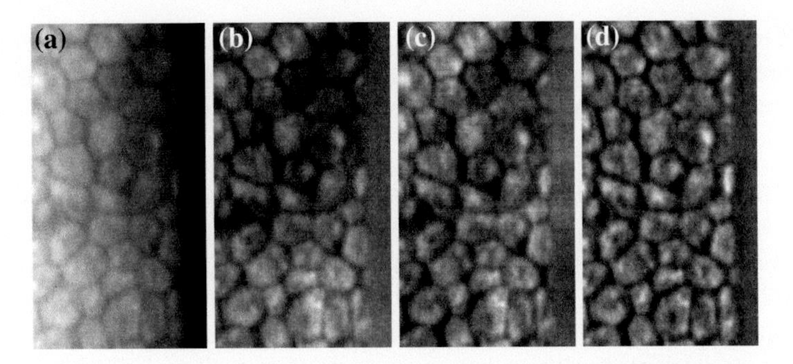

Fig. 4 Normalized and averaged (**a**) image of corneal endothelium with adjusted illumination along the columns (**b**) and in both directions (**c**) and in the round neighborhood using $r = 10$ (**d**)

for a round neighborhood of each pixel (9) (Fig. 3).

$$S = \frac{\sum_{i=1}^{x} \sum_{j=1}^{y} Pin_{ij}}{xy} \tag{1}$$

$$Sw_j = \sum_{i=1}^{x} Pin_{ij} \tag{2}$$

$$Sk_j = \sum_{j=1}^{y} Pin_{ij} \tag{3}$$

$$Pout_{ij} = \begin{Bmatrix} Pin_{ij} + S - Sw_j & for\ Sw_j \neq S \\ Pin_{ij} & for\ Sw_j = S \end{Bmatrix} \tag{4}$$

$$Pout_{ij} = \begin{Bmatrix} Pin_{ij} + S - Sk_i & for\ Sk_i \neq S \\ Pin_{ij} & for\ Sk_i = S \end{Bmatrix} \tag{5}$$

$$Pout_{ij} = \begin{Bmatrix} Pin_{ij} + S - Sw_j & for\ Sw_j > S \\ Pin_{ij} & for\ Sw_j = S \\ \frac{Pin_{ij} S}{Sw_j} & for\ Sw_j < S \end{Bmatrix} \tag{6}$$

$$Pout_{ij} = \begin{Bmatrix} Pin_{ij} + S - Sk_i & for\ Sk_i > S \\ Pin_{ij} & for\ Sk_i = S \\ \frac{Pin_{ij} S}{Sk_i} & for\ Sk_i < S \end{Bmatrix} \tag{7}$$

$$Pout_{ij} = \begin{Bmatrix} Pin_{ij} + S - Sk_i & for\ Sk_i > S \\ Pin_{ij} & for\ Sk_i = S \\ \frac{Pin_{ij} S}{Sk_i} & for\ Sk_i < S \end{Bmatrix} \quad (8)$$

$$Pout_{ij} = Pin_{ij} + (S - Sr_{ij}) \quad (9)$$

where:
Pin—input image
$Pout$—output image
S—arithmetic mean of brightness levels of all image pixels
Sw_j—arithmetic mean of brightness levels of pixels in j-row in the image
Sk_i—arithmetic mean for the levels of pixels brightness and in i-column in the image.
Sr_{ij}—arithmetic mean of brightness levels for all pixels in the distance equal or less than r from pixel i, j.

3 Methods of Binarization of Corneal Endothelium Cells

In order to obtain a particular and logical image of the corneal endothelium it is necessary to apply methods of binarization, as threshold binarization is not sufficient [14], even with proper noise removal and brightness level adjustment.

3.1 Other Binarization Approaches

There are a few approaches to corneal endothelium image binarization.

In [13] the author describes a binarization process that consists of the following stages: removal of the lowest frequency, gap closing, contrast enhancement, thresholding, skeletonization, and finally improvements to the output image.

An interesting approach which uses shape masks for convolution is presented in [5]. The authors designed three kinds of mask: 'tricorn', orthogonal (vertical and horizontal) and diagonal.

Another method is the use of pyramid methods in image processing [1]. In [4] the authors perform morphological operations to assess the thinned boundaries and next use wavelet pyramidal decomposition to carry out the binarization of corneal endothelial cells.

An interesting method is presented in [6]. The authors propose a scissoring operator which separates cells in the binary image, instead using a type of watershed algorithm.

A very promising approach to binarization is presented in [2]. The authors use the active contour technique to obtain the shape of each cell in the image.

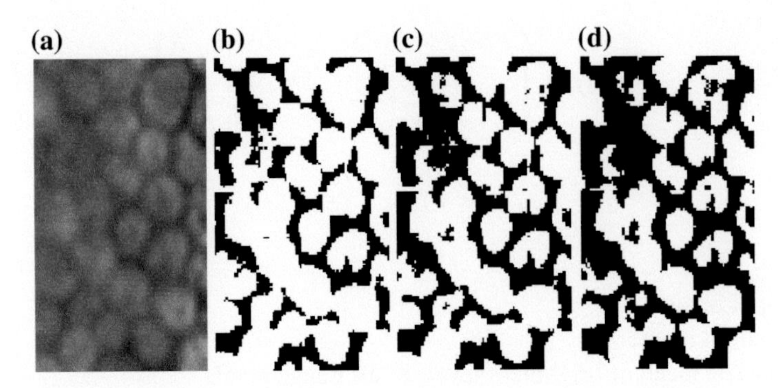

Fig. 5 Part of a normalized and averaged corneal endothelium image (**a**), binarized with adaptive method with low threshold (**b**), average threshold (**c**) and high threshold (**d**)

To assess cells contours a supervised classification scheme can be used [12]. The authors propose a multi-scale 2-dimensional matched filter to extract cells components. To obtain the final segmentation, a Support Vector Machine classifier is used.

Adaptive binarization Adaptive binarization is one of the easiest methods for the binarization of corneal endothelium images, however it is not very effective. This method moves masks with odd sizes (adjusted to the average size of cells in pixels) and determines a value of 0 or 1 for the central element on the basis of the mask context and taking into consideration a particular point. Figure 5 shows logical images which were the result of three different threshold adjustments. This method does not provide satisfying results. The vast majority of cells do not reflect their matrix shape, and in many places the cells either combine or are 'leaky'.

Top–Hat binarization algorithm Top-hat is a process which is performed by subtracting the opening result from the input image. This makes it possible to obtain an image with an adjusted background. The structural element used for the opening process needs to be large enough to successfully remove all corneal cells from the image. However, when it is too large it will cause incorrect background adjustment which further complicates binarization. Unfortunately, even if this method seems to be suitable for such processes as segmentation or binarization, it does not produce good results for corneal endothelium images. Figure 6 shows top-hat binarization results with the use of three different thresholds.

Binarization with the use of a 9 x 9 mask One of the suggested methods of binarization of corneal endothelium images uses a 9×9 size mask [8]. It was observed in corneal endothelium images that there is regular space between cells (7 pixels in average). With this is mind, a 9×9 size mask was created in order to distinguish intercellular boundaries. This fulfilled its aim relatively well because of its structure. The strongest weight was given to the central point which promoted the darkest pixels. If we invert all elements in the mask, then the mask detects corneal endothelium cells. Elements distribution is presented in (10). The filtering and binarization result from an output image with the use of this mask is presented in Fig. 7.

$$\begin{bmatrix} -3 & -3 & -2 & -2 & -2 & -2 & -2 & -3 & -3 \\ -3 & -2 & -1 & -1 & -1 & -1 & -1 & -2 & -3 \\ -2 & -1 & 0 & 0 & 0 & 0 & 0 & -1 & -2 \\ -2 & -1 & 0 & 6 & 12 & 6 & 0 & -1 & -2 \\ -2 & -1 & 0 & 12 & 32 & 12 & 0 & -1 & -2 \\ -2 & -1 & 0 & 6 & 12 & 6 & 0 & -1 & -2 \\ -2 & -1 & 0 & 0 & 0 & 0 & 0 & -1 & -2 \\ -3 & -2 & -1 & -1 & -1 & -1 & -1 & -2 & -3 \\ -3 & -3 & -2 & -2 & -2 & -2 & -2 & -3 & -3 \end{bmatrix} \tag{10}$$

Proposed binarization algorithm with the use of a 9 x 9 size directional masks (KH algorithm) The best binarization method so far (among methods presented above) is a 9×9 size filter mask. However, the mask is not perfect as it detects not only intercellular space, but also small objects that look like "holes" in cells, which are probably nuclei. A new method of four 9×9 direction mask emerged whilst taking into consideration filtering with a big mask and analyzing mistakes made during binarization conducted with the previous methods (Table 1, vertical mask and its transposition to horizontal, and, diagonal masks) [3]. The way these masks work is similar to the previous method in a sense that those masks emphasize not only one central point, but also a line in the middle of the mask which is directed at an angle of 0°, 45°, 90° or 135° against the vertical direction. In this way 4 output images were obtained (Fig. 8).

Images obtained with this method were binarized (assuming that during filtering there were some restrictions implemented and each value that exceeded the threshold of 255 was assigned a value equal to this threshold) and a logical value 1 was set for all pixels with value 255. Figure 9 shows images that were processed in this way.

In the binary images that emerged, it is visible that besides intercellular spaces, the mask has also detected small objects which look like "holes" in cells. With the use of indexation it was possible to get rid of elements smaller than 40 pixels. The is shown in Fig. 10. Binary images obtained in this way were bounded (binary addition), and

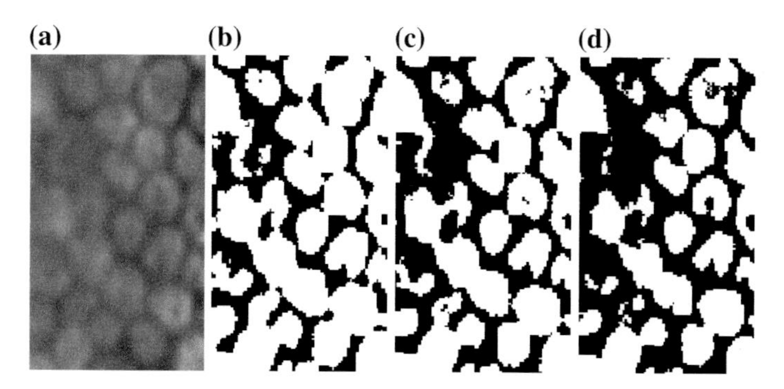

Fig. 6 Part of normalized and averaged image of corneal endothelium (**a**), binarized with the use of the top-hat algorithm with low threshold (**b**), average (**c**) and high (**d**)

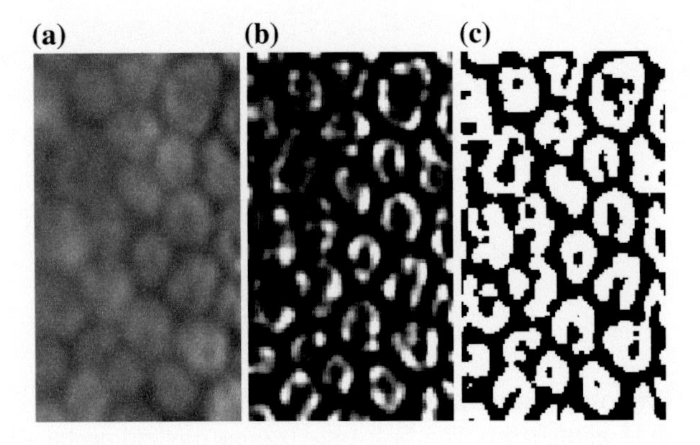

Fig. 7 Normalized and averaged (**a**) image of corneal endothelium with adjusted illumination along the columns (**b**) in both directions (**c**) in the round neighborhood using $r = 10$ (**d**)

Table 1 Four directional masks for cells' boundaries detection

$$\begin{bmatrix} 0 & 0 & 0 & 0 & 0 & 0 & 0 & 0 & 0 \\ 0 & 0 & 0 & 0 & 0 & 0 & 0 & 0 & 0 \\ 22 & -2 & -4 & -8 & -16 & -8 & -4 & -2 & 22 \\ 22 & -2 & -4 & -8 & -16 & -8 & -4 & -2 & 22 \\ 22 & -2 & -4 & -8 & -16 & -8 & -4 & -2 & 22 \\ 22 & -2 & -4 & -8 & -16 & -8 & -4 & -2 & 22 \\ 22 & -2 & -4 & -8 & -16 & -8 & -4 & -2 & 22 \\ 0 & 0 & 0 & 0 & 0 & 0 & 0 & 0 & 0 \\ 0 & 0 & 0 & 0 & 0 & 0 & 0 & 0 & 0 \end{bmatrix} \quad \begin{bmatrix} 0 & 0 & 22 & 22 & 22 & 22 & 22 & 0 & 0 \\ 0 & 0 & -2 & -2 & -2 & -2 & -2 & 0 & 0 \\ 0 & 0 & -4 & -4 & -4 & -4 & -4 & 0 & 0 \\ 0 & 0 & -8 & -8 & -8 & -8 & -8 & 0 & 0 \\ 0 & 0 & -16 & -16 & -16 & -16 & -16 & 0 & 0 \\ 0 & 0 & -8 & -8 & -8 & -8 & -8 & 0 & 0 \\ 0 & 0 & -4 & -4 & -4 & -4 & -4 & 0 & 0 \\ 0 & 0 & -2 & -2 & -2 & -2 & -2 & 0 & 0 \\ 0 & 0 & 22 & 22 & 22 & 22 & 22 & 0 & 0 \end{bmatrix}$$

$$\begin{bmatrix} 22 & 22 & 22 & 0 & 0 & 0 & 0 & 0 & 0 \\ 22 & -2 & -2 & -2 & 0 & 0 & 0 & 0 & 0 \\ 22 & -2 & -4 & -4 & -4 & -8 & -16 & 0 & 0 \\ 0 & -2 & -4 & -8 & -8 & -16 & -8 & 0 & 0 \\ 0 & 0 & -4 & -8 & -16 & -8 & -4 & 0 & 0 \\ 0 & 0 & -8 & -16 & -8 & -8 & -4 & -2 & 0 \\ 0 & 0 & -16 & -8 & -4 & -4 & -4 & -2 & 22 \\ 0 & 0 & 0 & 0 & 0 & -2 & -2 & -2 & 22 \\ 0 & 0 & 0 & 0 & 0 & 0 & 22 & 22 & 22 \end{bmatrix} \quad \begin{bmatrix} 0 & 0 & 0 & 0 & 0 & 0 & 22 & 22 & 22 \\ 0 & 0 & 0 & 0 & 0 & -2 & -2 & -2 & 22 \\ 0 & 0 & -16 & -8 & -4 & -4 & -4 & -2 & 22 \\ 0 & 0 & -8 & -16 & -8 & -8 & -4 & -2 & 0 \\ 0 & 0 & -4 & -8 & -16 & -8 & -4 & 0 & 0 \\ 0 & -2 & -4 & -8 & -8 & -16 & -8 & 0 & 0 \\ 22 & -2 & -4 & -4 & -4 & -8 & -16 & 0 & 0 \\ 22 & -2 & -2 & -2 & 0 & 0 & 0 & 0 & 0 \\ 22 & 22 & 22 & 0 & 0 & 0 & 0 & 0 & 0 \end{bmatrix}$$

then negations were created and the final result was obtained. Figure 11 presents the proposed algorithm schema.

4 Results and Comparison of Selected Binarization Methods

The algorithms presented in the research paper were compared according to the quality of binarization results. In Fig. 12 binarization results are presented with the use

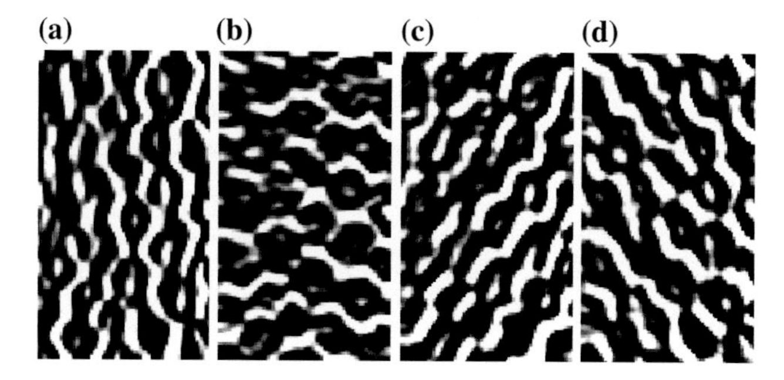

Fig. 8 Output images obtained after filtering with **a** *vertical* mask, **b** *horizontal* mask, **c** *left diagonal* mask , and **d** *right diagonal* mask

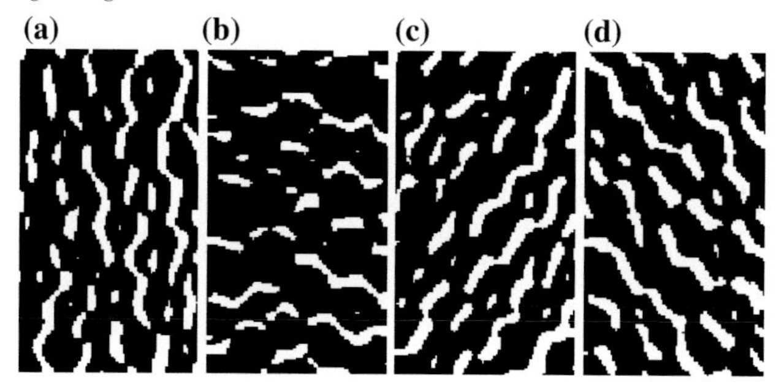

Fig. 9 Image binarization effects

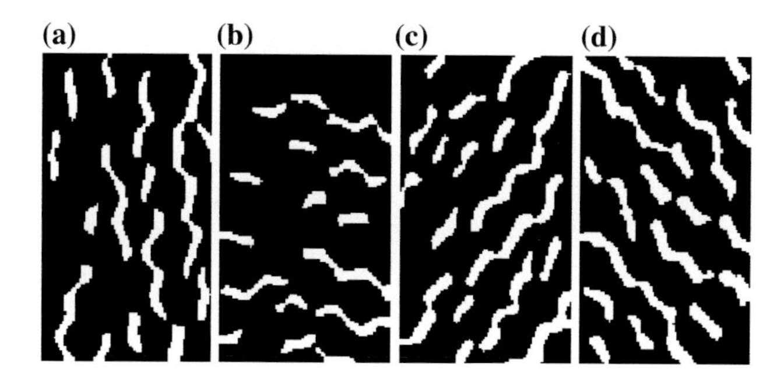

Fig. 10 Images excluding objects with an area less than 40 pixels

of all 4 algorithms. Based on the obtained results it is possible to subjectively evaluate the quality of output images. The proposed algorithm had the best performance in distinguishing cells by giving them a regular edge shape.

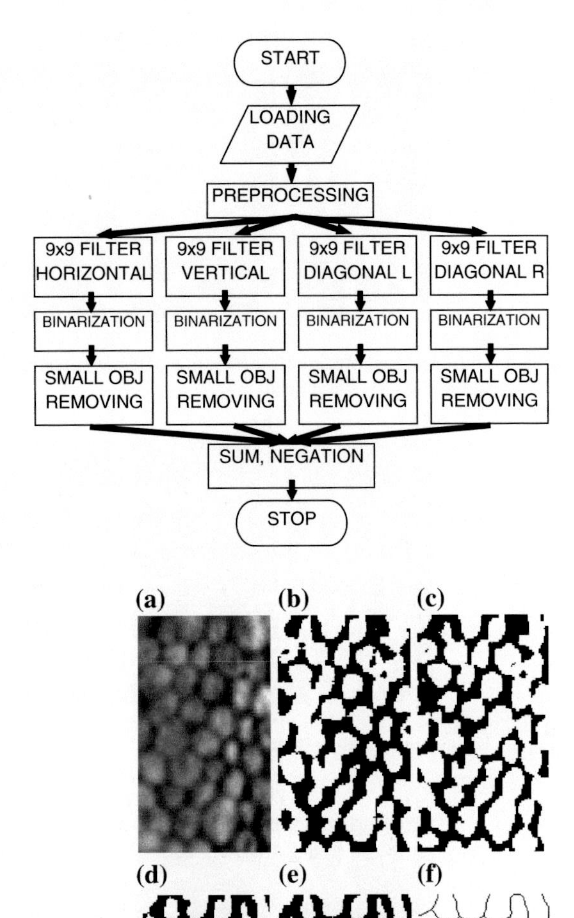

Fig. 11 The proposed algorithm schema

Fig. 12 Results of binarization of non-normalized image with adaptive method (**b**), with the use of top-hat (**c**), with the use of mask size 9×9 (**d**), proposed algorithm (using 4 direction masks size 9×9) (**e**) and final segmentation (**f**). Input image (**a**)—normalized

5 Summary

The described research presents an adaptable algorithm for the detection of objects that represent corneal endothelium cells. A detailed description of the methodology of clarifying blurry images and binarizing such images has been presented and its use has been compared to general and popular methods of image processing. The proposed methodology seems to be satisfactory from the point of view of the medical specialist. It is unfortunately sensitive to image resizing, which is the basis for the Authors to conduct further research.

Further work involves a research on methods of segmentation (e.g. precise segmentation [11]), a construction of a grid based on 'tripple points'. An interesting issue will be an adaptation of an algorithm for neighborhood map [10] for binarization of corneal endothelial microscopy images.

Acknowledgments This work was financed by the AGH—University of Science and Technology, Faculty of Geology, Geophysics and Environmental Protection as a part of statutory project.

References

1. Adelson, E.H., Anderson, C.H., Bergen, J.R., Burt, P.J., Ogden, J.M.: Pyramid methods in image processing. RCA Eng. **29**(6), 33–41 (1984)
2. Charłampowicz, K., Reska, D., Boldak, C.: Automatic segmentation of corneal endothelial cells using active contours. Adv. Comput. Sci. Res. **11**, 47–60 (2014)
3. Habrat, K.: Binarization of corneal endothelial digital images. Master's thesis, AGH University of Science and Technology (2012)
4. Khan, M.A.U., Niazi, M.K.K., Khan, M.A., Ibrahim, M.T.: Endothelial cell image enhancement using non-subsampled image pyramid. Inf. Technol. J. **6**(7), 1057–1062 (2007)
5. Mahzoun, M., Okazaki, K., Mitsumoto, H., Kawai, H., Sato, Y., Tamura, S., Kani, K.: Detection and complement of hexagonal borders in corneal endothelial cell image. Med. Imaging Technol. **14**(1), 56–69 (1996)
6. Nadachi, R., Nunokawa, K.: Automated corneal endothelial cell analysis. In: Fifth Annual IEEE Symposium on Computer-Based Medical Systems. Proceedings, pp. 450–457. IEEE (1992)
7. Oblak, E., Doughty, M., Oblak, L.: A semi-automated assessment of cell size and shape in monolayers, with optional adjustment for the cell-cell border width-application to human corneal endothelium. Tissue Cell **34**(4), 283–295 (2002)
8. Piorkowski, A., Gronkowska-Serafin, J.: Analysis of corneal endothelial image using classic image processing methods. In: KOWBAN—XVIII The Computer-Aided Scientific Research, The Works of Wroclaw Scientific Society, B, vol. 217, pp. 283–290. Wroclawskie Towarzystwo Naukowe (2011)
9. Piorkowski, A., Gronkowska-Serafin, J.: Selected issues of corneal endothelial image segmentation. J. Med. Inf. Technol. **17**, 239–245 (2011)
10. Piorkowski, A., Gronkowska-Serafin, J.: Towards automated cell segmentation in corneal endothelium images. In: Image Processing and Communications Challenges 6. Advances in Intelligent Systems and Computing, vol. 313, pp. 179–186. Springer (2015)
11. Piorkowski, A., Gronkowska-Serafin, J.: Towards precise segmentation of corneal endothelial cells. In: Bioinformatics and Biomedical Engineering. Lecture Notes in Computer Science, vol. 9043, pp. 240–249 (2015)
12. Poletti, E., Ruggeri, A.: Segmentation of corneal endothelial cells contour through classification of individual component signatures. In: XIII Mediterranean Conference on Medical and Biological Engineering and Computing 2013, pp. 411–414. Springer (2014)
13. Sanchez-Marin, F.: Automatic segmentation of contours of corneal cells. Comput. Biol. Med. **29**(4), 243–258 (1999)
14. Szostek, K., Gronkowska-Serafin, J., Piorkowski, A.: Problems of corneal endothelial image binarization. Schedae Informaticae **20**, 211–218 (2011)
15. Tadeusiewicz, R., Korohoda, P.: Komputerowa analiza i przetwarzanie obrazów (1997)

The Method of Probabilistic Nodes Combination in 2D Information Retrieval, Pattern Recognition and Biometric Modeling

Dariusz Jacek Jakóbczak

Abstract Proposed method, called Probabilistic Nodes Combination (PNC), is the method of 2D curve modeling and interpolation using the set of key points. Nodes are treated as characteristic points of unknown object for modeling and recognition. Identification of shapes or symbols need modeling and each model of the pattern is built by a choice of probability distribution function and nodes combination. PNC modeling via nodes combination and parameter γ as probability distribution function enables curve parameterization and interpolation for each specific object or symbol. Two-dimensional curve is modeled and interpolated via nodes combination and different functions as continuous probability distribution functions: polynomial, sine, cosine, tangent, cotangent, logarithm, exponent, arc sin, arc cos, arc tan, arc cot or power function.

Keywords Pattern recognition · Shape modeling · Curve interpolation · PNC method · Nodes combination · Probabilistic modeling

1 Introduction

The problem of curve modeling appears in many branches of science and industry. There are several methods to describe and calculate geometry of the curve. Also object identification and pattern recognition are still the open questions in artificial intelligence and computer vision. The paper is dealing with these problems via handwritten text identification and recognition. Handwriting based author recognition offers a huge number of significant implementations which make it an important research area in pattern recognition [18]. There are so many possibilities and applications of the recognition algorithms that implemented methods have to be concerned on a single problem. Handwriting and signature identification represents such

D.J. Jakóbczak (✉)
Department of Electronics and Computer Science, Technical University of Koszalin, Sniadeckich 2, 75-453 Koszalin, Poland
e-mail: dariusz.jakobczak@tu.koszalin.pl

R.S. Choraś (ed.), *Image Processing and Communications Challenges 7*,
Advances in Intelligent Systems and Computing 389,
DOI 10.1007/978-3-319-23814-2_15

a significant problem. In the case of biometric writer recognition, described in this paper, each person is represented by the set of modeled letters or symbols. The sketch of proposed method consists of three steps: first handwritten letter or symbol must be modeled by a curve, then compared with unknown letter and finally there is a decision of identification. Author recognition of handwriting and signature is based on the choice of key points and curve modeling. Reconstructed curve does not have to be smooth in the nodes because a writer does not think about smoothing during the handwriting. So curve interpolation in handwriting identification is not only a pure mathematical problem but important task in pattern recognition and artificial intelligence such as: biometric recognition [1, 7, 8], personalized handwriting recognition [14], automatic forensic document examination [20, 25], classification of ancient manuscripts [23]. Also writer recognition in monolingual handwritten texts is an extensive area of study and the methods independent from the language are well-seen. Proposed method represents language-independent and text-independent approach because it identifies the author via a single letter or symbol from the sample. So novel method is also applicable to short handwritten text.

Writer recognition methods in the recent years are going to various directions: writer recognition using multi-script handwritten texts [10], introduction of new features [2], combining different types of features [7], studying the sensitivity of character size on writer identification [15], investigating writer identification in multi-script environments [10], impact of ruling lines on writer identification [4], model perturbed handwriting [5], methods based on run-length features [7, 9], the edge-direction and edge-hinge features [1], a combination of codebook and visual features extracted from chain code and polygonized representation of contours [24], the autoregressive coefficients [10], codebook and efficient code extraction methods [11], texture analysis with Gabor filters and extracting features [22], using Hidden Markov Model [13, 17, 19] or Gaussian Mixture Model [18]. But no method is dealing with writer identification via curve modeling or interpolation and points comparing as it is presented in this paper. The paper wants to approach a problem of curve interpolation [3, 6, 16] and shape modeling [26] by characteristic points in handwriting identification. Proposed method relies on nodes combination and functional modeling of curve points situated between the basic set of key points. The functions that are used in calculations represent whole family of elementary functions with inverse functions: polynomials, trigonometric, cyclometric, logarithmic, exponential and power function. These functions are treated as probability distribution functions in the range [0; 1]. Nowadays methods apply mainly polynomial functions, for example Bernstein polynomials in Bezier curves, splines and NURBS [21]. But Bezier curves don't represent the interpolation method and cannot be used for example in signature and handwriting modeling with characteristic points (nodes). Numerical methods for data interpolation are based on polynomial or trigonometric functions, for example Lagrange, Newton, Aitken and Hermite methods. These methods have some weak sides [16] and are not sufficient for curve interpolation in the situations when the curve cannot be build by polynomials or trigonometric functions.

This paper presents novel Probabilistic Nodes Combination (PNC) method of curve interpolation and takes up PNC method of two-dimensional curve modeling.

The method of PNC requires minimal assumptions: the only information about a curve is the set of at least two nodes. Proposed PNC method is applied in handwriting identification via different coefficients: polynomial, sinusoidal, cosinusoidal, tangent, cotangent, logarithmic, exponential, arc sin, arc cos, arc tan, arc cot or power. Function for PNC calculations is chosen individually at each modeling and it represents probability distribution function of parameter $\alpha \in [0; 1]$ for every point situated between two successive interpolation knots. PNC method uses nodes of the curve $p_i = (x_i y_i) \in \boldsymbol{R}^2$, $i = 1, 2, \ldots n$: PNC needs 2 knots or more ($n \geq 2$); if first node and last node are the same ($p_1 = p_n$) then curve is closed (contour); for more precise modeling knots ought to be settled at key points of the curve, for example local minimum or maximum, the highest point of the curve in a particular orientation, convexity changing or curvature extrema. So this paper wants to answer the question: how to model a handwritten letter or symbol by a set of knots [12]?

2 Probabilistic Modeling

The method of PNC is computing points between two successive nodes of the curve: calculated points are interpolated and parameterized for real number $\alpha \in [0; 1]$ in the range of two successive nodes. PNC method uses the combinations of nodes $p_1 = (x_1, y_1), p_2 = (x_2, y_2), \ldots, p_n = (x_n, y_n)$ as $h(p_1, p_2, \ldots, p_m)$ and $m = 1, 2, \ldots, n$ to interpolate second coordinate y for first coordinate $c = \alpha \cdot x_i + (1 - \alpha) \cdot x_{i+1}$, $i = 1, 2, \ldots n - 1$:

$$\begin{gathered} y(c) = \gamma \cdot y_i + (1 - \gamma) y_{i+1} + \gamma(1 - \gamma) \cdot h(p_1, p_2, \ldots, p_m) \\ \alpha \in [0; 1], \gamma = F(\alpha) \in [0; 1] \end{gathered} \tag{1}$$

Two examples of h computed for MHR method [12] with good features because of orthogonal rows and columns at Hurwitz-Radon family of matrices:

$$h(p_1, p_2) = \frac{y_1}{x_1} x_2 + \frac{y_2}{x_2} x_1 \tag{2}$$

or

$$\begin{aligned} h(p_1, p_2, p_3, p_4) = & \tfrac{1}{x_1^2 + x_3^2}(x_1 x_2 y_1 + x_2 x_3 y_3 + x_3 x_4 y_1 - x_1 x_4 y_3) \\ & + \tfrac{1}{x_2^2 + x_4^2}(x_1 x_2 y_2 + x_1 x_4 y_4 + x_3 x_4 y_2 - x_2 x_3 y_4) \end{aligned}$$

The examples of other nodes combinations:

$$h(p_1, p_2) = \frac{y_1 x_2}{x_1 y_2} + \frac{y_2 x_1}{x_2 y_1}, \; h(p_1, p_2) = x_1 y_1 + x_2 y_2 \text{ or } h(p_1, p_2, \ldots, p_m) = 0 \tag{3}$$

Nodes combination is chosen individually for each curve. Formula (1) gives the infinite number of possibilities for curve calculations (determined by choice of F and h) as there is the infinite number of human signatures, handwritten letters and symbols. Nodes combination is the individual feature of each modeled curve (for example a handwritten letter or signature). Coefficient $\gamma = F(\alpha)$ and nodes combination h are key factors in PNC curve interpolation and shape modeling.

2.1 Distribution Functions in PNC Modeling

Points settled between the nodes are computed using PNC method. Each real number $c \in [a; b]$ is calculated by a convex combination $c = \alpha \cdot a + (1 - \alpha) \cdot b$ for $\alpha = (b - c)/(b - a) \in [0; 1]$. Key question is dealing with coefficient γ in (1). The simplest way of PNC calculation means $h = 0$ and $\gamma = \alpha$ (basic probability distribution). Then PNC represents a linear interpolation. MHR [12] is the example of PNC modeling. Each interpolation requires specific distribution of parameter α and γ (1) depends on parameter $\alpha \in [0; 1]$:

$$\gamma = F(\alpha),\ \ F : [0; 1] \rightarrow [0; 1],\ \ F(0) = 0, F(1) = 1$$

and F is strictly monotonic. Coefficient γ is calculated using appropriate function and choice of function is connected with initial requirements and curve specifications. Different values of coefficient γ are connected with applied functions $F(\alpha)$. These functions $\gamma = F(\alpha)$ represent the examples of probability distribution functions for random variable $\alpha \in [0; 1]$ and real number $s > 0$: $\gamma = \alpha^s$, $\gamma = \sin(\alpha^s \cdot \pi/2)$, $\gamma = \sin^s(\alpha \cdot \pi/2)$, $\gamma = 1 - \cos(\alpha^s \cdot \pi/2)$, $\gamma = 1 - \cos^s(\alpha \cdot \pi/2)$, $\gamma = \tan(\alpha^s \cdot \pi/4)$, $\gamma = \tan^s(\alpha \cdot \pi/4)$, $\gamma = \log_2(\alpha^s + 1)$, $\gamma = \log_2^s(\alpha + 1)$, $\gamma = (2^\alpha - 1)^s$, $\gamma = 2/\pi \cdot \arcsin(\alpha^s)$, $\gamma = (2/\pi \cdot \arcsin\alpha)^s$, $\gamma = 1 - 2/\pi \cdot \arccos(\alpha^s)$, $\gamma = 1 - (2/\pi \cdot \arccos\alpha)^s$, $\gamma = 4/\pi \cdot \arctan(\alpha^s)$, $\gamma = (4/\pi \cdot \arctan\alpha)^s$, $\gamma = \cot(\pi/2 - \alpha^s \cdot \pi/4)$, $\gamma = \cot^s(\pi/2 - \alpha \cdot \pi/4)$, $\gamma = 2 - 4/\pi \cdot \mathrm{arccot}(\alpha^s)$, $\gamma = (2 - 4/\pi \cdot \mathrm{arccot}\alpha)^s$.

Functions above are strictly monotonic for random variable $\alpha \in [0; 1]$ as $\gamma = F(\alpha)$ is probability distribution function. Choice of function and value s depends on curve specifications and individual requirements. What is very important in PNC method: two curves (for example a handwritten letter or signature) may have the same set of nodes but different h or γ results in different interpolations (Figs. 1, 2, 3 and 4).

Algorithm of PNC interpolation and modeling consists of five steps: first choice of knots p_i at key points, then choice of nodes combination $h(p_1, p_2, \ldots, p_m)$, choice of distribution $\gamma = F(\alpha)$, determining values of $\alpha : \alpha = 0.1, 0.2, \ldots, 0.9$ (nine points) or 0.01, 0.02, …, 0.99 (99 points) or others and finally the computations (1).

3 Biometric Modeling and Pattern Recognition

The process of biometric identification consists of three parts: pre-processing, feature extraction and comparison (the result). Pre-processing is a common stage for all methods with binarization, thinning, size standardization. Proposed approach is based on feature extraction and comparison. Feature extraction gives the key points (nodes) that are used in PNC curve reconstruction. PNC method enables signature and handwriting recognition, which is used for biometric purposes. The language does not matter because each symbol is treated as a curve. This process of recognition consists of three parts:

1. Before recognition—continual and never-ending building the data basis: patterns' modeling—choice of nodes combination and probabilistic distribution function (1) for known signature or handwritten letters of some persons in the basis;
2. Feature extraction: unknown author—choice of characteristic points (nodes) for unknown signature or handwritten words;
3. The result: recognition or identification—comparing the results of PNC interpolation for known patterns from the data basis with coordinates of unknown object.

3.1 *Modeling—The Basis of Patterns (Continually Being Larger)*

Letters or symbols of some persons ought to be modeled by the choice of nodes, determining specific nodes combination and characteristic probabilistic distribution function. As the example for recognition in Sect. 3.3, the signature "*rw*" looks different for person A (for example Richard Wright) and B (for example Rachel White). So how to model handwritten characters via PNC method? Each model has to be described (1) by the set of nodes, nodes combination h and a function $\gamma = F(\alpha)$ for each letter. Less complicated models can take $h(p_1, p_2, \ldots, p_m) = 0$ and then the formula of interpolation (1) looks as follows:

$$y(c) = \gamma \cdot y_i + (1 - \gamma)y_{i+1} \tag{4}$$

Formula (4) represents the simplest linear interpolation for basic probability distribution ($\gamma = \alpha$). So in this example, used for recognition in Sect. 3.3, how first letter "*r*" is modeled in two versions for nodes combination $h = 0$ (4) and $\alpha = 0.1, 0.2 \ldots 0.9$? Of course α is a random variable and $\alpha in[0; 1]$.

Person A Nodes (1;3), (3;1), (5;3), (7;2) and $\gamma = F(\alpha) = \alpha^2$:

Person B Nodes (1;3), (3;1), (5;3), (7;4) and $\gamma = F(\alpha) = \alpha^3$:

These two versions of letter "*r*" (Figs. 1, 2) with nodes combination $h = 0$ (4) differ at fourth node and probability distribution functions $\gamma = F(\alpha)$. Much more possibilities of modeling are connected with a choice of nodes combination $h(p_1, p_2, \ldots, p_m)$

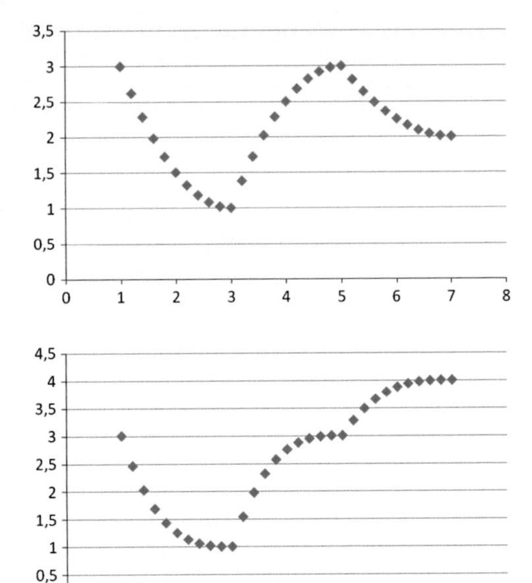

Fig. 1 PNC modeling of letter "*r*" with four nodes

Fig. 2 PNC modeling of handwritten letter "*r*"

as (2)–(3) or others. Now let us consider the example for letter "*w*" with nodes combination $h = 0$ (4), that is used for recognition in Sect. 3.3.

Person A Nodes (2;2), (3;1), (4;2), (5;1), (6;2) and $\gamma = F(\alpha) = (5^{\alpha} - 1)/4$:

Person B Nodes (2;2), (3;1), (4;2), (5;1), (6;2) and $\gamma = F(\alpha) = \sin^{3.5}(\alpha \cdot \pi/2)$:

These two versions of letter "*w*" (Figs. 3, 4) with nodes combination $h = 0$ (4) and the same nodes differ only at probability distribution functions $\gamma = F(\alpha)$. So persons A and B with the parameters of their signatures are allocated in the basis of patterns. The curve does not have to be smooth at the nodes because handwritten symbols are not smooth. The range of coefficients x has to be the same for all models because of comparing appropriate coordinates y. Every letter is modeled by PNC via three factors: the set of nodes, probability distribution function $\gamma = F(\alpha)$ and nodes combination h. These three factors are chosen individually for each letter, therefore this information about modeled letters seems to be enough for specific PNC curve interpolation, comparing and handwriting identification. What is very important—PNC modeling is independent of the language or a kind of symbol (letters, numbers or others). One person may have several patterns for one handwritten letter. Summarize: every person has the basis of patterns for each handwritten letter or symbol, described by the set of nodes, probability distribution function $\gamma = F(\alpha)$ and nodes combination h. Whole basis of patterns consists of models $\boldsymbol{S}_j$ for $j = 0, 1, 2, 3, \ldots, K$.

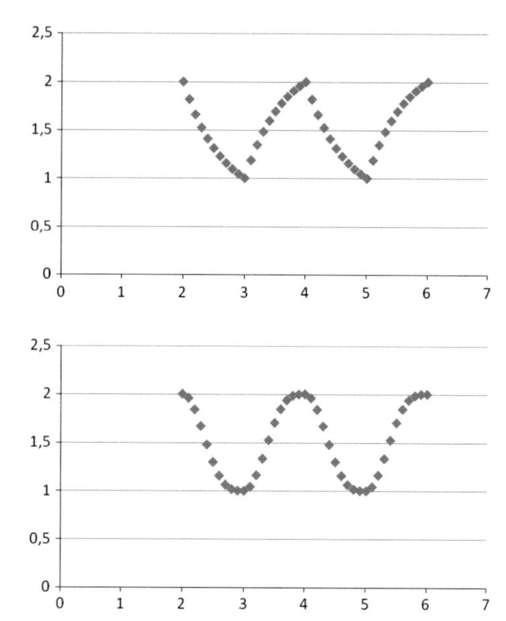

Fig. 3 PNC modeling for nine reconstructed points between nodes

Fig. 4 PNC modeling of handwritten letter "*w*"

3.2 Unknown Author—Key Points of the Curve (Feature Extraction)

After pre-processing (binarization, thinning, size standarization), feature extraction is second part of biometric identification. Choice of characteristic points (nodes) for unknown letter or handwritten symbol is a crucial factor in object recognition. The range of coefficients x has to be the same like the x range in the basis of patterns. When the nodes are fixed, each coordinate of every chosen point on the curve $(x_0^c, y_0^c), (x_1^c, y_1^c), \ldots, (x_M^c, y_M^c)$ is accessible to be used for comparing with the models. Then probability distribution function $\gamma = F(\alpha)$ and nodes combination h have to be taken from the basis of modeled letters to calculate appropriate second coordinates y_i^j of the pattern $\boldsymbol{S}_j$ for first coordinates x_i^c, $i = 0, 1, \ldots, M$. After interpolation it is possible to compare given handwritten symbol with a letter in the basis of patterns.

3.3 Decision of Recognition (The Result)

Comparing the results of PNC interpolation for required second coordinates of a model in the basis of patterns with points on the curve $(x_0^c, y_0^c), (x_1^c, y_1^c), \ldots, (x_M^c, y_M^c)$, one can say if the letter or symbol is written by person A, B or another. The comparison and decision of recognition [12] is done via minimal distance criterion. Curve points of unknown handwritten symbol are: $(x_0^c, y_0^c), (x_1^c, y_1^c), \ldots, (x_M^c, y_M^c)$.

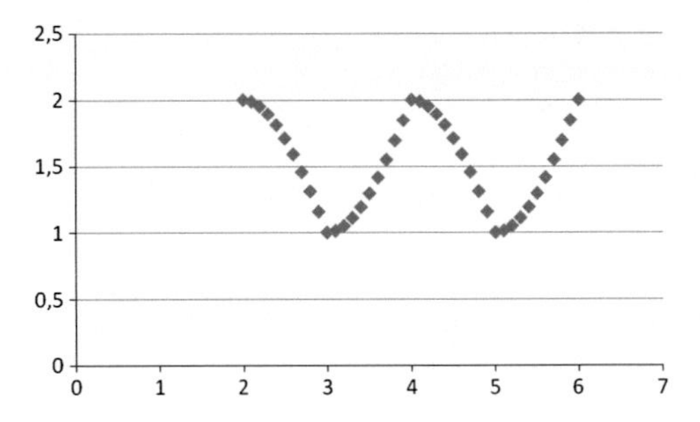

Fig. 5 Letter "*w*" by unknown author recognized as person A

The criterion of recognition for models $\boldsymbol{S}_j = (x_0^c, y_0^c), (x_1^c, y_1^c), \ldots, (x_M^c, y_M^c), j = \{1, 2, \ldots, K\}$ is given as:

$$\sum_{i=0}^{M} \left| y_i^c - y_i^{(j)} \right| \rightarrow \min \text{ or } \sqrt{\sum_{i=0}^{M} \left| y_i^c - y_i^{(j)} \right|^2} \rightarrow \min \tag{5}$$

Minimal distance criterion helps us to fix a candidate for unknown writer as a person from the model $\boldsymbol{S}_j$ in the basis. For example using (5), if above data basis consists of two persons A and B, the letter from Fig. 5 is recognized as person A.

4 Conclusions

The method of Probabilistic Nodes Combination (PNC) enables interpolation and modeling of two-dimensional curves using nodes combinations and different coefficients γ: polynomial, sinusoidal, cosinusoidal, tangent, cotangent, logarithmic, exponential, arc sin, arc cos, arc tan, arc cot or power function. Function for γ calculations is chosen individually at each curve modeling and it is treated as probability distribution function: γ depends on initial requirements and curve specifications. PNC method leads to curve interpolation as handwriting or signature identification via discrete set of fixed knots. So PNC makes possible the combination of two important problems: interpolation and modeling in a matter of writer identification. Main features of PNC method are: calculations for coordinates close to zero and near by extremum require more attention because of importance of these points; PNC interpolation develops a linear interpolation into other functions as probability distribution functions; PNC is a generalization of MHR method via different nodes combinations; interpolation of L points is connected with the computational cost of rank $O(L)$ as in MHR method; nodes combination and coefficient γ are crucial in the process of curve probabilistic parameterization and interpolation: they are computed individually for a single curve.

Future works are going to applications of PNC method in signature and handwriting biometric recognition: choice and features of nodes combinations h and coefficient γ.

References

1. Bulacu, M., Schomaker, L.: Text-independent writer identification and verification using textural and allographic features. IEEE Trans. Pattern Anal. Mach. Intell. **29**(4), 701–717 (2007)
2. Bulacu, M., Schomaker, L., Brink, A.: Text-independent writer identification and verification on off-line Arabic handwriting. In: International Conference on Document Analysis and Recognition, pp. 769–773 (2007)
3. Chapra, S.C.: Applied Numerical Methods. McGraw-Hill (2012)
4. Chen, J., Lopresti, D., Kavallieratou, E.: The impact of ruling lines on writer identification. In: International Conference on Frontiers in Handwriting Recognition, pp. 439–444 (2010)
5. Chen, J., Cheng, W., Lopresti, D.: Using perturbed handwriting to support writer identification in the presence of severe data constraints. In: Document Recognition and Retrieval, pp. 1–10 (2011)
6. Collins II, G.W.: Fundamental Numerical Methods and Data Analysis. Case Western Reserve University (2003)
7. Djeddi, C., Souici-Meslati, L.: A texture based approach for Arabic writer identification and verification. In: International Conference on Machine and Web Intelligence, pp. 115–120 (2010)
8. Djeddi, C., Souici-Meslati, L.: Artificial immune recognition system for Arabic writer identification. In: International Symposium on Innovation in Information and Communication Technology, pp. 159–165 (2011)
9. Galloway, M.M.: Texture analysis using gray level run lengths. Comput. Graph. Image Process. **4**(2), 172–179 (1975)
10. Garain, U., Paquet, T.: Off-line multi-script writer identification using AR coefficients. In: International Conference on Document Analysis and Recognition, pp. 991–995 (2009)
11. Ghiasi, G., Safabakhsh, R.: Offline text-independent writer identification using codebook and efficient code extraction methods. Image Vision Comput. **31**, 379–391 (2013)
12. Jakóbczak, D.J.: 2D Curve Modeling via the Method of Probabilistic Nodes Combination—Shape Representation, Object Modeling and Curve Interpolation-Extrapolation with the Applications. LAP Lambert Academic Publishing, Saarbrucken (2014)
13. Marti, U.-V., Bunke, H.: The IAM-database: an English sentence database for offline handwriting recognition. Int. J. Doc. Anal. Recognit. **5**, 39–46 (2002)
14. Nosary, A., Heutte, L., Paquet, T.: Unsupervised writer adaption applied to handwritten text recognition. Pattern Recogn. Lett. **37**(2), 385–388 (2004)
15. Ozaki, M., Adachi, Y., Ishii, N.: Examination of effects of character size on accuracy of writer recognition by new local arc method. In: International Conference on Knowledge- Based Intelligent Information and Engineering Systems, pp. 1170–1175 (2006)
16. Ralston, A., Rabinowitz, P.: A First Course in Numerical Analysis, 2nd edn. Dover Publications, New York (2001)
17. Schlapbach, A., Bunke, H.: A writer identification and verification system using HMM based recognizers. Pattern Anal. Appl. **10**, 33–43 (2007)
18. Schlapbach, A., Bunke, H.: Off-line writer identification using Gaussian mixture models. In: International Conference on Pattern Recognition, pp. 992–995 (2006)
19. Schlapbach, A., Bunke, H.: Using HMM based recognizers for writer identification and verification, 9th International Workshop on Frontiers in Handwriting Recognition, pp. 167–172 (2004)

20. Schomaker, L., Franke, K., Bulacu, M.: Using codebooks of fragmented connected—component contours in forensic and historic writer identification. Pattern Recogn. Lett. **28**(6), 719–727 (2007)
21. Schumaker, L.L.: Spline Functions: Basic Theory. Cambridge Mathematical Library (2007)
22. Shahabinejad, F., Rahmati, M.: A new method for writer identification and verification based on Farsi/Arabic handwritten texts. In: Ninth International Conference on Document Analysis and Recognition (ICDAR 2007), pp. 829–833 (2007)
23. Siddiqi, I., Cloppet, F., Vincent, N.: Contour based features for the classification of ancient manuscripts. In: Conference of the International Graphonomics Society, pp. 226–229 (2009)
24. Siddiqi, I., Vincent, N.: Text independent writer recognition using redundant writing patterns with contour-based orientation and curvature features. Pattern Recogn. Lett. **43**(11), 3853–3865 (2010)
25. Van, E.M., Vuurpijl, L., Franke, K., Schomaker, L.: The WANDA measurement tool for forensic document examination. J. Forensic Doc. Exam. **16**, 103–118 (2005)
26. Zhang, D., Lu, G.: Review of shape representation and description techniques. Pattern Recognit. **1**(37), 1–19 (2004)

3-D Reconstruction of Real Objects Using an Android Device

Krzysztof Kowalak, Łukasz Kamiński, Paweł Gardziński, Sławomir Maćkowiak and Radosław Hofman

Abstract In the paper, the autonomous system of reconstruction of 3-D model based on the matching characteristic features between the images for mobile devices with Android OS is proposed. Our method focuses on fully automated system with marker less calibration method. The experimental results show that although the reconstructed objects contain certain artifacts or loss, the end result can be successfully used by the average user.

1 Introduction

In recent years 3-D printing, 3-D visualization of a scene in everyday life and a video content creating have earned massive popularity. A large number of 3-D models is generated, put together in one scene, allowing viewers to navigate through it or print it and used it in prototyping process. People in the industry spent uncountable hours trying to model the world with detailed environment. This can be troublesome for most people. It would be nice if anyone can reconstruct any scene depicted in their photographs in a fairly accurate 3-D model.

This paper presents innovative approach towards 3-D reconstruction of objects, which is characterized by low-cost of the system. The solution is based on mobile

K. Kowalak (✉) · L. Kamiński · P. Gardziński · S. Maćkowiak
Poznań University of Technology, Poznań, Poland
e-mail: kkowalak@multimedia.edu.pl

L. Kamiński
e-mail: lkaminski@multimedia.edu.pl

P. Gardziński
e-mail: pgardzinski@multimedia.edu.pl

S. Maćkowiak
e-mail: smack@multimedia.edu.pl

R. Hofman
3D Vision Sp. z o.o., Warsaw, Poland
e-mail: radekh@3d-vision.com.pl

R.S. Choraś (ed.), *Image Processing and Communications Challenges 7*,
Advances in Intelligent Systems and Computing 389,
DOI 10.1007/978-3-319-23814-2_16

devices with Android operating system as an image acquisition system and an application which performs image processing operations and is running e.g. in the cloud systems.

The most popular methods of the 3-D model objects reconstruction based on images are: techniques based on the estimation of depth maps from two or more views of the object [7, 8], reconstructions based on the matching characteristic features between the images [9, 10] and volumetric reconstruction techniques from multiple camera views [11, 12].

The techniques based on the estimation of depth maps from many images are very complicated and time consuming. This process can be accelerated by the application of ToF cameras or cameras using structured light (e.g. an infrared light KinectOne) and generate directly a cloud of points which is the distance to the object (the depth of the scene). During the conducted research on the reconstruction of 3-D models of objects using volumetric reconstruction algorithm, the authors came across a problem with the accuracy of the internal and external camera parameters and the effect of these parameters on the quality of the estimated 3-D model. These camera parameters have a significant impact on the quality of the estimated model. Proposed solution uses characteristic feature points matching between two images to perform 3-D reconstruction. Authors decided to use marker less calibration method. This approach is dedicated for most people who want to create the 3-D models via mobile devices by creating the series of the photos of the object and then send these images to the server and nothing more. The details are presented in the next section.

This paper is organized as follows: next section describes the proposed system of the 3-D model reconstruction and the advances operations on images. Section 3 contains results obtained by running these apps. The Sect. 4 contains the conclusions and the future scope of this work.

2 System Description

Proposed system of 3-D model reconstruction is using an Android device. Scanned object must be captured from different sides in order to build a 3-D model. Captured images are used to reconstruct the spatial information about the scene. General diagram of proposed system is shown on Fig. 1.

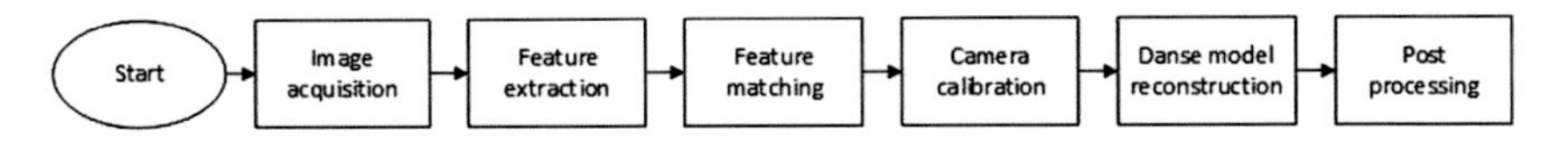

Fig. 1 General diagram of the proposed system

2.1 Image Acquisition

Image acquisition of scanned object is done using a specially prepared Android application. The main purpose of application is to support the user during scanning process. In order to build a model, user should take a series of images of object from different angles. It is worth mentioning that there is no calibration process of Android device. Intrinsic and extrinsic parameters of camera and lens distortion coefficients are estimated in the later stages of the algorithm. Besides the images of object, designed application also provides information about model and manufacturer of the device as well as focal length.

2.2 Feature Extraction

Spatial information about observing scene must be known in order to determinate camera coordinates for each image in one common coordinate system. Many camera calibration methods are well known in literature [1]. Generally, these methods can by divided into two groups. One group of them is using marker with known geometry, and the second group is marker less. Methods which are using a marker are less complex than marker less methods, and are giving accurate results. Unfortunately, the need to use marker carries some complications. The end user would have to have a marker in possession and know how to use it. Therefore authors decided to use marker less calibration method. In order to calibrate the camera without a marker it is needed to find some points of interest in all views. In proposed system SIFT [2] feature points are used. SIFT points are resistant and remain stable in case of scale, orientation and lighting change. Additionally for each SIFT point his own descriptor is calculated which is used in next step of algorithm.

2.3 Feature Matching

The next step of the algorithm is feature points matching between two views. Individual views are matched in each to each scheme. This is more complex solution than chain matching but provides resistance to interruption in case of impossibility of matching of adjacent pair of images. Feature point matching relies on descriptors comparison of these points. Brute Force method with L2 distance metric is used to point matching between two views. The result of this operation is a list of matched pairs among two images. It may happen that some points are matched incorrectly. For this reason the correction step is performed which is using a fundamental matrix. Fundamental Matrix is calculated between two images in RANSAC [3] scheme and describes the transformation of point set from first to second image.

Fig. 2 Example of reconstructed sparse model

2.4 Camera Calibration

The main goal of this step is to determine camera orientation and position in space for each image of scanned object. Calculated intrinsic and extrinsic parameters must be accurate that means reprojection error of 3-D point into image plane need to be as small as possible. The problem of camera parameters determination can be formulated as non-linear minimization of mean squared error and can be solved by Levenberg—Marquadt [4] algorithm.

The camera parameters estimation process starts with single pair of images. The initial pair should have high number of matched feature points and proper displacement so that the 3-D position of points can be identified well. Initialization of new view starts with using a DLT [5] (Direct Linear Transform) algorithm and RANSAC [3]. After determining the parameters of the camera, points which are observed by this camera are added to 3-D model of object. Example of sparse model reconstructed in this step is shown in Fig. 2.

2.5 Dense Model Reconstruction

Dense model algorithm relies on usage of SIFT characteristic features and projection matrices designated from previous step and increase model coverage by patches. Each characteristic point is a central point in patch with $\mu \times \mu$ size, where μ is equal 5 or 7. For single patch normal vector $n(p)$ is assigned, except central point $c(p)$. Normal vector is always directed on outside of the model. The next step is patch expansion. New patches are created in empty space between neighboring patches. When the expansion is finished, patches filtering process is performed. For each view, location of each patch is verifying. If some of them is in front of or behind set of patches plane, it is removed. Based on this dense model, surface reconstruction can be made (Fig. 3).

Fig. 3 Dense model reconstruction

2.6 Post Processing

The dense model which was obtained in previous step need some post processing. First of all in reconstructed model appear some points which do not belong to the scanned object. Additionally to build a complete 3-D model there is need to perform surface reconstruction step. Point cloud filtering is done by SOR [6] (Statistical Outlier Removal) filter. This filter is based on stochastic analysis of all points in the cloud and rejection of those which do not meet certain criteria. During the filtration process the distance between point and its neighborhood is calculated. Assuming that the distribution of these distance is Gaussian, point that do not match the distribution are removed. Example of point cloud before and after the filtration process is shown in Fig. 4.

Last step of algorithm is surface reconstruction. Surface reconstruction is complex problem because the quality of reconstructed point cloud is mostly low due to noise. In order to perform surface reconstruction Poisson [7] surface reconstruction algorithm was used. This algorithm treats model reconstruction problem as spatial Poisson problem. This method is resistant to noise in point cloud because is considering all points simultaneously while giving satisfactory result. The result of this method is shown of Fig. 5.

Fig. 4 Example of SOR filter application

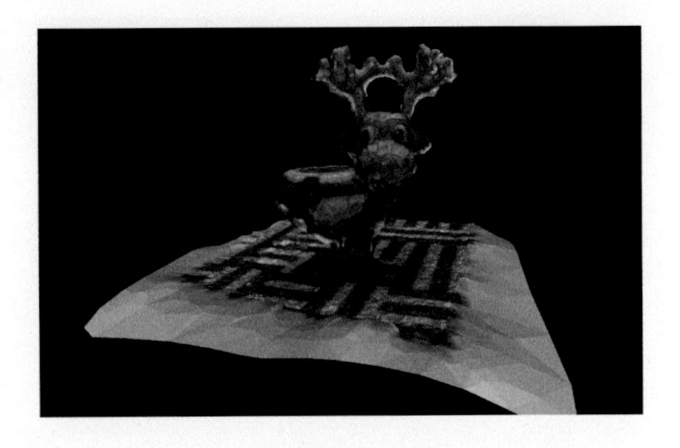

Fig. 5 Example of reconstructed surface from point cloud

3 Experimental Results

Studies conducted in this paper concerned the estimate subjective quality of reconstruction on the basis of prepared test sequences of real objects.

Data for the tests were recorded using a popular smartphone with Android OS. Four image sequences which vary in terms of difficulty of reconstruction were recorded. Two test sequences have low complexity and two others have a higher degree of complexity. All sequences were prepared in two versions: as a series of images and a video sequence.

Moreover, all sequences were prepared in different lengths, in order to verify the effectiveness of the reconstruction where there are plenty of images around the object (high density scan), and in the case of the small number of input images.

According to project assumptions, the solution presented in this paper is prepared for the end user who does not have the necessary knowledge to handle most of the currently available solutions. As already described in Sect. 2, the algorithm does not require any configuration by the user. In connection with this assumption, widely available data sets do not meet the prescribed requirements. These sequences are not recorded from the hand as it probably will do the average user. It was therefore necessary to record our own sequences that satisfy the above requirements.

The effectiveness of the reconstruction in the presented solution was evaluated based on a subjective assessment of quality of reconstruction by comparing the input images with model reprojection to the image plane. The experimental results of proposed approach is shown on Fig. 6.

The developed system of autonomous reconstruction of the 3-D model gives very good results. Although the reconstructed objects contain certain artifacts or loss, the end result can be successfully used by the average user. The reconstructed objects can easily be used to further work in graphic programs or 3-D printers. Computation time depends on number of images and images resolution. The presented technique is characterized by wide application, but needed further study in order to further improve the quality of reconstruction and reducing computation time.

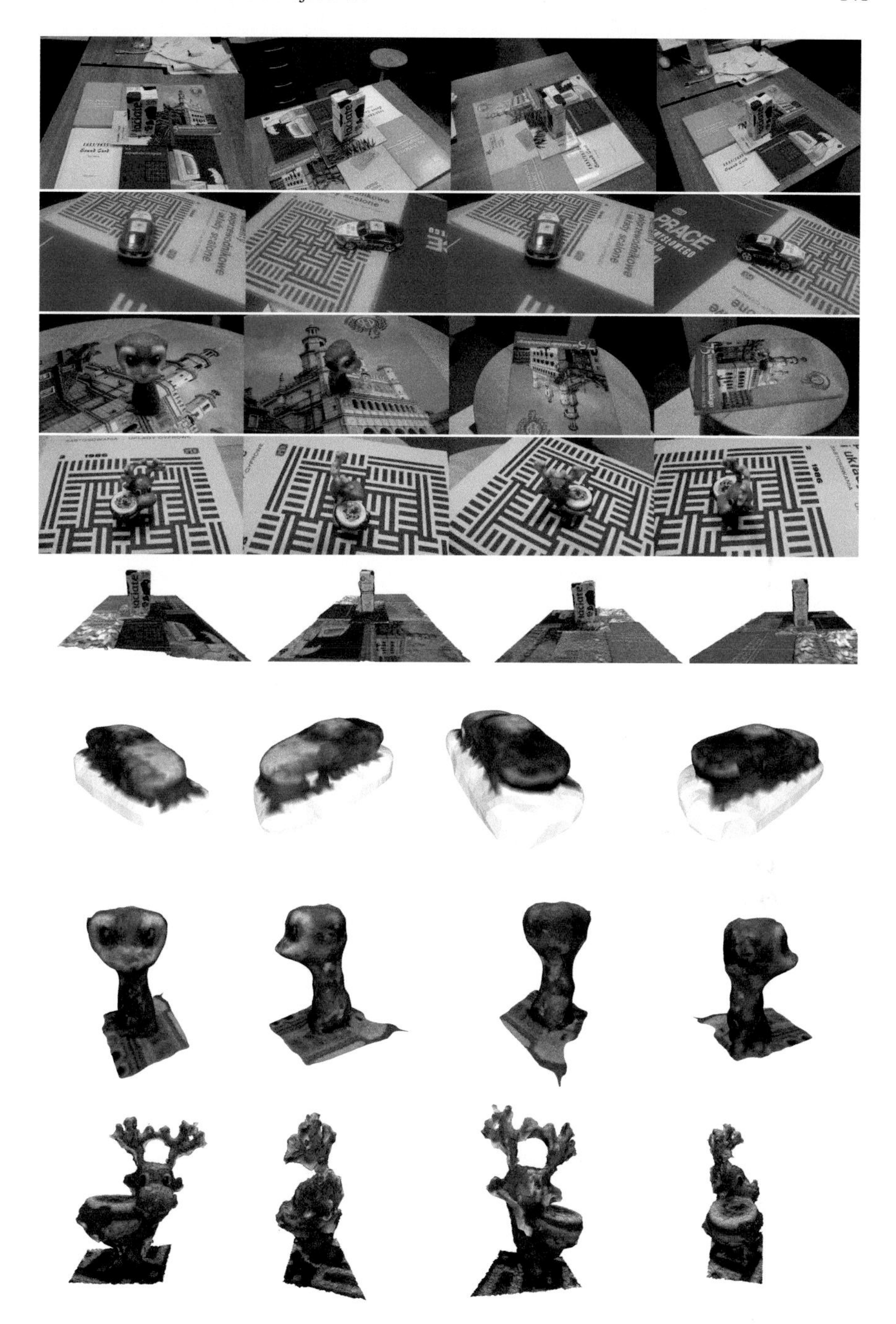

Fig. 6 Experimental results of the 3-D reconstruction of the test objects

4 Conclusions

In this paper, the novel system for 3-D model reconstruction is proposed. The proposed system is using only an Android device in order to capture images of scanned object. Special application for Android was designed to guide the user during scanning process. All calculations take place on the server where images were uploaded by application. The main advantage of proposed approach is that user does not need to have any special knowledge about scanning procedure. The marker is also unnecessary.

At the end of the algorithm the user receives complete 3-D model of scanned object. The experimental results shows that the quality of 3-D model is satisfactory. The model is saved in PLY file format which can be used by most graphics programs available on the market. Reconstructed model of object can be further processed or printed on 3-D printer.

Acknowledgments The project *Innovative stereoscopic webcam with dedicated software for image processing prototype construction* is the project co-founded by European Union (project number: UDA-POIG.01.04.00-30-018/10-03, realization period: 02.2012–06.2014). European funds for the development of innovative economy.

References

1. Baker, H.: Three-dimensional modelling. In: Proceedings of the 5th International Joint Conference on Artificial Intelligence, pp. 649–655. Cambridge, MA, USA (1977)
2. Cyganek, B.: Object Detection and Recognition in Digital Images: Theory and Practice. Wiley (2013)
3. Fischler M., Bolles R.: Random sample consensus: a paradigm for model fitting with applications to image analysis and automated cartography. In: Readings in Computer Vision: Issues, Problems, Principles and Paradigms, pp. 726–740 (1987)
4. Goldlcke, B., Magnor, M.: Real-time, free-viewpoint video rendering from volumetric geometry, In: Proceedings of Visual Computation and Image Processing (VCIP), vol. 5150, Issue. 2, pp. 1152–1158. Lugano, Switzerland (2003)
5. Hartley, R., Zisserman, A.: Multiple View Geometry in Computer Vision. Cambridge University Press (2003)
6. Lowe, D.G.: Distinctive image features from scale-invariant keypoints. Int. J. Comput. Vis. **60**(2), 91–110 (1997)
7. Nocedal, J., Wright, S.: Numerical Optimization. Springer, New York (1999)
8. Point Cloud Library (PCL) Documentation. http://dev.pointclouds.org/projects/pcl/wiki
9. Seitz, S.M., Curless, B., Diebel, J., Scharstein, D., Szeliski, R.: A comparison and evaluation of multi-view stereo re-construction algorithms. In: IEEE Conference on Computer Vision and Pattern Recognition, pp. 519–528 (2006)
10. Snavely, N., Seitz, S., Szeliski, R.: Photo tourism: exploring photo collections in 3D. In: ACM Transactions on Graphics (TOG), vol. 25, pp. 835–846. ACM (2006)
11. Snavely, N., Seitz, S., Szeliski, R.: Modeling the world from internet photo collections. Int. J. Comput. Vis. **80**(2), 189–210 (2008)
12. Xiao, J., Chen, J., Yeung, D.-Y., Quan, L.: Learning twoview stereo matching, In: Proceedings of the 10th Eu-ropean Conference on Computer Vision: Part III, ECCV '08, pp. 15–27. Springer, Heidelberg (2008)

Methods of Natural Image Preprocessing Supporting the Automatic Text Recognition Using the OCR Algorithms

Piotr Lech and Krzysztof Okarma

Abstract Reading text from natural images is much more difficult than from scanned text documents since the text may appear in all colors, different sizes and types, often with distorted geometry or textures applied. The paper presents the idea of high-speed image preprocessing algorithms utilizing the quasi-local histogram based methods such as binarization, ROI filtering, line and corners detection, etc. which can be helpful for this task. Their low computational cost is provided by a reduction of the amount of processed information carried out by means of a simple random sampling. The approach presented in the paper allows to minimize some problems with the implementation of the OCR algorithms operating on natural images on devices with low computing power (e.g. mobile or embedded). Due to relatively small computational effort it is possible to test multiple hypotheses e.g. related to the possible location of the text in the image. Their verification can be based on the analysis of images in various color spaces. An additional advantage of the discussed algorithms is their construction allowing an efficient parallel implementation further reducing the computation time.

Keywords Image binarization · Natural images · OCR

P. Lech (✉) · K. Okarma
West Pomeranian University of Technology, Szczecin, Poland
e-mail: piotr.lech@zut.edu.pl

K. Okarma
e-mail: krzysztof.okarma@zut.edu.pl

P. Lech · K. Okarma
Faculty of Electrical Engineering, Department of Signal Processing
and Multimedia Engineering, 26. Kwietnia 10, 71-126 Szczecin, Poland

R.S. Choraś (ed.), *Image Processing and Communications Challenges 7*,
Advances in Intelligent Systems and Computing 389,
DOI 10.1007/978-3-319-23814-2_17

1 Introduction

Currently much attention in the image recognition applications is paid to the issues related to the text recognition from natural images. Increasingly, the electronic devices have some functions implemented allowing the real-time recognition of text from images e.g. in driver assistance systems or mobile applications supporting the visually impaired people making their life easier.

Nevertheless, well-known Optical Character Recognition (OCR) methods which can be efficiently applied for scanned text documents may not always lead to satisfactory results for natural images captured in open air environment with differing lighting conditions. Proper recognition of such text is also dependent on the geometrical distortions which may be present on the acquired images as well as the presence of textures, different types of fonts as well as their colors. All those elements should be considered during image preprocessing operations which may include color to grayscale conversion as well as further image binarization, extraction of Regions Of Interest (ROI), detection of lines, corners of different features etc. Quite similar operations together with additional shape analysis can also be applied e.g. for stamps detection on document images [7].

In many cases some of these operations should be conducted locally or in an adaptive way. Since most of modern electronic devices used for real-time text recognition have some limitations related to available amount of memory and processing speed, it is necessary to use relatively fast image preprocessing algorithms in order to increase the accuracy of text recognition during further steps of analysis.

Typical natural images captured by cameras may contain text data which is located almost anywhere on the image plane so a proper extraction of Regions Of Interest seems to be one of the most relevant issues as it allows significant savings in calculations conducted in further steps of image analysis.

A good example can be the images acquired by the cameras mounted inside the vehicles which can be helpful e.g. for supporting the detection of speed limit signs as well as road information which can be useful for supporting the navigation e.g. during temporary closing of some roads and necessity of detours. Useful text information in such images is located on relatively small area of image and therefore an important preprocessing operation which may influence the overall speed and accuracy of the OCR algorithms is the ROI extraction which may be based on the object boundaries detection often conducted using the binary image.

Recently many attempts to natural image OCR problem have been made using various methods. One of the most popular, included e.g. in MATLAB's Computer Vision System Toolbox, is based on detection of Maximally Stable Extremal Regions (MSER) [5]. Some other ideas utilize also some other features such as convolutional co-occurrence Histogram of Oriented Gradients (HOG) [17] or some others (Shape Contexts, Geometric Blur, Scale Invariant Feature Transform—SIFT, Spin image, Maximum Response of filters and Patch descriptor) compared in the paper [4]. The application of Stroke Width Transform (SWT) for color reduction has been proposed by Ikica and Peer [10] which can be helpful as a preprocessing step. A similar idea has also been investigated by Epshtein et al. [6]. Nevertheless, the computational complexity of such approaches remains relatively high.

An interesting attempt to integrate text detection and its recognition into one process has been investigated in the paper [15], similarly as in the paper [12] where the evaluation of 12 different binarization methods has also been done. A complete OCR system for natural images known as PhotoOCR has been recently presented by Bissacco et al. [3] which utilizes quite complex segmentation approach based on a combination of HOG and Weighted Direction Code Histogram (WDCH) features.

Nevertheless, the ROI identification in many cases may require the analysis of the whole image which can be time consuming. In order to avoid such necessity statistical random sampling may be applied which decreases significantly the amount of processed data leading to very similar results.

Exemplary algorithms which can be easily applied using the statistical sampling are the histogram based image binarization methods such as Kapur et al. [11] and Otsu's [14] algorithms. It is worth to notice that binarization and further processing of binary images may also be considered as one of the methods for significant increase of the overall OCR process. An additional useful operation may be the detection of horizontal and vertical lines which may be helpful for corrections of perspective projection or geometrical image distortions.

In the remaining part of the paper some experimental results obtained applying the fast statistical sampling approach in order to support the automatic text recognition from natural images are presented together with the description of the proposed approach. Experimental testing of the algorithms has been conducted using high-resolution natural images containing information characteristic for the surroundings of roads, such as banners, traffic signs, etc.

The verification of obtained results has been done using the popular Tesseract OCR engine [16] supported by Google with the use of the images from NEOCR database [13] which contains a set of natural text images together with meta-descriptions with text boundaries on the test images as well as using some other available images.

2 Relevance of Color Conversion and Binarization

In order to illustrate the influence of the color to grayscale conversion method and further binarization algorithm, an exemplary original image containing "natural text" together with the result of its binarization using Otsu's thresholding are presented in Fig. 1. Before thresholding the original color image has been converted to grayscale using the YUV and CIE LAB color spaces. As can be easily noticed, a proper recognition of text from the right part of the image (for YUV luminance channel) turned out to be impossible.

Apart from the problem of the proper choice of color to grayscale conversion method and thresholding algorithm for such obtained image, an important issue is the variability of attributes of the "natural text" used by different creators of graphic elements which can be found on the information signs or billboards. It is

Fig. 1 An exemplary natural image with results of its Otsu binarization for CIE LAB and YUV luminance channels (from *left* to *right* respectively)

worth noticing that depending on the chosen color model the luminance of pixels can be expressed as a linear of nonlinear combination of the RGB channels. The most popular linear combinations are used in so called "television color models" according to recommendations of the International Telecommunication Union ITU BT-601.7 [2] for SDTV or ITU BT.709-5 [1] for HDTV signals, as well as in the HSV color model where the Value (V) is just the average of the RGB channels.

Another possibility is the choice of the maximum of three RGB values according to the luminance calculation in the HLS color model or a nonlinear combination suggested by Commission Internationale de l'Eclairage (CIE) in CIE LAB and CIE LUV color spaces.

Some other quite popular methods of color to grayscale conversion, which are not considered in this paper mainly due to their computational demands, are Color2Gray proposed by Gooch et al. [8] and Grundland and Dodgson [9].

After the conversion to grayscale further image thresholding is typically made in order to obtain the binary image which can be analyzed. The most popular are histogram based methods such as Kapur et al. [11] and Otsu's [14] thresholding leading to the choice of threshold value separating the objects from the background. In Otsu's thresholding the optimal value is obtained by minimizing the intra-class variance (or maximizing the variance between the classes i.e. inter-class one) whereas Kapur method utilizes the probability distributions of object and background pixels.

Since the calculation of the histogram for the whole image may be time consuming, this procedure may be significantly accelerated by probabilistic histogram estimation.

3 Proposed Fast Natural Image Preprocessing

Acceleration of the procedure of the histogram calculation can be achieved by the use of a simplified representation of image obtained by a simple statistical experiment. Assuming that the specified image pixels reflect the population, the histogram estimation can be based on a randomly chosen sample being a fragment of a population.

Nevertheless, it is important that the random sample (new representation of the image) should be possibly the most similar to the whole population being a representative statistical miniature of the whole image. In this case, you can expect the results

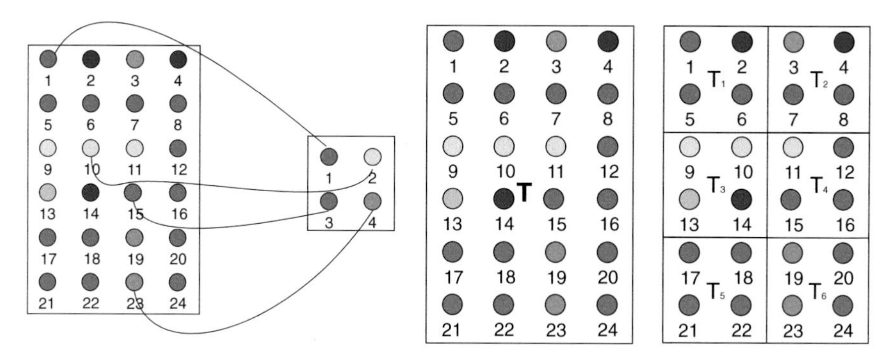

Fig. 2 The idea of the randomly chosen miniature representation of an image assuming 4 draws from 24 pixels (*left*) and the idea of the quasi-local binarization (*right*)

of the histogram calculation obtained for the random sample (i.e. new ultra-small image) to be identical or at least very similar to those obtained for the whole population (the original image). The idea of the miniature representation of the original image is illustrated in the left part of Fig. 2.

It should be noted that when making e.g. 10,000 draws of pixels in comparison to the original High Definition image (i.e. resolution of 1920 × 1080 elements and therefore containing 2,073,600 pixels) approximately 200-fold decrease of the number of performed operations can be achieved.

The idea of global binarization is based on the assumption that the image histogram represents two classes related to objects and background, however it may be not always fulfilled, especially for images containing relatively small luminance differences where even small lighting irregularities may cause serious distortions in the resulting binary image. The influence of lighting irregularity on the results of binarization can be observed in Fig. 1. In fact, most of natural images typically acquired in outdoor open space conditions may be troublesome for analysis in view of irregular lighting and similar phenomena.

The solution which can be applied for a partial elimination of those disadvantages may be the division of an image into smaller parts and calculation of local histograms for them. They can be further used for a quasi-local binarization which idea is illustrated in the right part of Fig. 2 where T denotes the global threshold and $T_1 - T_6$ are the local ones. A disadvantage of the method is the necessity of choosing the proper size of the block dividing the image into smaller fragments.

Both considered approaches can be combined, especially for high resolution images where the chosen block size is relatively large, and the local threshold values can also be determined using the statistical sampling within each block.

Improper results of binarization may also influence the results of edge detection which is often conducted in order to determine the shape of individual characters. Popular gradient based edge detection algorithms, such as Sobel, Prewitt or Roberts filters, are very sensitive to uneven lighting of the scene, especially in their classical applications for grayscale images. Nevertheless, such edge detectors can also be

Result of global Otsubinarization

Result of fast quasi-local Otsubinarization with marked text boundaries

Result of global Kapur binarization

Result of fast quasi-local Kapur binarization with marked text boundaries

Fig. 3 Comparison of the results of global and fast quasi-local binarization for an exemplary image from the NEOCR dataset

applied successfully for binary images obtained using the proposed fast quasi-local binarization method with the use of Otsu method for determining the local threshold values. In such case the edges in the binary image can be considered as closer to the ideal representation of a sharp edge of an object.

In the next step a contour tracing algorithm (in our case a popular Moore Neighbor Tracing method has been used which is available e.g. in MATLAB as *bwboundaries* function) can be applied in order to find the fragments of an image containing possible text areas further processed using the OCR engine. The comparison of results obtained for an exemplary representative test image from the NEOCR dataset is shown in Fig. 3. It can be easily noticed that the proposed quasi-local binarization significantly improves text detection for Kapur thresholding. The results of boundary analysis obtained for global and quasi-local Otsu method are very similar and therefore they have been shown only for the images obtained after the quasi-local binarization.

Figure 4 illustrates the comparison of the binarization results obtained using the proposed statistical quasi-local approach and the popular method based on the MSER features. It can be easily noticed that the similar readability of individual characters using the MSER based approach requires additional segmentation whereas comparable results may be obtained much faster using the proposed approach.

Further application of the Tesseract OCR engine for both images has led to similar results as well. Applying the OCR algorithm to the binary image obtained as the result

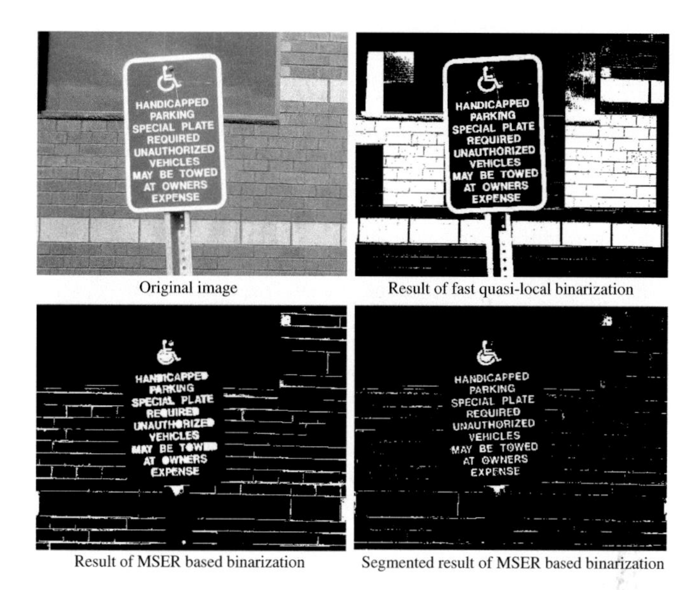

Fig. 4 Comparison of the quasi-local binarization with MSER based before and after additional segmentation

of the statistical quasi-local binarization, only one of the words remains unrecognized what can be considered as a very good result.

4 Conclusions and Future Work

Obtained experimental results have confirmed the validity of the proposed approach which may be helpful for a fast preprocessing of natural images subjected to further Optical Character Recognition. Nevertheless, our method requires a further research in order to improve its universality by an automatic choice of the number of randomly drawn samples as well as the proper block size depending on the resolution of the image and rough estimation of its contents.

Further experiments can also be directed towards a fast correction of lighting differences, which may be present in natural images, based on the obtained map of local threshold values.

References

1. International Telecommunication Union recommendation BT.709-5—parameter values for the HDTV standards for production and international programme exchange (2001)
2. International Telecommunication Union recommendation BT.601-7—studio encoding parameters of digital television for standard 4:3 and wide-screen 16:9 aspect ratios (2011)

3. Bissacco, A., Cummins, M., Netzer, Y., Neven, H.: PhotoOCR: Reading text in uncontrolled conditions. In: Proceedings of IEEE International Conference on Computer Vision (ICCV), pp. 785–792 (2013)
4. de Campos, T.E., Babu, B.R., Varma, M.: Character recognition in natural images. In: Proceedings of the International Conference on Computer Vision Theory and Applications (2009)
5. Chen, H., Tsai, S., Schroth, G., Chen, D., Grzeszczuk, R., Girod, B.: Robust text detection in natural images with edge-enhanced Maximally Stable Extremal Regions. In: Proceedings of the 18th IEEE International Conference on Image Processing (ICIP), pp. 2609–2612 (2011)
6. Epshtein, B., Ofek, E., Wexler, Y.: Detecting text in natural scenes with stroke width transform. In: Proceedings of the IEEE Conference on Computer Vision and Pattern Recognition (CVPR), pp. 2963–2970 (2010)
7. Forczmański, P., Frejlichowski, D.: Robust stamps detection and classification by means of general shape analysis. In: Bolc, L., Tadeusiewicz, R., Chmielewski, L., Wojciechowski, K. (eds.) Computer Vision and Graphics. Lecture Notes in Computer Science, vol. 6374, pp. 360–367. Springer, Berlin (2010)
8. Gooch, A.A., Olsen, S.C., Tumblin, J., Gooch, B.: Color2Gray: salience-preserving color removal. ACM Trans. Graph. **24**(3), 634–639 (2005)
9. Grundland, M., Dodgson, N.A.: Decolorize: fast, contrast enhancing, color to grayscale conversion. Pattern Recogn. **40**(11), 2891–2896 (2007)
10. Ikica, A., Peer, P.: Swt voting-based color reduction for text detection in natural scene images. EURASIP J. Adv. Sig. Process. **2013**(1), Article ID 95 (2013)
11. Kapur, J., Sahoo, P., Wong, A.: A new method for gray-level picture thresholding using the entropy of the histogram. Comput. Vis. Graph. Image Process. **29**(3), 273–285 (1985)
12. Milyaev, S., Barinova, O., Novikova, T., Kohli, P., Lempitsky, V.: Image binarization for end-to-end text understanding in natural images. In: Proceedings of the 12th International Conference on Document Analysis and Recognition (ICDAR), pp. 128–132 (2013)
13. Nagy, R., Dicker, A., Meyer-Wegener, K.: NEOCR: A configurable dataset for natural image text recognition. In: Iwamura, M., Shafait, F. (eds.) Camera-Based Document Analysis and Recognition. Lecture Notes in Computer Science, vol. 7139, pp. 150–163. Springer, Berlin (2012)
14. Otsu, N.: A threshold selection method from gray-level histograms. IEEE Trans. Syst. Man Cybern. **9**(1), 62–66 (1979)
15. Roubtsova, N.S., Wijnhoven, R.G.J., de With, P.H.N.: Integrated text detection and recognition in natural images. In: Image Processing: Algorithms and Systems X and Parallel Processing for Imaging Applications II. Proceedings of SPIE, vol. 8295, pp. 829507–829521 (2012)
16. Smith, R.: An overview of the Tesseract OCR engine. In: Proceedings of the 9th International Conference on Document Analysis and Recognition (ICDAR), vol. 2, pp. 629–633 (2007)
17. Su, B., Lu, S., Tian, S., Lim, J.H., Tan, C.L.: Character recognition in natural scenes using convolutional co-occurrence HOG. In: Proceedings of 22nd International Conference on Pattern Recognition (ICPR), pp. 2926–2931 (2014)

Fast Machine Vision Line Detection for Mobile Robot Navigation in Dark Environments

Piotr Lech, Krzysztof Okarma and Jarosław Fastowicz

Abstract Navigation of mobile robots based on video analysis becomes one of the most popular application areas of machine vision in automation and robotics. Recently growing popularity of Unmanned Aerial Vehicles (drones) as well as some other types of autonomous mobile robots leads to rapid increase of their application possibilities e.g. related to exploration of some areas hardly accessible for people, such as caves, underground corridors, bunkers etc. However, such places are specific in view of lighting conditions so many classical image analysis algorithms cannot be applied effectively for navigation of mobile robots in such environments. In order to utilize the image data for robot navigation in such places some modified machine vision algorithms should be applied such as fast line detection based on statistical binarization discussed in this paper.

Keywords Machine vision · Robot navigation · Edge detection

1 Introduction

Fully visual universal navigation of mobile robots is still considered rather as a future trend than a working solution. Such situation is caused partially by relatively high price of high resolution cameras in comparison to widely available "classical" sensors e.g. infrared, ultraviolet, laser, sonar, etc. Another reason is the necessity to develop

P. Lech (✉) · K. Okarma · J. Fastowicz
West Pomeranian University of Technology, Szczecin, Poland
e-mail: piotr.lech@zut.edu.pl

K. Okarma
e-mail: krzysztof.okarma@zut.edu.pl

J. Fastowicz
e-mail: jaroslaw.fastowicz@zut.edu.pl

P. Lech · K. Okarma · J. Fastowicz
Faculty of Electrical Engineering, Department of Signal Processing and Multimedia Engineering, 26. Kwietnia 10, 71-126 Szczecin, Poland

R.S. Choraś (ed.), *Image Processing and Communications Challenges 7*,
Advances in Intelligent Systems and Computing 389,
DOI 10.1007/978-3-319-23814-2_18

the appropriate image analysis algorithms as the amount of data present on images is much higher than can be collected using many other types of robot sensors. For this reason image data is often considered as supplementary for signals acquired by some other sensors [4].

Nevertheless, during recent years the availability of relatively cheap high quality cameras has become much higher as well as the development of many embedded systems and platforms which can be effectively applied in mobile robotics, including drones and underwater vehicles [7, 8], allowing to perform much more computations useful for real-time robot navigation purposes [1, 3]. The applications of machine vision are becoming much more common not only in mobile robotics but also in mechatronics and industrial automation [5] or transport, especially in view of development of Intelligent Transportation Systems [12].

Despite the necessity of processing of relatively big amount of image data, sensitivity of cameras to changing lighting conditions and computational demands, machine vision applications in mobile robotics have several advantages allowing to gain much more useful information than obtained from classical sensors. A good example can be the application of cameras for line following robots [13] allowing the prediction of line's direction as well as navigation along the interrupted line.

Some of the most relevant issues of machine vision in robotics are related to applications of Visual Self-Localization And Mapping (SLAM) [15, 16] as well as visual odometry [11]. One of the most recent trends is the fusion of local image features for robot visual localization discussed in the paper [2].

Since one of the challenges for machine vision applications is still related to their applications in varying lighting conditions, the idea of fast statistical line detection in dark images is presented in the paper together with results of some experiments related to its applicability for the visual navigation of mobile robots in dark corridors.

2 Line Detection for Robot Navigation

Visual navigation of mobile robots in corridors is typically based on detection of lines representing the elements of known environment such as wall corners, doors and windows. Assuming good lighting conditions, the navigation task is quite easy and the necessary lines can be determined by one of the classical edge detectors such as Sobel, Prewitt, Roberts or Canny filters, well-known in image processing. In case of high accuracy demands some of subpixel edge detection algorithms can be applied, which are useful also in some other application areas [6]. However, in most mobile robotic applications, the subpixel accuracy is not necessary as the time constraints usually become more critical.

A reliable line detection based on edge filters requires images with relatively high contrast so in case of dark images some problems occur as the gradient values become relatively small and the presence of noise plays an important role.

It is worth noticing that an image containing high number of detected lines or edges may be useless for robot's navigation as the most desired situation is the presence of

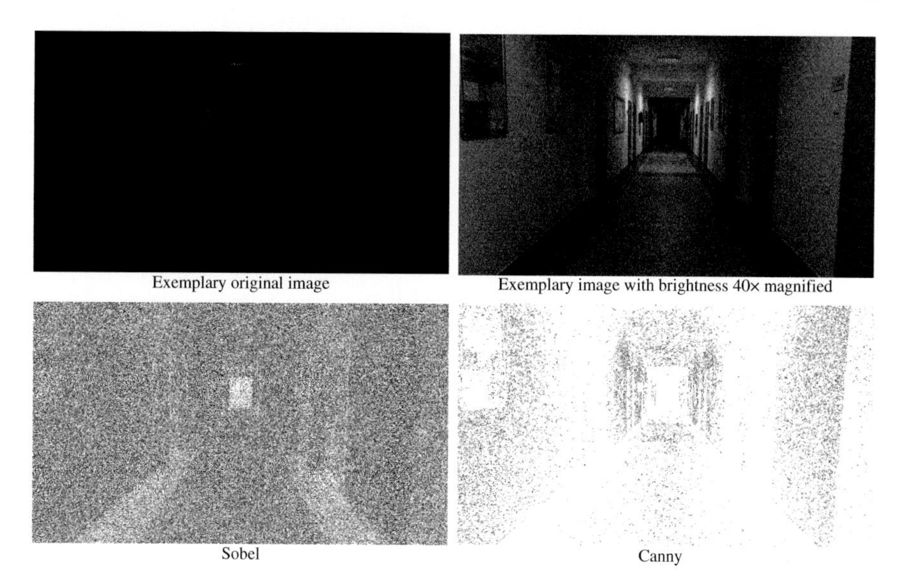

Fig. 1 Exemplary results of line detection for a high resolution *dark image*

a limited number of lines in the image, useful for navigation purposes, representing the boundaries of real objects. On the other hand, assuming the real-time navigation, application of some time-consuming image analysis algorithms for high resolution (e.g. FullHD or higher) images can be troublesome so decrease of image resolution should be made in order to increase the performance of image analysis.

Some examples of edge detection applied for high resolution dark images are presented in Fig. 1 where the high number of detected edges can be noticed both for Sobel and Canny filters (due to high resolution neighboring edges have been averaged and look like greyscale). Application of some other popular edge detection filters leads to similar results.

Although the computational complexity of Canny edge detector is much higher than using Sobel filter, in both cases obtained processing time may be too high for real-time robot navigation purposes. The average processing time for 4608×2592 pixels images during our experiments in MATLAB environment with Image Processing Toolbox using the built-in *edge* function is about 4.3 s for Canny and 0.35 s for Sobel filter. It should be stated that those values should not be considered as the absolute values in view of real-time constraints as more efficient implementation is possible. Nevertheless treating them as the reference values further comparison of the processing speed may be conducted using the same environment apart from specific implementation issues.

It is worth to notice that denoising filters (e.g. low-pass or median), which could be potentially useful for removing any artifacts from images obtained as the result of edge detection, are considered as too complex from computational point of view and therefore their application has not been further analyzed in our experiments.

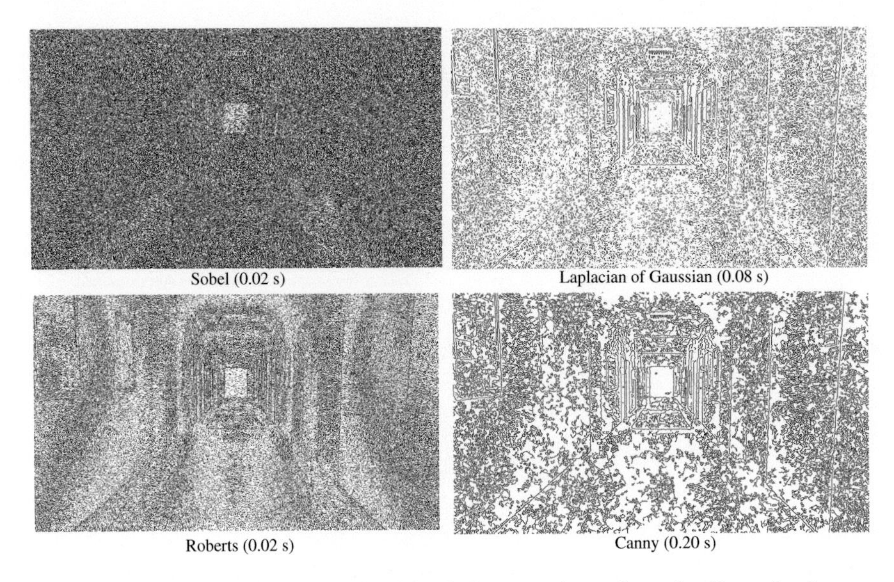

Fig. 2 Exemplary results of line detection for a *dark image* using various edge filters after decrease of image resolution with comparison of processing time

Downsizing the dark images should not be conducted using classical fast nearest neighbor approach since the noise present in the image would be mixed together with useful information representing the edges. Relatively good results can be obtained using the bicubic or bilinear interpolation kernels as well as simple averaging, depending on the specific edge filter applied for downsized greyscale image. An illustration of such dependence is shown in Fig. 2 where the results of edge detection in such downsized images, obtained using *imresize* function using box (averaging) kernel, are presented together with average processing time for comparison purposes.

As can be noticed although the lines are almost invisible in the original image, the most useful results (however far from ideal) can be obtained by Canny filter and the effects of applying much faster Sobel or Roberts filters, or even Laplacian of Gaussian (LoG) method, cannot be used effectively for robot's navigation. Nevertheless, obtained binary images require further processing e.g. using morphological filters.

3 Proposed Approach

Since the computation speed plays an important role in many robotic applications, one of typical solutions in this area is the application of binary image analysis. Regardless of the fact that image thresholding decreases the amount of image data which can be used in further processing, such generalization of an image can be considered positively, especially for images containing many unnecessary details or contaminated by noise.

Considering the fact that an input images for edge detection filters can be both greyscale and binary, where the latter one represents rapid changes of luminance associated with sharp edges, the image binarization step can be included into the edge detection procedure before applying the edge filter. Since the most popular universal image binarization method has been proposed by Otsu [14], it has been chosen for the verification of the validity of the idea proposed in this paper. In order to compare the results and the processing speed, two possible solutions, differing by the order of consecutive operations, have been investigated:

- downsizing of greyscale image, binarization and edge detection,
- binarization, downsizing of binary image and edge detection.

The comparison of the obtained experimental results is presented in Fig. 3. Since edge detection results achieved for two analyzed approaches are very similar, only

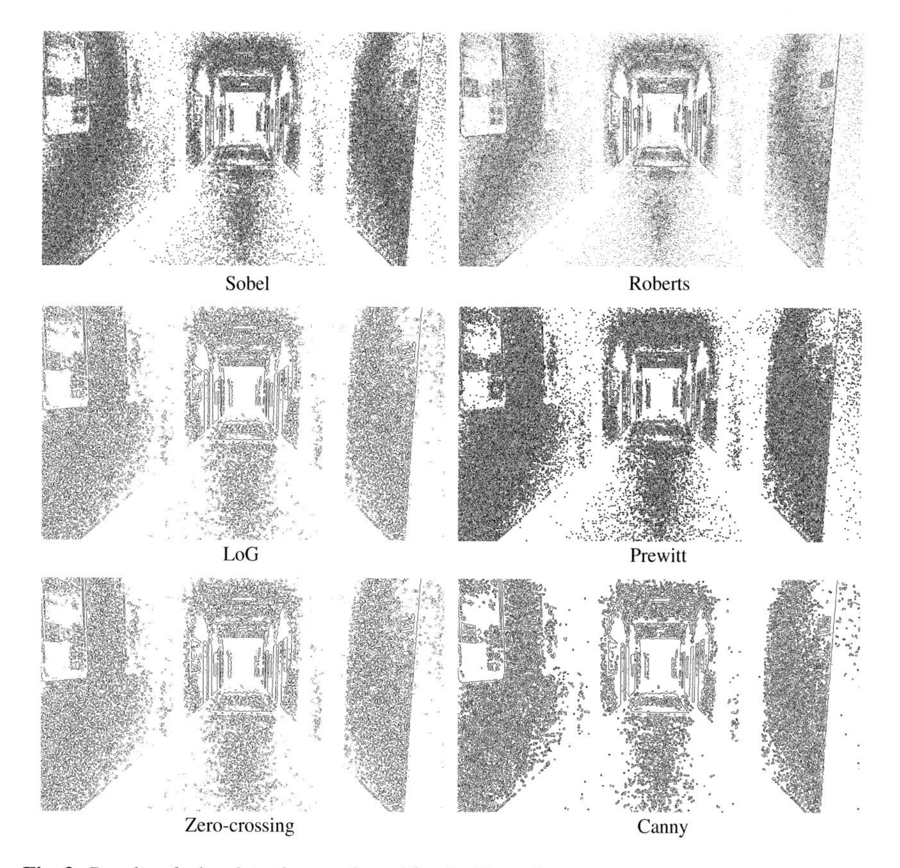

Fig. 3 Results of edge detection conducted for the binary images

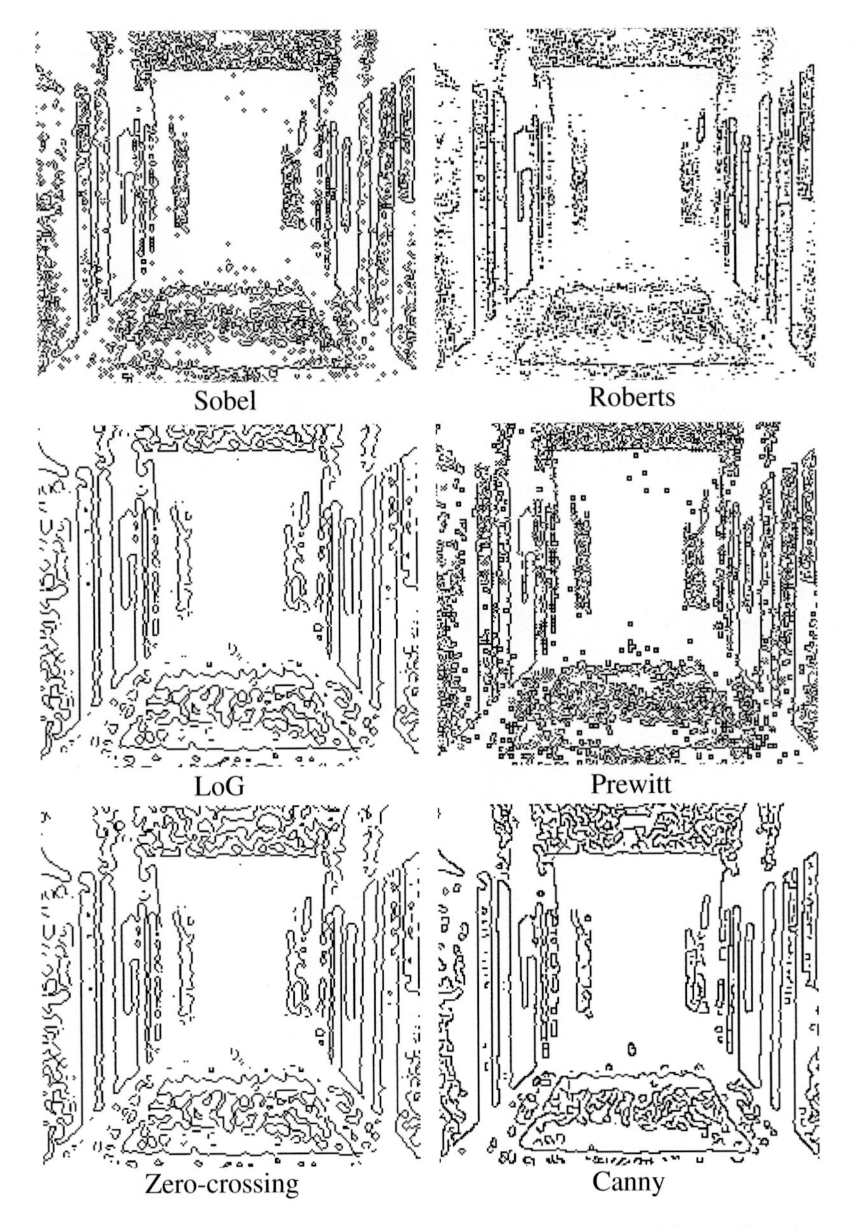

Fig. 4 Zoomed fragments of images obtained by edge detection conducted for the binary images

one set of images is presented in order to illustrate the differences among various edge filters, which has been obtained using the first processing scheme.

Analyzing the zoomed images presented in Fig. 4 an advantage of relatively fast Roberts filter can be noticed as the resulting image contains lines and isolated small groups of pixels which are relatively easy to remove using morphological operations. The artifacts introduced by the other edge detection methods, resulting from the noise present in the original dark image, can be much harder to remove leading to further increase of total computational cost.

Considering the overall computation time for the proposed processing schemes a significant increase of performance can be noticed in comparison to edge detection conducted using the full resolution greyscale image (leading to unsatisfactory results as shown in Fig. 1). Using the same experimental environment the average total processing time for the first scheme is about 0.028 s whereas for the second scheme is about 0.051 s.

In both cases image downsizing (by the factor of 4.8 leading to 960×540 pixels resolution) lasts about 0.011 s and edge detection using Roberts or Sobel filter takes about 0.016 s. The difference is related to Otsu's binarization which requires the computation of image histogram which is much faster for smaller images—in our experiments we have obtained 0.025 s for full resolution image and 0.0015 s for the downsized one.

Since one of the bottlenecks of the proposed processing scheme is the binarization step, especially conducted for the full resolution image, further increase of the processing speed should be related with faster image thresholding. This can be achieved using the statistical sampling proposed in the paper [10] allowing to estimate the image histogram using strongly limited number of samples (e.g. using only 100 samples from the whole image) with the accuracy which is enough for proper calculation of the threshold using e.g. Otsu's method. Applying the proposed fast statistical histogram estimation for Otsu's binarization about 30 % of the computation time has been saved during our experiments obtaining the same binarization results.

4 Concluding Remarks and Future Work

Another possibility of increasing the processing speed is the application of the statistical sampling for image downsizing procedure removing the necessity of averaging of all pixels in the block which is represented by a single pixel in the destination image. Although such approach decreases the quality of the resulting image, it can still be useful for line detection [9]. Nevertheless, an efficient application of such approach can be considered mainly for significant change of image size. Its usefulness for dark images combined with the approach discussed in this paper will be verified in our further research.

Proposed method of the fast edge detection on the dark images using the statistical histogram estimation applied for Otsu's thresholding before the use of Roberts filter, allows an efficient processing of images which can be used for mobile robot navigation purposes in hard lighting conditions. Our future experiments will concentrate on the implementation and optimization of the algorithm on the physical mobile robot's board. Some simulation experiments e.g. using the Simbad robot simulating environment are planned as well.

References

1. Bonon-Font, F., Ortiz, A., Oliver, G.: Visual navigation for mobile robots: a survey. J. Intel. Rob. Syst. **53**(3), 263–296 (2008)
2. Campos, F.M., Correia, L., Calado, J.: Robot visual localization through local feature fusion: an evaluation of multiple classifiers combination approaches. J. Intel. Rob. Syst. **77**(2), 377–390 (2015)
3. Chatterjee, A., Rakshit, A., Singh, N.N.: Vision Based Autonomous Robot Navigation—Algorithms and Implementations. Studies in Computational Intelligence, vol. 455. Springer, Berlin (2013)
4. Desouza, G., Kak, A.: Vision for mobile robot navigation: a survey. IEEE Trans. Pattern Anal. Mach. Intel. **24**(2), 237–267 (2002)
5. Domek, S., Dworak, P., Grudziński, M., Okarma, K.: Calibration of cameras and fringe pattern projectors in the vision system for positioning of workpieces on the CNC machines. In: Gosiewski, Z., Kulesza, Z. (eds.) Mechatronic Systems and Materials V, Solid State Phenomena, vol. 199, pp. 229–234. Trans Tech Publications (2013)
6. Fabijańska, A.: A survey of subpixel edge detection methods on images of heat-emitting metal specimens. Int. J. Appl. Math. Comput. Sci. **22**(3), 695–710 (2012)
7. Garcia, R., Nicosevici, T., Ridao, P., Ribas, D.: Towards a real-time vision-based navigation system for a small-class UUV. In: Proceedings of the IEEE/RSJ International Conference on Intelligent Robots and Systems (IROS 2003), vol. 1, pp. 818–823 (2003)
8. Horgan, J., Toal, D.: Computer vision applications in the navigation of unmanned underwater vehicles. In: Inzartsev, A.V. (ed.) Underwater Vehicles, pp. 195–214. InTech (2009)
9. Lech, P.: The detection of horizontal lines based on the monte carlo reduced esolution images. In: Chmielewski, L., Kozera, R., Shin, B.S., Wojciechowski, K. (eds.) Computer Vision and Graphics, Lecture Notes in Computer Science, vol. 8671, pp. 374–381. Springer International Publishing (2014)
10. Lech, P., Okarma, K.: Optimization of the fast image binarization method based on the Monte Carlo approach. Elektron. Ir Elektrotechnika **20**(4), 63–66 (2014)
11. Maimone, M., Cheng, Y., Matthies, L.: Two years of visual odometry on the Mars exploration rovers. J. Field Robot. **24**(3), 169–186 (2007)
12. Mazurek, P., Okarma, K.: Vehicle tracking using a multi-scale Bayesian algorithm for a perspective image of a road. In: Mikulski, J. (ed.) Modern Transport Telematics. Communications in Computer and Information Science, vol. 239, pp. 346–353. Springer, Berlin (2011)
13. Okarma, K., Lech, P.: A fast image analysis technique for the line tracking robots. In: Rutkowski, L., Scherer, R., Tadeusiewicz, R., Zadeh, L.A., Zurada, J.M. (eds.) Artificial Intelligence and Soft Computing. Lecture Notes in Computer Science, vol. 6114, pp. 329–336. Springer, Berlin (2010)
14. Otsu, N.: A threshold selection method from gray-level histograms. IEEE Trans. Syst. Man. Cybern. **9**(1), 62–66 (1979)
15. Se, S., Lowe, D., Little, J.: Vision-based global localization and mapping for mobile robots. IEEE Trans. Robot. **21**(3), 364–375 (2005)
16. Steder, B., Grisetti, G., Stachniss, C., Burgard, W.: Visual slam for flying vehicles. IEEE Trans. Robot. **24**(5), 1088–1093 (2008)

Adjustment of Viterbi Algorithm for Line Following Robots

Grzegorz Matczak and Przemysław Mazurek

Abstract Line following robots are used in numerous application areas. The tracking of weak line is challenging, especially if SNR is high, so application of Track–Before–Detect algorithm is necessary. The Viterbi algorithm is assumed in this paper and the possibilities of optimization are considered. Two metric are applied in Monte Carlo tests—the direct metric and proposed boundary metric. The optimization of Viterbi algorithm is based on non single row movements of moving window.

Keywords Track-Before-Detect · Tracking · Viterbi algorithm · Line following robots

1 Introduction

Robot navigation could be based on satellite navigation (GPS), inertial navigation or vision systems. Vision based navigation requires sophisticated algorithms for general case, but the simplification for well known or defined working area is possible. Line following robots use intentionally added line that defines possible routes. Many line following robots are proposed and used, because vision system could be very simple. Simplest systems uses two light reflective sensors for the measurement of reflected light from the line and the background. Fixed value for both measurements is obtained if the sensor is over the line. Different values occurs if the sensor is misaligned and the error could be used by the control algorithm. Rapid movements and high velocities require more advanced sensors, so linear light sensors are applied also. Forward looking cameras (Fig. 1) allow the estimation of line before robot

G. Matczak̦ (✉) · P. Mazurek
Department of Signal Processing and Multimedia Engineering, West Pomeranian University of Technology Szczecin, 26. Kwietnia 10 St., 71126 Szczecin, Poland
e-mail: grzegorz.matczak@gmail.com

P. Mazurek
e-mail: przemyslaw.mazurek@zut.edu.pl

R.S. Choraś (ed.), *Image Processing and Communications Challenges 7*,
Advances in Intelligent Systems and Computing 389,
DOI 10.1007/978-3-319-23814-2_19

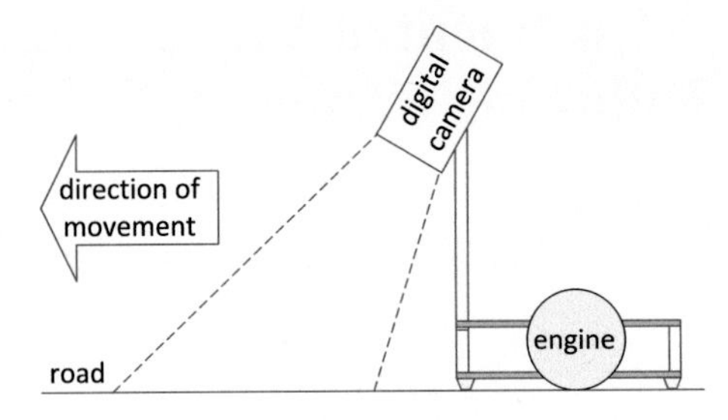

Fig. 1 Schematic of line following robot with forward looking camera

Fig. 2 Example line following robot with forward looking camera

(Fig. 2) at far distance also and this is the best solution, but requires sophisticated algorithms. Systems with artificially added lines are simple to design and the desired high background to line contrast could be achieved.

There are numerous application where the line is natural or artificial but significantly degraded. The quality of the line is low in such cases, so the contrast between the background and the line is not sufficient for the application of simple vision systems and algorithms. This paper assumes such cases, that are important for numerous applications of line following robots.

1.1 Related Works

Navigation using lines, that are highly deteriorated are considered in numerous Lane Departure Warning (LDW) systems. The traffic safety could be improved if LDW system is used [14]. The very interesting research area is the combined navigation and inspection of power lines by UAVs [2]. There are numerous application of line following robots related to the harvesting, trash compacting [12].

Line estimation could be improved by the application of Track–Before–Detect algorithms. Such class of tracking algorithm [1, 13] allows the line estimation for $SNR < 1$ cases also. Viterbi algorithm [15] for line following robots is proposed in [8]. Lines that are noise pattern only, could be processed by the approach considered in [10]. Directional filtering of lines improves estimation also [7]. Local approach for the detection of disturbed lines is considered in [4] for example.

1.2 Content of the Paper

Two metrics for quality of estimation are considered in Sect. 2 (direct and new proposed boundary metric). The selection of metric influences the results and it is shown in Sect. 2. The optimization of Viterbi algorithm could be based on different parameters and two of them are selected and considered in Sect. 3 due to mobile robot applications. Monte Carlo tests are provided in Sect. 4. The results and discussion are provided in Sect. 4. Final conclusions are described in Sect. 5.

2 Metrics for Horizontal Error

The direct metric (E_i) for the horizontal error value could be defined as a difference between true X_i and estimated $\hat{X}_i$ positions:

$$E_i = X_i - \hat{X}_i. \quad (1)$$

Alternative approaches are possible and another metric is proposed (E^B) that considers width of line, so left (X_i^L) and right (X_i^R) boundaries are used. The error value in horizontal direction is typically non–zero for direct metric, but the error value should be zero if the estimated line position is located between line boundaries:

$$E_i^B = \begin{cases} 0 & : X_i^L \le \hat{X}_i \le X_i^R \\ X_i^L - \hat{X}_i & : \hat{X}_i \le X_i^L \\ \hat{X}_i - X_i^R & : X_i^R \le \hat{X}_i \end{cases}. \quad (2)$$

Estimated line position (Fig. 3 left), with pixel accuracy, that is inside boundaries gives zero value error for direct metric. The application of boundary metric allows the acceptation of all position between boundaries (Fig. 3 right).

New boundary metric is proposed, because the selection of metric has impact on results. The example of line tracking for wide line is presented in Fig. 4 and direct metric and boundary metrics are compared. The cumulative error value is significantly reduced which shows impact of the metric. The results of Monte Carlo test for variable noise are shown in Fig. 5 and the difference between both metrics is well visible.

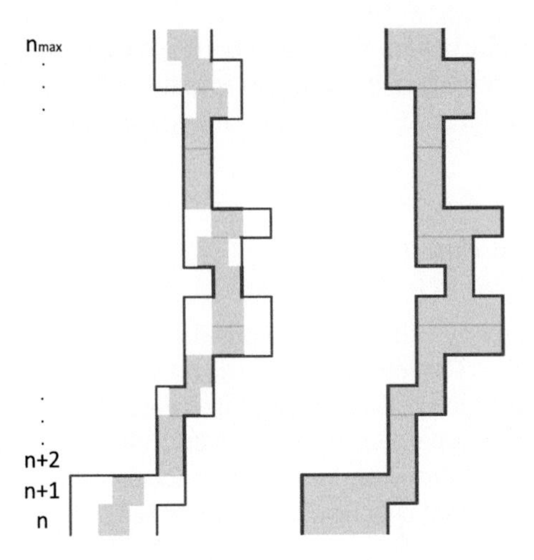

Fig. 3 Direct (*left*) and boundary metrics (*right*) regions of zero value error (*gray*)

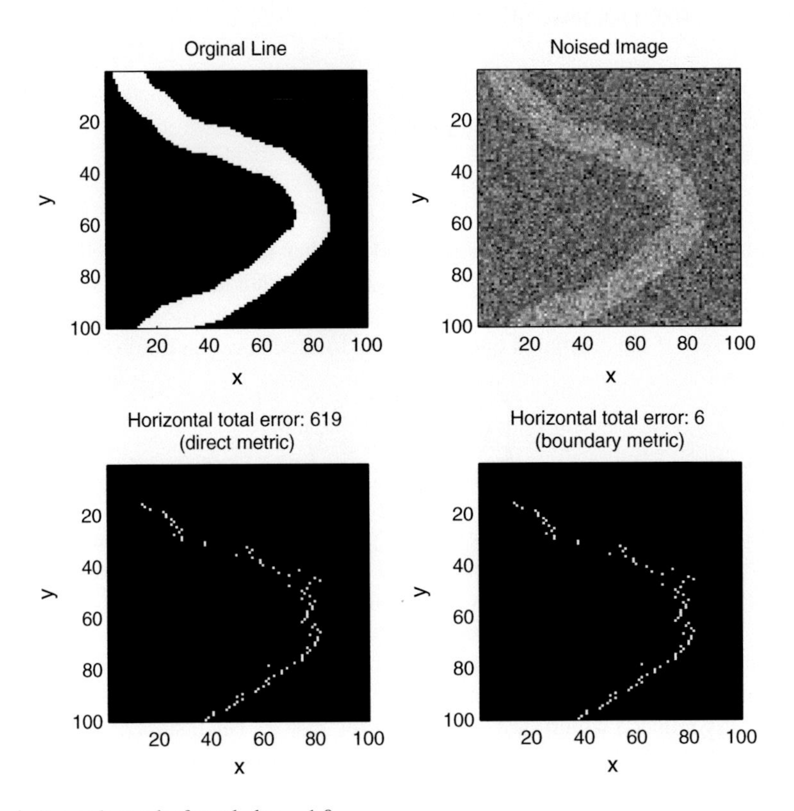

Fig. 4 Example results for std. dev. = 1.0

3 Adjustment of Viterbi Algorithm for Mobile Robots

Moving windows approach has a high degree of similarity to driving mobile robot with digital camera looking down on route. Mobile robots very often moves at different speed and directions. Windows approach ensure possibility of moving mobile robot simulation in the same direction with different speed.

Windows approach from previous publications [7, 8, 10] defines shifting windows in image. Window has dimensions width as image width and window height as *depth* parameter. Window in the image was started from $n = 1$ row and finish on $n = n_{max}$, where n_{max} is known from value of *depth* parameter. Next processing of window starts from $n = 2$ row and finish on $n = n_{max} + 1$ row. The shift of window is equal one pixel in this case. This is the smallest possible move of the window in simulation. Added ability to change value of shifted window contributed to model mobile robot speed. If we parametrize this adding part we get $(n + step)$, where the *step* is mentioned parameter of shifted window. Markings shown in Fig. 5.

The selection of *depth* and *step* parameters depends on several factors. First is resolution of image. The image from digital camera can be in different resolution. When resolution is low, the computation is fast but with less accuracy. When resolution increases, the accuracy is improved and the number of computation is growing. When the same image is captured with differs resolution, the depth parameter is various, because for higher resolution gained greater number of pixel. Second factor is height of the digital camera relative to ground, because when camera is on higher position it can register more route. Last factor is *FPS* (Frames Per Second) value of

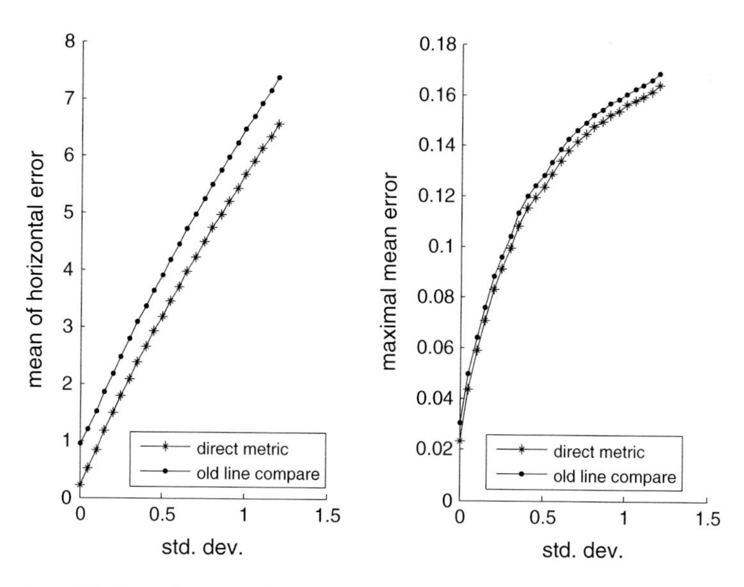

Fig. 5 Mean of horizontal error (*left*) and maximal mean error (*right*). Monte Carlo test for std. dev. = 0.13 and line direction change probability 0.25

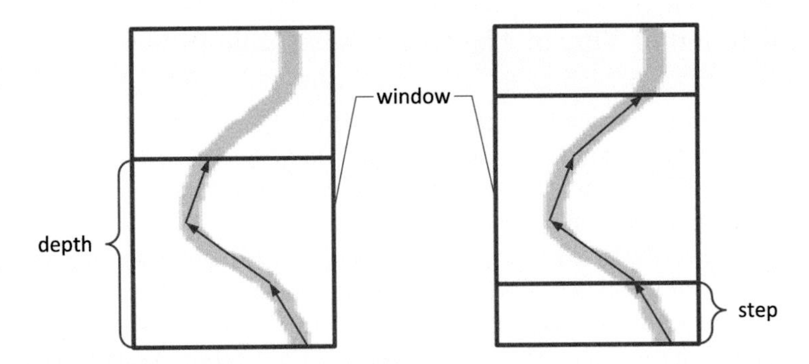

Fig. 6 Shifted window approach

digital camera. Higher number *FPS* provided better accuracy with following line, because faster moving of mobile robot can be register (Fig. 6).

4 Results and Discussion

The same set of generated line was made on the image with 200×60 pixel resolution. After generated line, disturbing lines with standard deviation from Gaussian noise were added. Next to the image Gaussian noise was added. Each set has 300 test scenario for specified Gaussian noise in std. dev. range (0–1.3) every 0.05 value.

Two values of parameters were studied in this paper: $depth = (16, 32)$, and three value of parameter $step = (1, 0.25 \cdot depth, 0.5 \cdot depth)$. For parameter $depth = 16$ and $step = (1, 4, 8)$, results are shown in Fig. 7. The results for parameter $depth = 32$ and $step = (1, 8, 16)$ are shown in Fig. 8.

The change of parameter *step* does not aggravate the algorithm, so algorithm will work with high speed of mobile robot. The result is better for higher noise level. The number of computation changes with step parameter. The algorithm finds faster for higher step parameter. The comparison between different depth parameter (Figs. 7 and 8) shown that parameter have small impact on the results.

The algorithm is resistant to change speed of mobile robot that is related to *depth* parameter. If we want our robot to move quickly we need to have as much as possible scope of the road. It can be achieved in several ways: by increasing the field–of–view of camera, or by increasing of height from the ground. The camera with more *FPS* could be applied for the better accuracy.

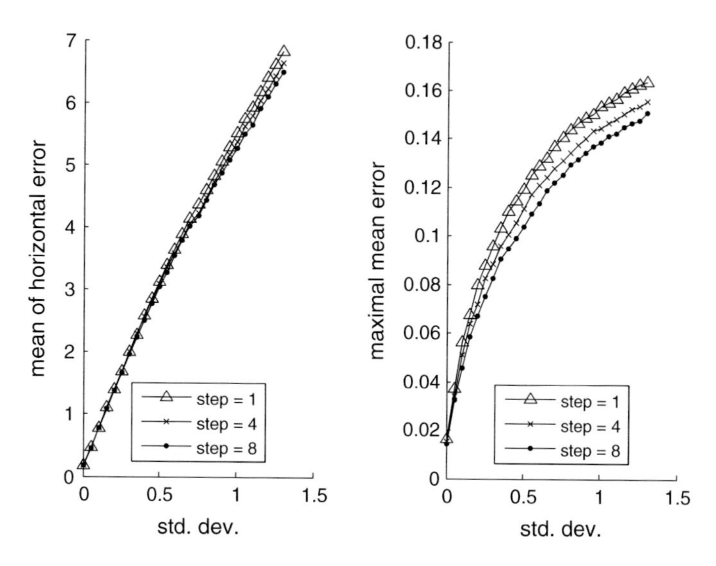

Fig. 7 Mean error and maximal mean error for different *step* parameter, (Monte Carlo test for std. dev. and *line direction* change probability 0.25 and *depth* = 16)

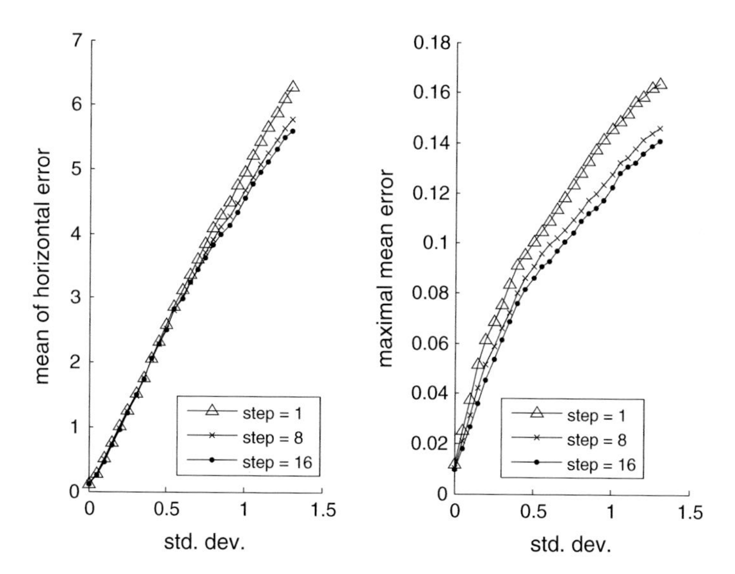

Fig. 8 Mean error and maximal mean error for different *step* parameter, (Monte Carlo test for std. dev. and *line direction* change probability 0.25 and *depth* = 32)

5 Conclusions

The computational cost of considered algorithm is acceptable for modern microcontrollers for low–resolution images. Proposed approach allows the reduction of computation cost by the application of larger step of moving window. High resolution images need TBD processing using multiple processors [9] or GPGPUs [5, 6].

Presented example assumes additive Gaussian noise disturbance. The removal of non–line object from the image could be based on concept presented in [3]. Texture related disturbances could be also identified [11].

References

1. Blackman, S., Popoli, R.: Design and Analysis of Modern Tracking Systems. Artech House (1999)
2. Chen, Z., Ellis, T.: Automatic lane detection from vehicle motion trajectories. In: Workshop on Vehicle Retrieval in Surveillance (VRS) in conjunction with 2013 10th IEEE International Conference on Advanced Video and Signal Based Surveillance, pp. 466–471 (2013)
3. Frejlichowski, D., Gościewska, K., Forczmański, P., Nowosielski, A., Hofman, R.: The removal of false detections from foreground regions extracted using adaptive background modelling for a visual surveillance system. Computer Information Systems and Industrial Management. Lecture Notes in Computer Science, vol. 8104, pp. 253–264. Springer, Berlin (2013)
4. Marchewka, A.: Crack detection on asphalt surface image using local minimum analysis. Adv. Intel. Soft Comput. **84**, 353–359 (2010)
5. Mazurek, P.: Optimization of bayesian track-before-detect algorithms for GPGPUs implementations. Electr. Rev. R. **86**(7), 187–189 (2010)
6. Mazurek, P.: Code reordering using local random extraction and insertion (LREI) operator for GPGPU-based track-before-detect systems. Soft Comput. **18**(6), 1095–1106 (2013)
7. Mazurek, P.: Directional filter and the viterbi algorithm for line following robots, Computer Vision and Graphics, ICCVG 2014, LNCS, vol. 8671, pp. 428–435. Springer (2014)
8. Mazurek, P.: Line estimation using the viterbi algorithm and track-before-detect approach for line following mobile robots. pp. 788–793 (2014)
9. Mazurek, P.: Parallel distributed downsampled spatio-temporal track-before-detect algorithm. pp. 119–124 (2014)
10. Mazurek, P.: Viterbi algorithm for noise line following robots. Adv. Intel. Syst. Comput. AISC **313**, 111–118 (2015)
11. Okarma, K., Frejlichowski, D., Czapiewski, P., ForczmańĎski, P., Hofman, R.: Similarity estimation of textile materials based on image quality assessment methods. Lect. Notes Comput. Sci. **8671**, 478–485 (2014)
12. Ollis, M.: Perception Algorithms for a Harvesting Robot. CMU-RI-TR-97-43, Carnegie Mellon University (1997)
13. Stone, L., Barlow, C., Corwin, T.: Bayesian Multiple Target Tracking. Artech House (1999)
14. Taubel, G., Yang, J.S.: A lane departure warning system based on the integration of the optical flow and Hough transform methods. In: 2013 10th IEEE International Conference on Control and Automation (ICCA) Hangzhou, China, June 12–14, pp. 1352–1357 (2013)
15. Viterbi, A.: Error bounds for convolutional codes and an asymptotically optimum decoding algorithm. IEEE Trans. Inf. Theory **13**(2), 260–269 (1967)

Studentized Range for Spatio–Temporal Track–Before–Detect Algorithm

Przemysław Mazurek

Abstract Two preprocessing approaches, dedicated to the tracking of low SNR objects, are compared—the variance and studentized range. Both approaches are applied using sliding window for noised signal for improving the detection of weak object. Influences of window size are compared and studentized range shows improvement over variance. Both approaches are compared using the Monte Carlo test with numerous tracking scenarios using ST–TBD (Spatio–Temporal Track–Before–Detect) algorithm. The results shows better performance of the studentized range for the detection and tracking weak point objects.

Keywords Track-Before-Detect · Tracking · Studentized range

1 Introduction

Tracking systems are applied for numerous applications [1, 10] and the tracking of weak objects ($SNR < 1$) is important especially for rapid and robust detection of moving objects. Applications of advanced detection techniques together with proper motion model allow the detection and tracking of weak signals. The conventional detection and tracking scheme is not sufficient for really weak objects, due to limitations of the detection part, unfortunately. The detection algorithm converts raw to binary data, so the part of information about object is lost. Even the application of motion estimation algorithms cannot improve final results. The alternative approach based on the tracking before detection allows raw data processing and is well suited for such class of objects. Signals hidden in the background noise floor could be processed and the object could be tracked successfully. The main limitation of this

P. Mazurek (✉)
West–Pomeranian University of Technology, Szczecin, Poland
e-mail: przemyslaw.mazurek@zut.edu.pl

P. Mazurek
Department of Signal Processing and Multimedia Engineering, 26. Kwietnia 10 St., 71126 Szczecin, Poland

R.S. Choraś (ed.), *Image Processing and Communications Challenges 7*,
Advances in Intelligent Systems and Computing 389,
DOI 10.1007/978-3-319-23814-2_20

approach (TBD–Track–Before–Detect) is the giant computation cost, because all possible trajectories should be processed even if no one object is in the range.

1.1 Related Works

Advantages of TBD approach could be achieved only if the signal is preprocessed to the expected value range by the particular TBD algorithm. The typical TBD algorithm improves SNR only if the signal values are over the mean value of the noise. Numerous techniques are proposed for conversion (preprocessing) of specific signals to the desired signal representation. In [4] is presented solution for amplitude modulated signals using hierarchical approach. In [5] are considered noise signals (object position is characterized by the local noise only), so the noise that is below the background noise floor is tracked. The application of autocovariance allows the preprocessing [8] repetitive object patterns also. Filter banks with multiple window of analysis are considered in [7]. The considered preprocessing techniques were applied to the class of extended objects. This class is related to the object that occupies multiple pixels of signal. Point object with single pixel excitation in observed signal is assumed in this paper. Moreover the observed excitation is not directly related to the linear trajectory, but the position is perturbed by the uniform noise.

1.2 Content and Contribution of the Paper

There are numerous TBD algorithms and the selected one is the ST–TBD (Spatio–Temporal Track–Before–Detect) algorithm that could model a few another algorithms also, e.g. TBD Velocity Filters. This algorithm works as a kind of multidimensional filter, that estimates the position and velocity for all possible trajectories assumed by the motion model.

The detection of noise objects cannot be based on the threshold oriented algorithm, because observed signal is noise only. The preprocessing of measured signal allows the estimation of noise parameters and the application of the detection. Such preprocessing is not sufficient for really weak signals because estimated value is noised also, but long time measurements desired by TBD algorithm allows the estimation of the trajectory.

The model of object and two preprocessing techniques are considered in Sect. 2. ST–TBD algorithm is considered briefly in Sect. 3. The performance of preprocessing algorithms is estimated using Monte Carlo test in Sect. 4. The tracking performance using preprocessing techniques and ST–TBD algorithm together is presented in Sect. 5. The discussion is provided in Sect. 6 and the final conclusion in Sect. 7.

2 Variance and Studentized Range Preprocessing

The tracking of the noise signals needs preprocessing techniques but numerous of them are possible. Additive Gaussian noise of background floor and zero mean value are assumed. The signal is assumed as point object.

The example tracking scenario with multiple 1D measurements is presented in the Fig. 1. The trajectory is linear but the position of observed excitation is perturbed by the uniform noise:

$$x(n) = x(0) + nV + Wu, \tag{1}$$

where $x(n)$—position of object at time n, $x(0)$—starting position, V—velocity of object, and W magnitude of noise u.

The value of excited pixel is modeled as:

$$x^*_{pix} = x_{pix} e_{coef} \tag{2}$$

and this pixel is from W–range area. Such model assumes fixed moving window assigned to ideal trajectory. The position of excited pixel is selected from this window as a pixel with highest value. Excitation coefficient e_{coef} increases the value of this pixel and during the evaluation of preprocessing algorithms $e_{coef} \in \langle 1 - 2 \rangle$ is assumed. Exemplary reference and after excitation signals are shown in Fig. 2.

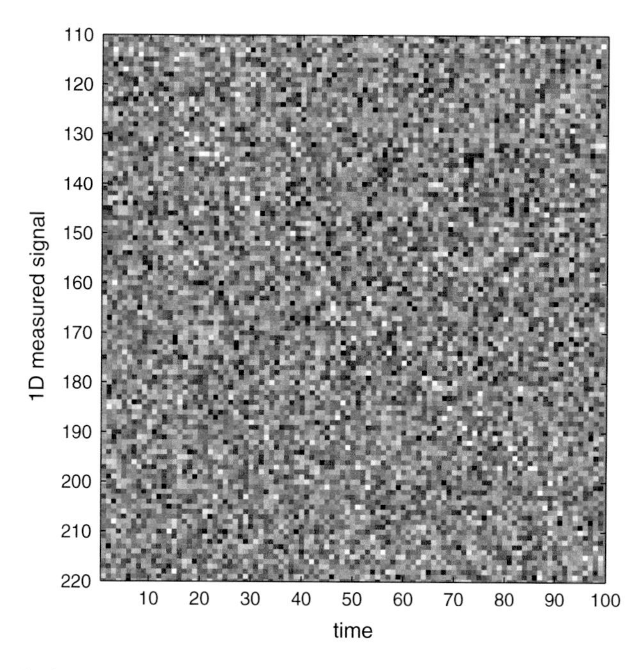

Fig. 1 Example signal—object movement from *top—left* to *bottom—right corner*

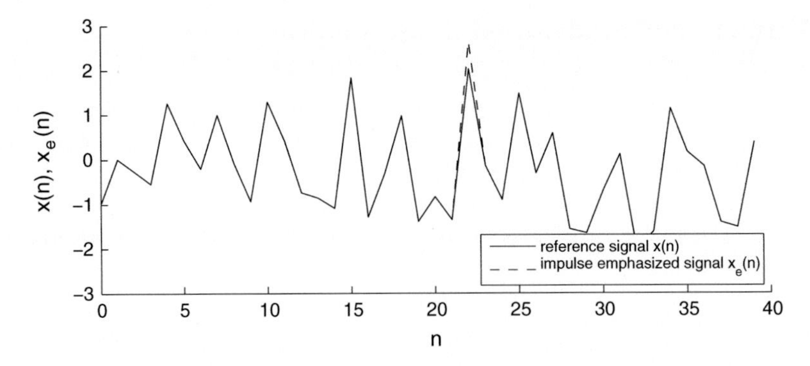

Fig. 2 Reference signal and signal after excitation of single value

Increased value of single pixel (maximal one) changes mean and variance. During the evaluation of preprocessing algorithm the mean changes were rejected due to very small effects.

The preprocessing based on the variance uses moving window approach:

$$std(x)^2 = \frac{1}{N-1} \sum_{i=1}^{N} (x_i - \bar{x})^2 . \tag{3}$$

Alternative preprocessing approach, that is especially important for the detection of outliers, is the studentized range [2, 9]. It is difference between the largest and smallest data in a sample measured in units of sample standard deviations:

$$q(x) = \frac{\max\{x_1, x_2, \ldots, x_N\} - \min\{x_1, x_2, \ldots, x_N\}}{std(x)} . \tag{4}$$

Window of analysis N of the preprocessing algorithm is not identical to the window of dispersion of the signal model R.

3 Spatio–Temporal Track–Before–Detect Algorithm

ST–TBD algorithm is a kind of the multidimensional recursive filter [3]. The following pseudocode shows processing details:

Start

$$P(k = 0, s) = 0 : Initial \tag{5}$$

For $k \geq 1$ and $s \in S$

$$P^{-}(k,s) = \int_{S} q_k(s|s_{k-1}) P(k-1, s_{k-1}) ds_{k-1} : MotionUpdate \quad (6)$$

$$P(k,s) = \alpha P^{-}(k,s) + (1-\alpha) X(k) : InformationUpdate \quad (7)$$

EndFor

End

where: S—state space: e.g. 2D position and motion vectors, s—state (spatial and velocity components), k—time moment, α—smoothing coefficient $\alpha \in (0, 1)$, $X(k)$—measurements, $P(k, s)$—estimated value of objects, $P^{-}(k, s)$—predicted value of objects, $q_k(s|s_{k-1})$ —state transitions (Markov matrix).

There are two processes inside ST–TBD: smoothing of the state–space due to Markov transition between velocities and sharpening due to Information Update formula. Smoothing coefficient is responsible for the balance between both processes. High valued smoothing coefficient reduces the influence of incoming noised data, improves SNR and increases time response of algorithm. The Information Update formula is a kind of exponential smoothing filter for state–space.

4 Performance of Preprocessing Algorithms

The assumed model of signal synthesis allows the determination of the properties of preprocessing algorithm using Monte Carlo test. Such test gives unbiased estimates of algorithms. The comparison of two signal processing algorithm is based on the defined ratios. The result of particular algorithm for emphasized signal is compared to the reference one, so this ratio is related to the properties of the detection. Moreover, ratios are important, because allow the comparison between two different algorithms. They are defined as:

$$R_{s^2} = \frac{s_e^2}{s^2}, \; R_q = \frac{q_e}{q} \quad (8)$$

and for variable values of e_{coef} is obtained result show in Fig. 3. Monte Carlo approach is applied and mean value of ratios is calculated using 1000 tests.

There are two important results—studentized range gives better estimation, for more important, small values range of e_{coef}, and results depends on the window size N. Variance estimator is preferred for small window size (e.g. $N = 11$) and studentized range is preferred for larger window size (e.g. $N = 21$ or $N = 31$).

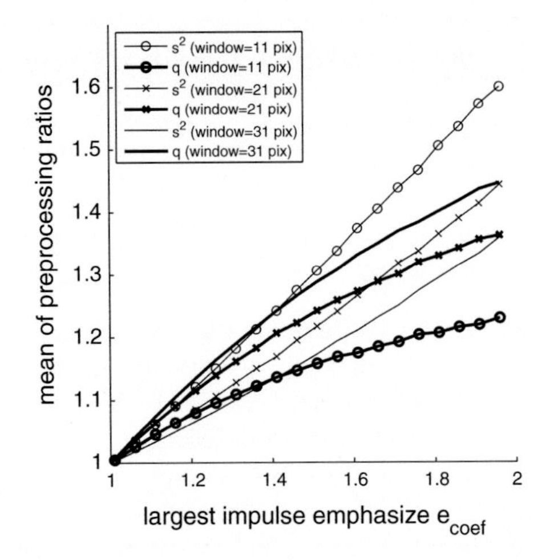

Fig. 3 Comparison of preprocessing algorithms based on mean R_{s^2} and R_q ratios

5 Tracking System Performance

Tracking system allows the detection and the tracking of weak signals, so even small e_{coef} values could be detected using numerous measurements. Such cases for assumed motion model are undetectable for humans. The Monte Carlo test is applied and 7000 cases of tracking scenarios are processed for every window of analysis size and smoothing coefficient. Such test requires 2 days of computation using 2.4 GHz CPU and Matlab code (all tests require 24 days of computations).

Tracking scenarios assume 1D tracking and 11 of motion vectors (velocities: 0–10). The length of the signal is 1400 pixels. Maximal value of the state–space after 100 processing steps is detected for the estimation of position and velocity. Mean position error is calculated for similar e_{coef}. The results are shown in Figs. 4, 5 and 6 for windows of analysis: 11, 21 and 31 respectively. The velocity of object (excluding noise) is linear and is fitted to the one of values from velocity set.

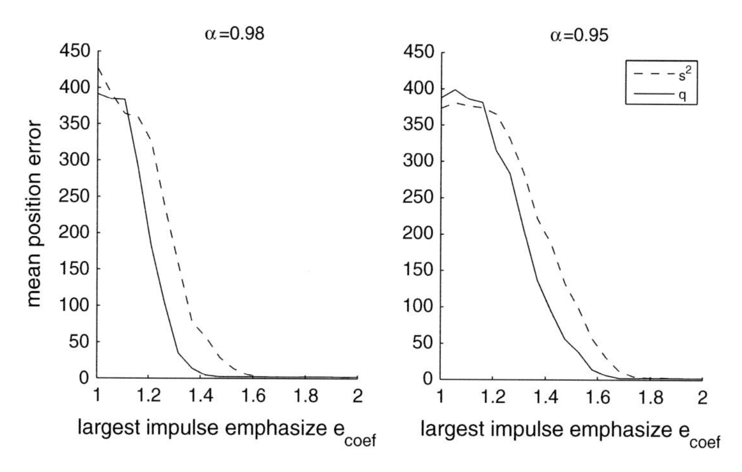

Fig. 4 Comparison of variance and studentized range for variable emphasize of object impulse and two different smoothing coefficient values of ST–TBD algorithm. Window analysis size is 11 pixels

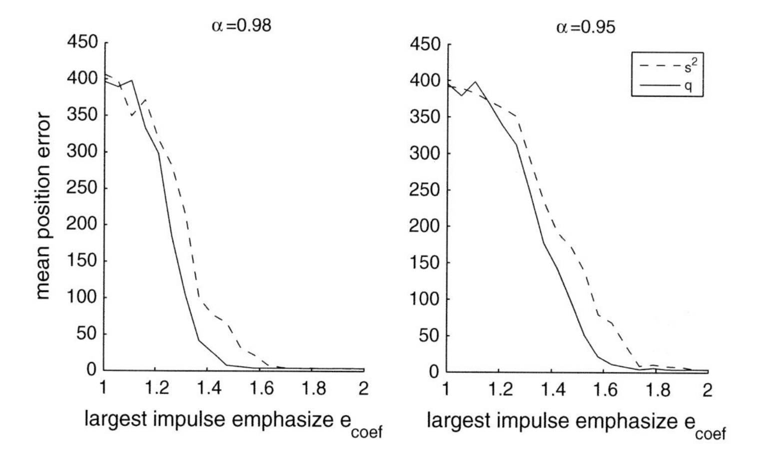

Fig. 5 Comparison of variance and studentized range for variable emphasize of object impulse and two different smoothing coefficient values of ST–TBD algorithm. Window analysis size is 21 pixels

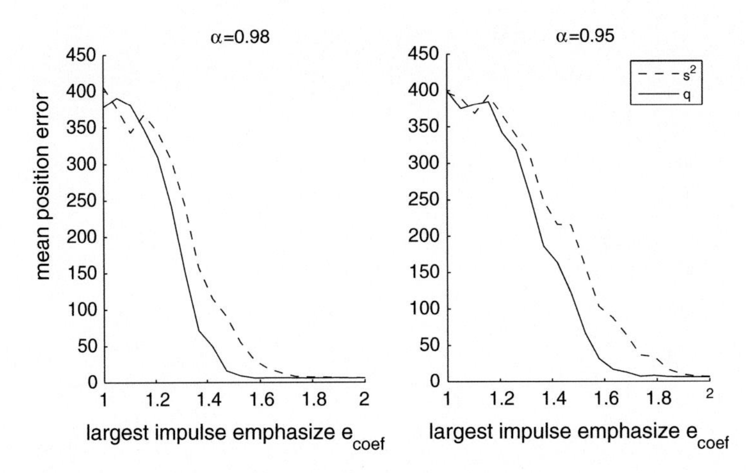

Fig. 6 Comparison of variance and studentized range for variable emphasize of object impulse and two different smoothing coefficient values of ST–TBD algorithm. Window analysis size is 31 pixels

6 Discussion

The smoothing coefficient influences the results. The same assumed fixed error value (mean position error) is obtained for higher values of e_{coef}. Studentized range estimator allows better detection (smaller mean position error) objects if the peak pulse is emphasized more then 1.1–1.2 ($e_{coef} = 1.1 - 1.2$). The improvement depends on the window of analysis. The lower boundary of this range is obtained for smaller smoothing coefficient. Increased window of analysis changes the lower boundary and smaller window size reduces value of this boundary.

Variance estimator gives comparable results to studentized range for $e_{coef} < 1.1$, that is expected, because studentized range formula used maximal and minimal value from window of analysis. Peak that is not observed as an emphasized does not increases value of the studentized range.

7 Conclusions

The proposed approach based on the preprocessing of input signal using studentized range shows important advantages over the variance. Studentized range applied for the preprocessing of signals for the ST–TBD algorithm gives smaller mean position errors or gives better sensitivity on weaker pulses (5–20 %).

Better detection of emphasized pulses in noise is possible using sliding window approach and TBD algorithm. The signal processing using raw signals uses all available information about the object. Peaks related to the object could be observed even for non–point object. Extended object, that occupies a few pixels, could be detected if some parts reflect or emit signal in random order. The proposed preprocessing algorithm is simple and the main computation cost is related to ST–TBD [3, 6], but efficient processing is possible using modern CPU, GPGPU and FPGA chips. Moreover, parallel processing of 1D, 2D and more dimensional signals using TBD algorithms is possible without additional costs.

References

1. Blackman, S., Popoli, R.: Design and Analysis of Modern Tracking Systems. Artech House (1999)
2. Keuls, M.: The use of the "studentized range" in connection with an analysis of variance. Euphytica **1**(1), 112–122 (1952)
3. Mazurek, P.: Optimization of bayesian track-before-detect algorithms for GPGPUs implementations. Electr. Rev. R. **86**(7), 187–189 (2010)
4. Mazurek, P.: Hierarchical track-before-detect algorithm for tracking of amplitude modulated signals. Adv. Intel. Soft Comput. **102**, 511–518 (2011)
5. Mazurek, P.: Chi-square statistic for noise object tracking in track-before-detect systems. Poznań Univ. Technol. Acad. J. Electri. Eng. **71**, 177–184 (2012)
6. Mazurek, P.: Code reordering using local random extraction and insertion (LREI) operator for GPGPU-based track-before-detect systems. Soft Comput. **18**(6), 1095–1106 (2013)
7. Mazurek, P.: Track-before-detect filter banks for noise object tracking. Int. J. Electr. Telecommun. **59**(4), 325–330 (2013)
8. Mazurek, P.: Preprocessing using maximal autocovariance for spatio-temporal track-before-detect algorithm. Adv. Intel. Syst. Comput. **233**, 45–54 (2014)
9. Raftery, J.: M.L., A., J.P., B.: Multiple comparison methods for means. SIAM Rev. **44**(2), 259–278 (2002)
10. Stone, L., Barlow, C., Corwin, T.: Bayesian Multiple Target Tracking. Artech House (1999)

Real-Time US Image Enhancement by Forward-Backward Diffusion Using GPU

Mariusz Nieniewski and Paweł Zajączkowski

Abstract The forward-backward diffusion (FBD) can be used for edge improvement in any image. The case considered in this paper is that of ultrasound (US) images. The US images are imperfect since the useful signal representing anatomical details is obscured by speckles inherent in this approach. A method such as the nonlinear coherent diffusion (NCD) can remove speckles to a large extent. However, the NCD-processed image can be further enhanced by the FBD. This paper describes the details of the proposed GPU implementation of the FBD algorithm. The obtained execution time of one iteration of the FBD algorithm is less than 2.2 ms for 1 MB image.

Keywords Forward-backward diffusion · US image · Parallel computing

1 Introduction

The US imaging is a common diagnostic tool used all over the world. However, the US images have some limitations due to the fact that they are covered with bright speckles. There exist many algorithms for speckle removal as described in the book [4] and review paper [2]. One of the best despeckling methods is the NCD [1, 5]. However, the careful examination of effects of this method reveals that the quality of the filtered images might be further improved [6] and this can be done with FBD algorithm that is inherently faster than the NCD.

The specific aim of the current paper is a development of the fast FBD algorithm executable in a GPU. The FBD can be used jointly with the NCD or independently.

M. Nieniewski (✉) · P. Zajączkowski
Department of Mathematics and Informatics, University of Lodz, ul. Banacha 22, 90-238 Lodz, Poland
e-mail: mnieniew@math.uni.lodz.pl

P. Zajączkowski
e-mail: zajaczkowski.pawel@math.uni.lodz.pl

R.S. Choraś (ed.), *Image Processing and Communications Challenges 7*,
Advances in Intelligent Systems and Computing 389,
DOI 10.1007/978-3-319-23814-2_21

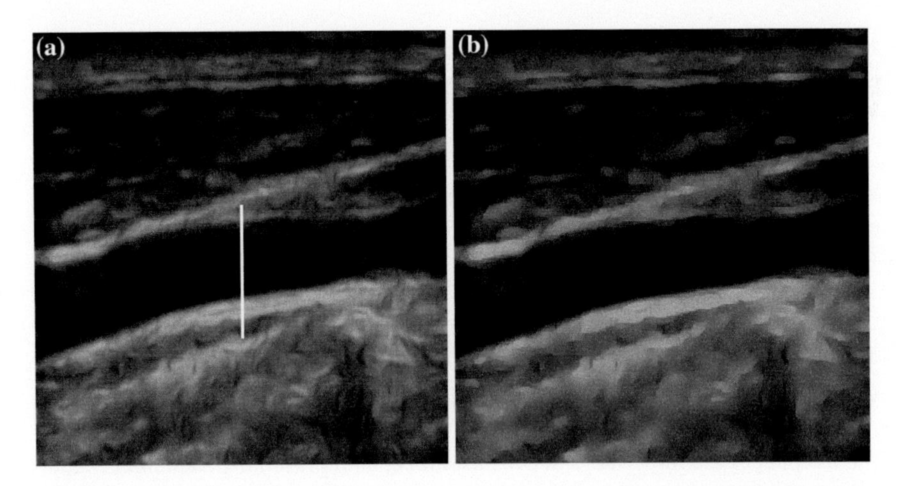

Fig. 1 Carotid artery. **a** Result of three iterations of the NCD. **b** Result of subsequent 10 iterations of the FBD

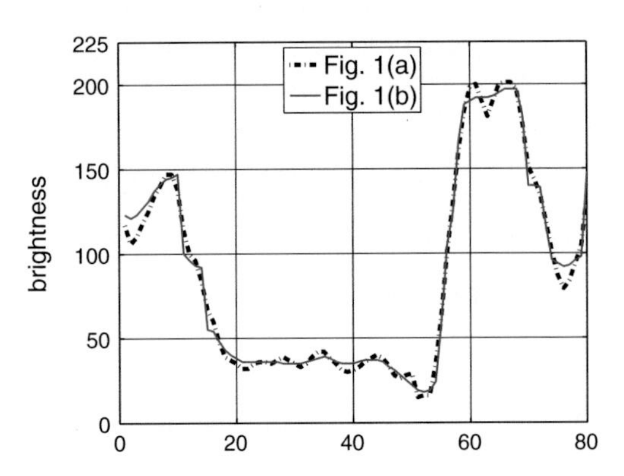

Fig. 2 Change of brightness distribution along the *vertical line* segment in Fig. 1a after 10 FBD iterations. The parameters $\alpha = 0.1$, $\lambda = 0.2$ (to be explained below)

An example of the NCD-despeckled US image of a carotid artery is shown in Fig. 1a. The size of image is 260×256 pixels, and the number of brightness levels is 256. The result of FBD iterations (Fig. 1b) can be appreciated by considering the brightness distribution along the straight line segment superposed on the image in Fig. 1a. Figure 2 indicates some sharpening of the edges and the smoothing of flatter areas along this line segment.

Another example of a carotid artery, not related to the NCD-despeckling, is shown in Fig. 3a, obtained from [9]. This image obviously has been despeckled in some way. The size of images in Fig. 3 is 369×446 pixels. The interesting thing about Fig. 3 is that it shows the upper and lower intima media. Particularly the upper intima media usually is hardly visible, and it is desirable to improve its visibility for purposes of image segmentation and facilitating medical diagnosing. As illustrated by Figs. 3b and 4, the FBD sharpens the cross section near the ends of the straight line segments where intima media is visible as a local maximum of intermediate height.

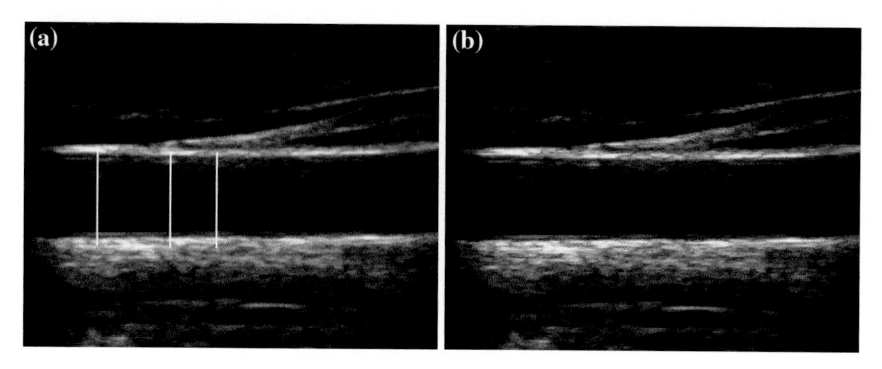

Fig. 3 Carotid artery with visible intima media. **a** Original image. **b** Result of five iterations of the FBD

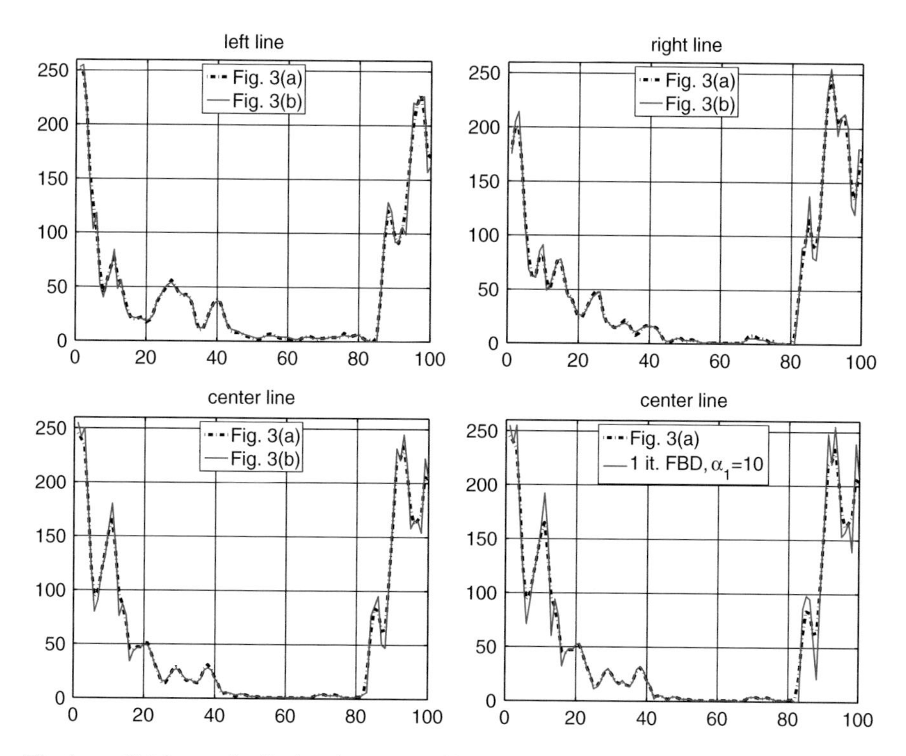

Fig. 4 **a–c** Brightness distribution along *vertical line*segments in Fig. 3. The parameters $\alpha = 1$, $\lambda = 0.1$. **d** Brightness along the middle segment obtained for one FBD iteration. The actual image is not shown. The parameters $\alpha_1 = 10$, $\lambda = 0.1$

2 Forward-Backward Diffusion

The FBD in general follows developments in [6, 7]. The difference scheme for one iteration of the FBD is as follows. The new brightness $I_{i,j}^{k+1}$ of the pixel with coordinates i, j, in iteration $k+1$ is calculated as a function of pixels in iteration k

$$I_{i,j}^{k+1} = I_{i,j}^{k} + \lambda[c_N \nabla_N I + c_S \nabla_S I + c_E \nabla_E I + c_W \nabla_W I]_{i,j}^{k} \tag{1}$$

The delta increments in the above equation are defined in "geographical" N, S, E, and W directions

$$\nabla_N I_{i,j} = I_{i-1,j} - I_{i,j} \qquad \nabla_S I_{i,j} = I_{i+1,j} - I_{i,j} \tag{2}$$

$$\nabla_E I_{i,j} = I_{i,j+1} - I_{i,j} \qquad \nabla_W I_{i,j} = I_{i,j-1} - I_{i,j} \tag{3}$$

The λ coefficient in Eq. (1) should lie in the range [0, 0.25] for numerical stability [7]. The diffusion coefficients $c_N, \ldots, c_W$ are calculated at every iteration as a function of the magnitude of the brightness gradient $\|\nabla'.\|$ in the respective directions

$$c_{N_{i,j}}^{k} = c(\|\nabla_N' I_{i-1/2,j}^{k}\|) \quad c_{S_{i,j}}^{k} = c(\|\nabla_S' I_{i+1/2,j}^{k}\|) \tag{4}$$

$$c_{E_{i,j}}^{k} = c(\|\nabla_E' I_{i,j+1/2}^{k}\|) \quad c_{W_{i,j}}^{k} = c(\|\nabla_W' I_{i,j-1/2}^{k}\|) \tag{5}$$

The function $c(.)$ in the above equations is specific to the FBD, and in accordance with [3, 6] it has two components, the first being responsible for the forward diffusion, and the second for the backward diffusion

$$c(s_1) = \frac{1}{1 + (s_1/k_f)^n} - \frac{\alpha_1}{1 + ((s_1 - k_b)/w)^{2m}} \tag{6}$$

In Eq. (6), s_1 denotes the magnitude of the gradient component, and the parameters k_f, k_b, w, n, m, and α_1 define the shape of the $c(s_1)$ function. In our experiments the following values were assumed: $k_f = 10$, $k_b = 40$, $w = 20$, $n = 4$, $m = 1$, $\alpha_1 = 0.1$, and $\lambda = 0.1$ unless otherwise specified. The gradients $\nabla_N', \ldots, \nabla_W'$ in Eqs. (4) and (5) are replaced by respective delta increments in Eqs. (2) and (3). The parameters specified above were obtained experimentally and can be used for a relatively wide class of US images.

3 Implementation of the NCD Algorithm Using the GPU

The principle of operation of the FBD algorithm is explained by means of Fig. 5, together with the List of Symbols. The main diffusion equation Eq. (1) can be rewritten in the form

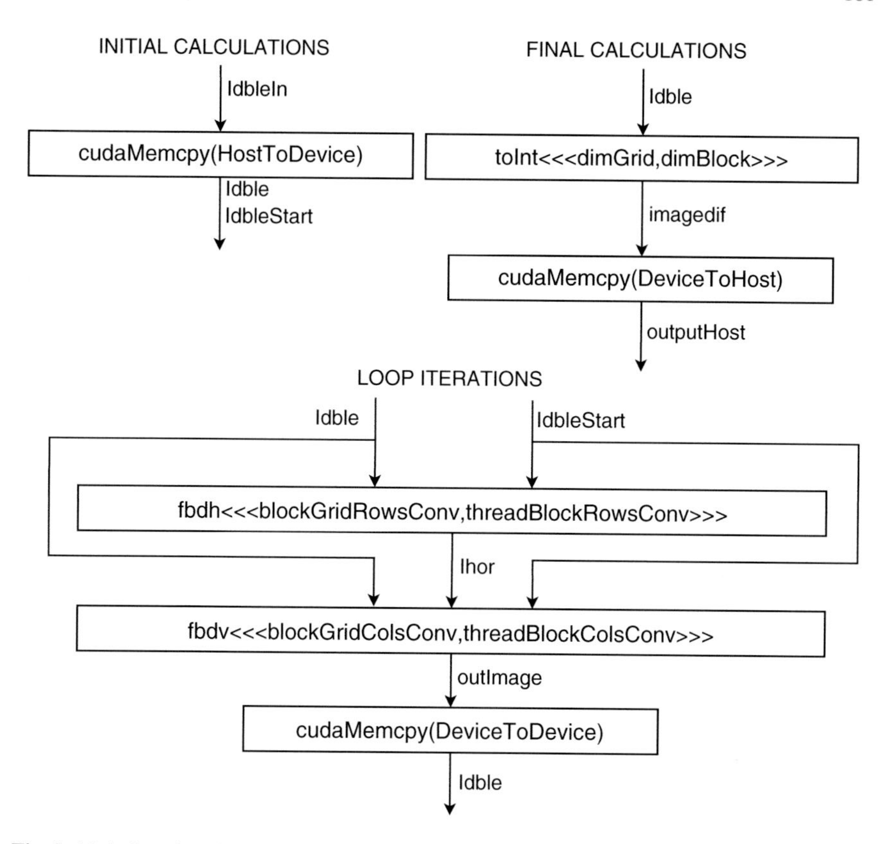

Fig. 5 Main function: initial calculations, loop iterations, and final calculations

$$I_{i,j}^{k+1} = I_{i,j}^{k} + \underbrace{\lambda[c_N \nabla_N I + c_S \nabla_S I]_{i,j}^{k}}_{vertical} + \underbrace{\lambda[c_E \nabla_E I + c_W \nabla_W I]_{i,j}^{k}}_{horizontal} \tag{7}$$

where two components corresponding to the vertical and horizontal diffusion are recognized. These components are calculated by two GPU kernel functions `fbdh` and `fbdv` to be shown in simplified form below.

The main function, executed in the CPU, takes the input image `IdbleIn` in double format from the CPU memory and transfers it to the GPU global memory. `Idblein` is maintained as `Idblestart` without changes through all the iterations, and its copy `Idble` is modified in each iteration. The main function executes the main iteration loop a given number of times. In each iteration the input image is `Idble` from the previous iteration and the pixels needed for padding, that is extending the image outside the `Idble`, are taken from `IdbleStart`. The main loop successively calls the `fbdh` and `fbdv` kernels. The `fbdh` outputs the auxiliary array `Ihor`. The `fbdv` uses the input image `Idble` as well as `Ihor`. The `Ihor` is necessary for summation of the horizontal and vertical diffusion components. The output image outImage from the fbdv is sent to the global GPU memory. The image Idble obtained after the final execution of the loop is sent to the additional kernel toInt for converting it into imagedif in integer format. The imagedif is transferred from the GPU global memory to the CPU memory. Each of the fbdh and fbdv kernels transfers data from the global memory to the shared memory, calculates the values of the function gFunc implementing Eq. (6) and sends the results to the global memory. If a certain block is to generate some portion of the output image then it must have access to a larger window of the input data and uses a so called apron around the required output [8]. Furthermore, when processing the pixels on the edge of the image the window sticks out of the image and then image padding is used. The fbdh uses a window of size equal to the image portion to be generated plus one column to the left and one to the right and transfers pixel values to the data array. Similarly, the fbdv uses a window including one row below and one row above the image portion to be generated and transfers the results to the data.

Consider now the fbdh (Algorithm 1) for an image of size 1024×1024 pixels, for which the critical parameters are specified after the symbol $\mapsto$ in several lines of the List of Symbols. The width of the block is 192 pixels, and each warp reads one pixel value. Line 20 of the algorithm calculates $\nabla_W I^k_{i,j}$, whereas line 21 calculates $\lambda[c_W \nabla_W I]^k_{i,j}$ (Eq. 3) and sends the result to dataNeighbor. Line 25 calculates the intermediate value $I^k_{i,j} + \lambda[c_E \nabla_E I + c_W \nabla_W I]^k_{i,j}$ and stores it in the array Ihor. When calculating this value, the following relations obtained from Eqs. (2) and (3) are used

$$\nabla_W I_{i,j} = -\nabla_E I_{i,j-1} \qquad \nabla_S I_{i,j} = -\nabla_N I_{i+1,j} \tag{8}$$

List of Symbols

width, height (width and height of the image), e.g. (1024, 1024)

ROW_TILE_W = 190 (width of a tile used in row processing)

KERN_RAD = 1 (radius of a 3×3 kernel)

COL_TILE_W = 32 (width of a tile used in column processing)

COL_TILE_H = 80 (height of a tile used in column processing)

blS = 16 (block size)

$\lambda = 0.1$, n = 4, m = 1, kf = 10, kb = 40, w = 20, $\alpha_1 = 0.1$ (diffusion parameters)

dim3 blockGridRowsConv(⌈width/Row_TILE_W⌉, height) ↦ (6, 1024)

dim3 blockGridColsConv(⌈width/COL_TILE_W⌉, ⌈height/COL_TILE_H⌉) ↦ (32, 13)

dim3 threadBlockRowsConv(ROW_TILE_W + 2 × KERN_RAD) ↦ 192

dim3 threadBlockColsConv(COL_TILE_W, 32) ↦ (32, 32)

dim3 dimBlock(blS, blS) ↦ (16, 16)

dim3 dimGrid(⌈(width + blS − 1)/bS⌉, ⌈(height + blS − 1)/blS⌉) ↦ (64, 64)

Turning now to the fbdv (Algorithm 2), the size of the block for fbdv is COL_TILE_W×32 = 32×32, and the size of the tile is COL_TILE_W × (2 × KER_RAD +COL_TILE_H) = 32 × 82. The tile is larger than the block, and each warp has to process several pixels. As a result, the Algorithm 2 has appropriate iterative `for` loops. The loop in lines 11–18 transfers pixel values to the `data`. The loop in lines 22–26 first calculates the differences according to Eq. (2), and then $\lambda[c_N \nabla_N I]^k_{i,j}$. The loop in lines 35–39 calculates the sum of the result from the kernel `fbdh` and $\lambda[c_N \nabla_N I + c_S \nabla_S I]^k_{i,j}$. Using the values given above each warp executes at most three times any of the `for` loops.

The function `gFunc` implements Eq. (6) with the exception that the values k^n_f and α_1/w^{2m} are calculated in advance and then used in each call of the function (Fig. 6).

Algorithm 1 `fbdh`

```
1: input: Idble in global memory, IdbleStart in global memory;
2: output: Ihor in global memory;
3: __shared__data[ROW_TILE_W + 2 × KERN_RAD];
4: __shared__dataNeighbor[ROW_TILE_W + 2 × KERN_RAD];
5: tileStart ← IMUL(blockIdx.x, ROW_TILE_W);
6: loadPos ← tileStart − KERN_RAD + threadIdx.x;
7: rowStart ← IMUL(blockIdx.y, width);
8: if loadPos >= tileStart − KERN_RAD then
9:     smemPos ← loadPos − tileStart + KERN_RAD;
10:     if max(tileStart − 1, 0) <= loadPos <= min(tileStart + ROW_TILE_W,
            width − 1) then
11:         data[smemPos] ← pixel value from Idble;
12:     else
13:         data[smemPos] ← pixel value from the first or last column of
                IdbleStart
14:     end if
15: end if
16: writePos ← tileStart + threadIdx.x;
17: smemPos ← writePos − tileStart + KERN_RAD;
18: if writePos <= min(tileStart + ROW_TILE_W − 1, width − 1) + 1 then
19:     v ← data[smemPos − 1] − data[smemPos];
20:     dataNeighbor[smemPos − 1] ← ˘ × gFunc (v, . . .) × v;
21: end if
22: if writePos <= min(tileStart + ROW_TILE_W − 1, width − 1) then
23:     Ihor[rowStart + writePos] ← data[smemPos]
            +dataNeighbor[smemPos − 1] − dataNeighbor[smemPos];
24: end if
```

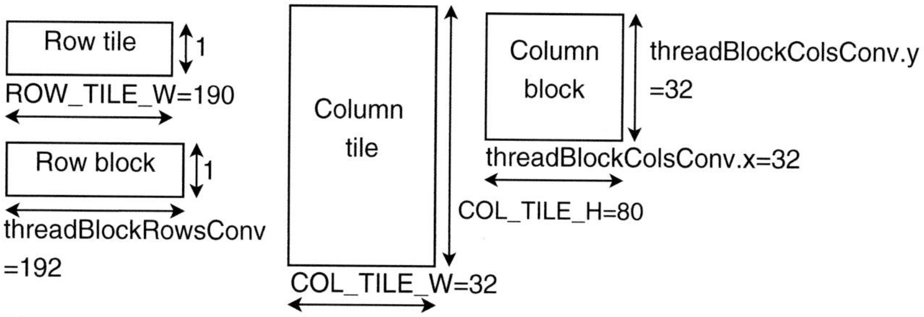

Fig. 6 Blocks and tiles

4 Experimental Results

The experiments conducted with the proposed FBD algorithm consisted in modifying dimensions of the grid and block and measuring the execution times for the largest image considered. The use of the NSight allowed us to analyze the occupancy of registers, the operation of warps and of shared memory. In particular, by changing

Algorithm 2 fbdv

```
1: input: Ihor, IdbleStart, Idble in global memory
2: output: outImage in global memory;
3: smemStride ← COL_TILE_W × 32;
4: smemSize ← COL_TILE_W × (2 × KERN_RAD + COL_TILE_H);
5: __shared__data[smemSize];
6: __shared__dataNeighbor[smemSize];
7: tileStart ← IMUL(blockIdx.y, COL_TILE_H);
8: columnStart ← IMUL(blockIdx.x, COL_TILE_W) + threadIdx.x;
9: if columnStart < width then
10:    smemPos ← IMUL(threadIdx.y, COL_TILE_W) + threadIdx.x;
11:    for i = tileStart + threadIdx.y − KERN_RAD : blockDim.y : tileStart +
         COL_TILE_H do
12:       if max(tileStart−1, 0)<= i<=min(tileStart+COL_TILE_H, height −
            1) then
13:          data[smemPos] ← pixel value from Idble;
14:       else
15:          data[smemPos] ← pixel value from the first or last row of IdbleIn;
16:       end if
17:       smemPos ← smemPos + smemStride;
18:    end for
19:    smemPos ← IMUL(threadIdx.y + 1, COL_TILE_W) + threadIdx.x;
20:    tileEndClamped ← min(tileStart + COL_TILE_H − KERN_RAD,
             height − 1);
21:    for i = tileStart + threadIdx.y : blockDim.y : tileEndClamped + 1 do
22:       v ← data[smemPos − COL_TILE_W] − data[smemPos];
23:       dataNeighbour[smemPos − COL_TILE_W] ← ˘ × gFunc (v, . . .) × v;
24:       smemPos ← smemPos + smemStride;
25:    end for
26:    smemPos ← IMUL(threadIdx.y + 1, COL_TILE_W) + threadIdx.x;
27:    gmemPos ← IMUL(threadIdx.y + tileStart, width) + columnStart;
28:    if mod(width,32) ==0 then
29:       gmemStride = width;
30:    else
31:       gmemStride = width − mod(width, 32) + 32;
32:    end if
33:    for i = tileStart + threadIdx.y : blockDim.y : tileEndClamped do
34:       outImage[gmemPos] ← Ihor[gmemPos] + dataNeighbor[smemPos−
                COL_TILE_W]−dataNeighbor[smemPos];
35:       smemPos ← smemPos + smemStride;
36:       gmemPos ← gmemPos + 32 × gmemStride
37:    end for
38: end if
```

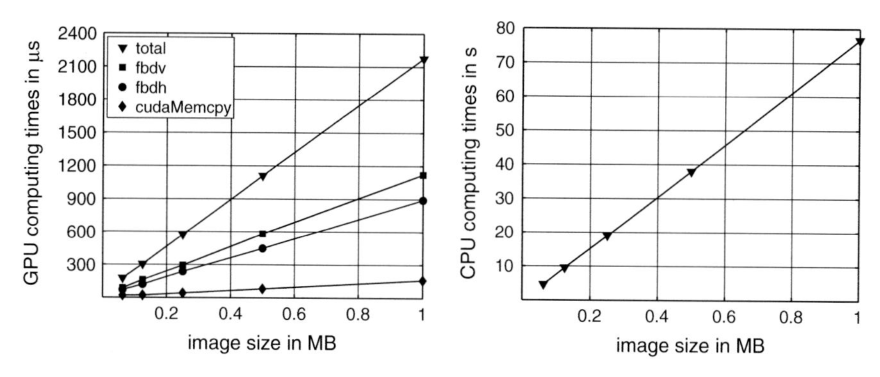

Fig. 7 Computing times for the FBD algorithm executed in GPU (*left*) and Matlab (*right*). Height × width of test images are: `256 × 256`, `256 × 512`, `512 × 512`, `512 × 1024`, `1024 × 1024`

the row tile width and column tile height we found that the shortest total execution time was obtained for `ROW_TILE_W=190`, `COL_TILE_H=80`, although the differences were rather small for a wide range tested dimensions. We used the GeForce GTX 560 graphics card, the Intel Celeron 1.8 GHz CPU with 2 GB RAM, and Visual Studio 2010 with 64-bit Windows 7. As can be checked in Fig. 7 the best total execution time of a single iteration of the FBD algorithm was less than 2.2 ms.

Execution time of one iteration of diffusion algorithm implementing directly Eq. (7) was on the order of 30 ms for 1 MB image. Taking advantage of Eq. (8) reduced the execution time to about 15 ms. Assuming constant values $m = 1$, $n = 4$, $w = 20$, and replacing raising to the power by multiplication in Eq. (6) reduced the execution time further down to about 2.6 ms, when measured using functions inserted in the main code. Figure 7 illustrates the times reported by the CUDA Launches Summary obtained by the Nsight. These times are the averages of five iterations, measured for the release version of the program. The difference between execution times of release and debug versions was less that 1 μs. Apart from the times specified in Fig. 7 there is a time connected with the transfer of the input image from the CPU memory to the global memory and the transfer from the global memory to the CPU memory. These times are similar to `cudaMemcpy` times shown in Fig. 7.

5 Conclusion

The execution time of the FBD algorithm in the GPU is on the order of 2.2 ms for one iteration for 1 MB image using a moderate graphics card, which is acceptable in real-time US applications. In the case of an equivalent Matlab program with similar acceleration measures as for the GPU version the time obtained for 1 MB image is 77 s. This means that Matlab program is out of the question for real-time applications, where ten or more frames per second have to be processed.

There are two possible scenarios of the use of the FBD. One is a general sharpening of the image. The weight of the backward diffusion α_1 can then be set to lower value. The other case is that when one wants to facilitate extraction of a particular component, say the carotid artery, even at the price of the deterioration of the other parts of the image. In this case α_1 may be set to a higher value. An illustration of these two cases of α_1 is given Fig. 4c, d. Finally, it is worth noting that the FBD algorithm is by no means restricted to US images, and enhancement of a variety of other images is possible.

References

1. Abd-Elmoniem, K.Z., Youssef, A.-B.M., Kadah, Y.M.: Real-time speckle reduction and coherence enhancement in ultrasound imaging via nonlinear anisotropic diffusion. IEEE Trans. Biomed. Eng. **49**(9), 997–1014 (2002)
2. Contreras Ortiz, S.H., Chiu, T., Fox, M.D.: Ultrasound image enhancement: a review. Biomed. Sign. Proc. Control **7**, 419–428 (2012)
3. Gilboa, G., Sochen, N., Zeevi, Y.Y.: Forward-and-backward diffusion processes for adaptive image enhancement and denoising. IEEE Trans. Image Process. **11**(7), 689–703 (2002)
4. Loizou, C.P., Pattichis, C.S.: Despeckle Filtering Algorithms and Software for Ultrasound Imaging. Morgan & Claypool (2008)
5. Nieniewski, M., Zajączkowski, P.: Real-time speckle reduction in ultrasound images by means of nonlinear coherent diffusion using GPU. In: ICCVG 2014. LNCS, vol. 8671, pp. 446–453. Springer (2014)
6. Nieniewski, M.: Enhancement of despeckled ultrasound images by forward-backward diffusion. In: ICCVG 2014. LNCS, vol. 8671, pp. 454–461. Springer (2014)
7. Perona, P., Malik, J.: Scale-space and edge detection using anisotropic diffusion. IEEE Trans. Patt. Anal. Mach. Intell. **12**(7), 629–639 (1990)
8. Podlozhnyuk, V.: Image Convolution with CUDA. Nvidia (Jun 2007). http://docs.nvidia.com/cuda/samples/3_Imaging/convolutionSeparable/doc/convolutionSeparable.pdf
9. Zukal, M., et al.: Ultrasound Image Database. http://splab.cz/en/download/databaze/ultrasound

Face-Based Distributed Visitor Identification System

Adam Nowosielski

Abstract In this paper the problem of the expansion of the simple face recognition system using the distributed system architecture is considered. The basic principle of the simple real-life face recognition system assumes the operating in specific localization using a single camera. People's faces are detected first and then recognized. Taking as a basis a primary face recognition system for the task of visitor identification the extension to additional localizations is analyzed. In order to maintain the simplicity of the components the final Distributed Visitor Identification System integrates the results from the individual subsystems and provides the data exchange. As a result face-based people re-identification is obtained. The resultant solution, by linking the information between separate localizations, enables new applications like for example the reported people flow analysis.

Keywords Face recognition · Face detection · Distributed face recognition · Re-identification · People flow

1 Introduction

The problem of visitor identification encompasses many applications where the essence is to recognize a human. The recognition process may be performed using variety of techniques and clues but biometrics emerges here as the most important and suitable technology. With the face recognition the process of visitor identification becomes non-invasive and not disturbing.

Potential applications of visitor identification include the recognition of [8, 13]: residents or their guests (visitors) in smart home, a frequent buyer in the shop, a missed or suspected person in a specific location etc. The foundation of the functioning of this class of systems is local-area usage with relatively small and dynamic

A. Nowosielski (✉)
Faculty of Computer Science, West Pomeranian University of Technology,
Żołnierska 52, 71-210 Szczecin, Poland
e-mail: anowosielski@wi.zut.edu.pl

R.S. Choraś (ed.), *Image Processing and Communications Challenges 7*,
Advances in Intelligent Systems and Computing 389,
DOI 10.1007/978-3-319-23814-2_22

face database [8, 13]. The system is directly associated with the specific camera and the operating time should be real-time.

The extension of the face recognition system for the visitor identification task to additional localizations and cameras is analyzed in the paper. Using the distributed system architecture individual face recognition systems are combined to form a Distributed Visitor Identification System. In addition to increasing the area of operation the association of the information between the separate localizations enables new applications. The resultant solution has a great potential in people flow analysis. The improvements in transfer conditions, schedule optimization and route planning can be achieved with pedestrian traffic distribution analysis in big public transport interchanges [11]. With the distributed face recognition this task is feasible and reliable.

Since the local visitor identification is limited to a specific location comprising private property or premises with the expansion process some privacy issues may emerge. Whereas the expansion within the closed area of the private or company property should not raise objections the concerns may arise with the application in public spaces. There is a possibility to overcome the issue in some applications with the anonymous recognition. For example in the problem of traffic distribution analysis there is no need of name or other privacy sensitive data storage [11].

The rest of the paper is structured as follows. In Sect. 2 a brief discussion on people re-identification and face recognition is provided. Then, in Sect. 3 face recognition system structures for local visitor identification are reported. The structure of the Distributed Visitor Identification System is proposed in Sect. 4 and an example implementation is outlined in Sect. 5. The paper ends with a summary.

2 Re-identification as a Face Recognition Problem

Identification and identity verification are the two classical tasks in face recognition [5, 8, 15]. The identification tries to specify the identity of a person using the database of known individuals. In the second task the claimed identity is verified on the base of comparison of provided biometrics with the stored template. The recognition here is one-to-one whereas the identification process is based on one-to-many comparisons. In both cases some thresholds values ensure the security of the system.

In recent years in the literature devoted to computer vision a new problem of people re-identification appeared. It is defined as [4]: recognizing an individual in diverse locations over different non-overlapping camera views, considering a large set of candidates. A person is tracked across multiple cameras which requires the ability to re-identification on leaving and reappearing between the field-of-view of different cameras [2]. Scene observation with multiple cameras is of particular interest for [17]: video surveillance, sports coverage and healthcare. The association of objects across views is one of the most important tasks here. Another issue is the dynamically evolving gallery and probe sets associated with the open set scenario [2].

Many researchers indicate that for such conditions the classical biometrics like face or gait may not be applicable (due to low resolution or frame rate [2, 4]) and they propose to use the entire body appearance. A person wears the same clothes and looks similar between subsequent observation points [4]. Re-identification based only on visual information is considered as appearance-based [2, 4]. The most commonly used features are color and texture information [17]. These features model the static appearance of a person [17]. Many solutions utilize additional spatial information about the location of cameras and predictable temporal reasoning [17]. Since the human body is an example of articulated objects it is willingly modeled by segments for which features of different kind are extracted [1, 2]. What is more, a person may be modeled with single-shot or multiple-shot (several frames) strategy [1, 2].

The task of re-identification is considered challenging since many unfavorable factors exists: varying pose, illumination and viewpoint diversity. By providing the appropriate imaging conditions: high resolution cameras and the appropriate camera placement, it is possible to employ the face recognition for the re-identification problem. The adequate system architecture should be provided and the appropriate set-independent face features should be used. The dynamically evolving gallery and probe set exclude the possibility of using entire dataset calculated features (e.g. classical PCA methods [5, 8, 15]).

3 Visitor Identification System Structures

The face-based Visitor Identification System is a specific case of a face recognition system and more specifically—the face identification system. It consists of the following fundamental blocks [8, 13]: face detector, feature extractor, database and classifier. Since the Visitor Identification System operates on real-life video stream, the face detection block is compulsory. Its task is to analyze input image for the presence of a face and provide the detected and properly cropped face image to the next stage. The actual process of face recognition starts at the stage of feature extraction which aims at describing face using a set of features. From the input face image only relevant information are preserved in the resultant feature vector. These feature vectors are stored in face database and compared in the classifier. The whole process is depicted on the scheme in Fig. 1. There are two types of connections between the blocks. The dotted line denotes the registration mode—the stage when users are enrolled to the database. The regular line reflects the operation of the system in the identification mode.

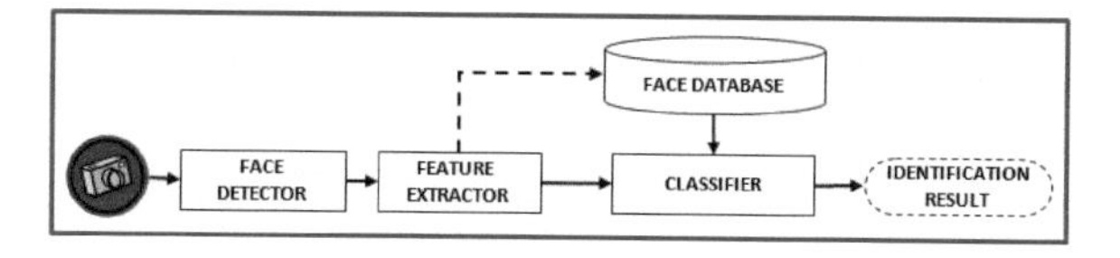

Fig. 1 Typical face recognition system structure

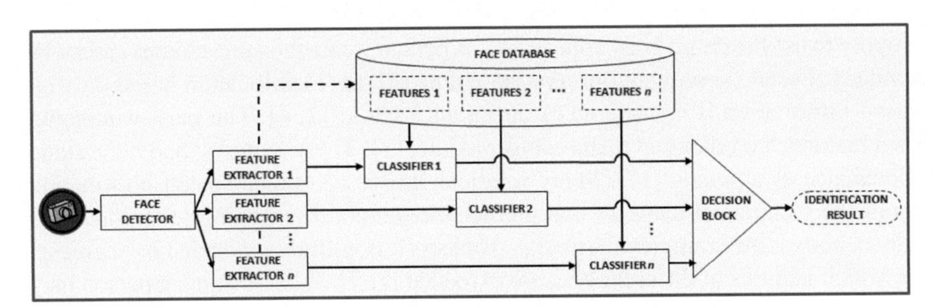

Fig. 2 Local visitor identification system in parallel architecture (based on [8])

The classic face recognition methods have been divided into two categories [8]: feature-based and appearance-based (holistic) approaches. Those categories are presently refereed as intensity based approaches [5]. According to the data acquisition two other techniques are referenced [5]: video based and special sensor based (like 3D approaches or infrared face models). Numerous attempts have been made to build a hybrid solution with different techniques fusion (e.g. [5–8, 10, 13]). This leads to a duplication of feature extractor blocks and classifiers. The repetition of the functional blocks is achieved using the parallel or serial architecture [8]. Both structures are provided on the the schemes in Figs. 2 and 3 [7, 8]. Additionally, the recognition result in the parallel system organization is formulated in a decision block on the base of some certain logical rules or in a voting fashion [8].

Another possibility of the expansion of the base structure of the Visitor Identification System (Fig. 1) is by the integration of additional function blocks. The capabilities of the system can be improved and some of characteristics can be strengthened. An example of the extension of the holistic face recognition system is provided in [14] where the following additional function blocks are incorporated in the base structure: virtual face image generator, light compensator, masking block, face tracker, feature space reducer and template selector. These additional function blocks implement some expansion strategies like [14]: improvement of the face database representativeness, stabilization of the results, reduction in computational complexity, increase in recognition efficiency, increase of the capabilities of the system.

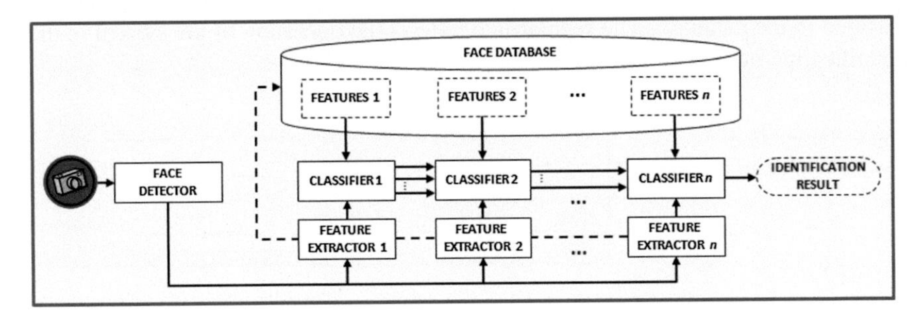

Fig. 3 Local visitor identification system in cascade architecture (based on [8])

The above quoted extensions of additional function blocks, the serial and parallel architecture, all allow improvements in the base face recognition system and they may be integrated successfully in the visitor identification task. On the other hand, all these enhancements improve the efficiency of the individual instance of of the system. The high expectations and new possibilities are associated with the integration of several instances of the Visitor Identification System in the distributed structure.

4 Distributed Visitor Identification System Structure

The local visitor identification task assumes that the face recognition is performed in a specific location. The system has a dedicated camera which video stream is directly analyzed. For many locations many individual systems operates independently. With the information exchange between these systems a distributed system for visitor identification can be constructed. Registered in one location the face can be searched for in other locations. It is of great importance for the high security places (e.g. buildings with some restricted areas) or for the people flow analysis. The Distributed Visitor Identification System may also be used for the task of loitering detection. The prolonged loitering behavior may indicate suspicious intentions [3].

The distributed system may operate in an open or closed space scenario. The second case is straightforward in the database management context. On the entry to some protected area (a building) a person is registered, tracked while moving inside the closed space, and finally removed from the database on leaving. With the open space scenario the person registered in one place may not appear in other locations. The automatic removal of individuals should be ensured in order to secure the database from overloading. A time of registration should be associated with a new face and the time window should be used for the validity inspection.

The information exchange between individual Visitor Identification Systems should be provided with the emphasis on the privacy issues. The local system should only be limited to send or receive data required for the process of recognition. In this context, a coded face image information, i.e. a feature vector, is the only data required.

Based on the above discussion the structure of the face-based Distributed Visitor Identification System can be proposed. The appropriate scheme is presented in Fig. 4. The system named GlobVIS encompasses a number of local visitor identification subsystems LocVIS. The only connecting element of individual subsystems is the face database block which must be considered in a broader sense. It acquires feature vectors, manages the feature vectors and distributes them to local systems. There is a feedback loop from each local classifier to the database which indicates the information transfer of the recognition results. Recognized at a time in one location a person cannot be considered for recognition in other places. Some localizations may also be associated unambiguously with the exit from the observed area and with the feedback information the removal from the database may be triggered without regard for the time window.

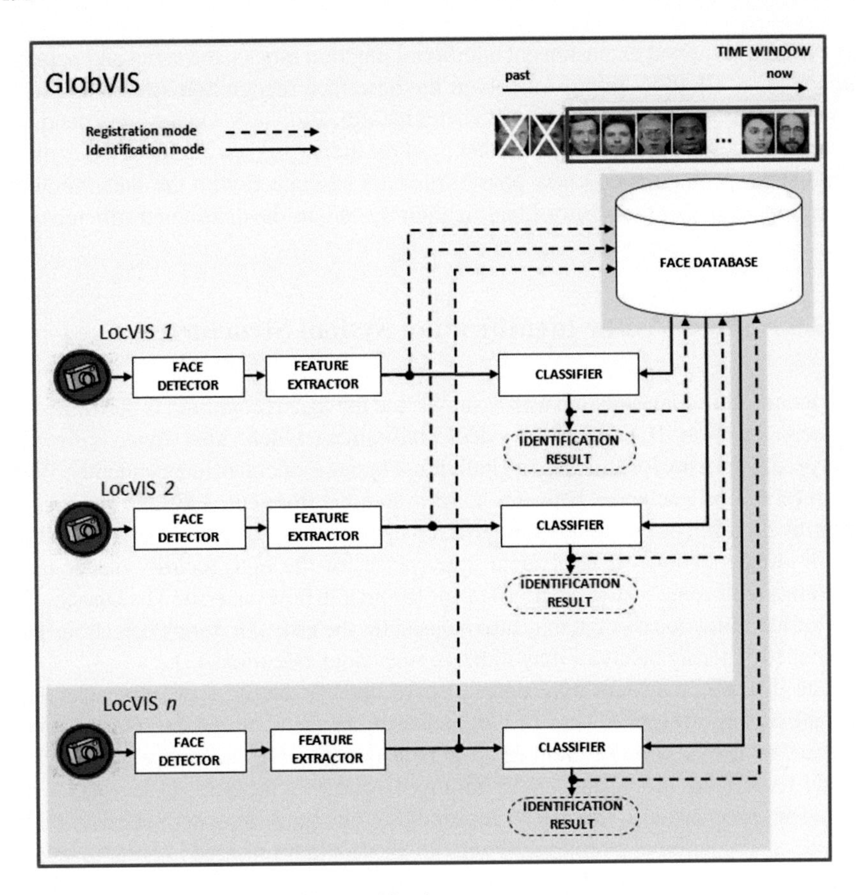

Fig. 4 Face-based distributed visitor identification system structure

5 Evaluation and the Application in Pedestrian Traffic Distribution Analysis Task

The system structured on the presented concept has been built for the task of pedestrian traffic distribution analysis. This kind of analysis is performed for public transport schedule optimization and route planning. For these tasks the traditional survey of people flow is based on counting and interviewing of travelers and the utilization of face recognition techniques is a new solution [11].

The concept of the distributed face recognition system has been utilized in the examination of Bemowo-Ratusz interchange in Warsaw. Pedestrian traffic distribution analysis was needed for the planned redevelopment of tram lines. The interchange consisted of three bus stops and two tram stops with 25 different lines. The survey based on face recognition techniques was accompanied with the traditional interview-based examination. The results of the comparison and the outcome of the appropriate survey have already been published in [11]. The results of traditional and face recognition based studies demonstrated good correlation. The current analysis concentrates on the aspects of the distributed face recognition.

In the Bemowo-Ratusz interchange case the Distributed Visitor Identification System consisted of 10 local subsystems. Each local Visitor Identification System was constructed on the base of cascade face detector and parallel feature extraction module. The Viola Jones method [16] was used at first stage of the serial face detector. The results were verified by the proprietary Verilook algorithm [12] at the second stage. For the classification task, the global approaches based on appearance have been combined with the local feature extractor LIFE (Light Invariant Feature Extractor) [9]. The integration was made using the weighted voting procedure. The global approaches included simple features [7]: scale, random, histogram, spectrumDFT and spectrumDCT.

The material was gathered during the morning and evening rush hours (the period of 3 h each). The time window of 10-min has been adopted here. For the face detection stage the already reported results equal to 83 % for the morning session and 79 % for the evening session [11]. In the respective sessions about 500 and 570 thousands separate face images have been detected. For the recognition stage the value of 75 % have been obtained. For more details of the examination and the comparison with traditional (survey based) approaches the reader is referenced to [11]. Since the examination of conducted people flow was a pilot study, the system prototype operated on the material gathered simultaneously by 10 cameras and processed later offline. Nonetheless, these results were achieved for the real-life data.

6 Conclusions

The extension of the face recognition system for the visitor identification task to additional localizations and cameras was analyzed in the paper. Using the distributed system architecture individual face recognition systems were combined to form the Distributed Visitor Identification System. The resultant solution enables face-based re-identification of people under dynamically evolving gallery and probe sets associated with the open set scenario.

References

1. Aziz, K.E., Merad, D., Fertil, B.: People re-identification across multiple non-overlapping cameras system by appearance classification and silhouette part segmentation. In: 8th IEEE International Conference on Advanced Video and Signal-Based Surveillance (AVSS), pp. 303–308, (Aug 2011)
2. Bedagkar-Gala, A., Shah, S.: Multiple person re-identification using part based spatio-temporal color appearance model. In: IEEE International Conference on Computer Vision Workshops (ICCV Workshops), pp. 1721–1728, (Nov 2011)

3. Bird, N., Masoud, O., Papanikolopoulos, N., Isaacs, A.: Detection of loitering individuals in public transportation areas. IEEE Trans. Intell. Transp. Syst. **6**(2), 167–177 (June 2005)
4. Farenzena, M., Bazzani, L., Perina, A., Murino, V., Cristani, M.: Person re-identification by symmetry-driven accumulation of local features. In: IEEE Conference on Computer Vision and Pattern Recognition (CVPR), pp. 2360–2367, (June 2010)
5. Jafri, R., Arabnia, H.R.: A survey of face recognition techniques. J. Inf. Process. Syst. **5**(2), 41–68 (2009)
6. Kamencay, P., Hudec, R., Benco, M., Zachariasova, M.: 2d–3d face recognition method based on a modified ccapca algorithm. Int. J. Adv. Rob. Syst. **11:36** (2014)
7. Kukharev, G., Forczmański, P., Nowosielski, A.: Simple Facial Features Extractors Utilization in Hardware-Software Biometric Systems (2006)
8. Kukharev, G., Kuzminski, A., Nowosielski, A.: Structure and characteristics of face recognition systems. Comput. Multimedia Intell. Tech. Special issue on Live Biometrics and Security **1**(1), 111–124 (2005)
9. Kukharev, G., Mikłasz, M., Sabuda, R., Kawka, G.: Feature Extraction Method Directed on Hardware Realization for Image Recognition Tasks (2009)
10. Liau, H.F., Seng, K.P., Ang, L.M., Chin, S.W.: New parallel models for face recognition. In: Delac, K., Grgic, M., Bartlett, M.S. (eds.) Recent Advances in Face Recognition, pp. 15–26. InTech (2008), http://www.intechopen.com/books/recent_advances_in_face_recognition/new_parallel_models_for_face_recognition
11. Mikłasz, M., Olszewski, P., Nowosielski, A., Kawka, G.: Pedestrian traffic distribution analysis using face recognition technology. In: Mikulski, J. (ed.) Activities of Transport Telematics, Communications in Computer and Information Science, vol. 395, pp. 303–312. Springer, Berlin, (2013). http://dx.doi.org/10.1007/978-3-642-41647-7_37
12. Neurotechnology: Verilook sdk (2009). http://www.neurotechnology.com/verilook.html
13. Nowosielski, A.: Three stage face recognition algorithm for visitor identification system. In: Pejaś, J., Piegat, A. (eds.) Enhanced Methods in Computer Security, Biometric and Artificial Intelligence Systems, pp. 177–184. Springer, US (2005). http://dx.doi.org/10.1007/0-387-23484-5_17
14. Nowosielski, A.: Mechanisms for increasing the efficiency of holistic face recognition systems. Przegl. Elektrotechniczny (Electr. Rev.) (R. 91 NR 2/2015), 51–55 (2015)
15. Smiatacz, M.: Eigenfaces, fisherfaces, laplacianfaces, marginfaces - how to face the face verification task. In: Burduk, R., Jackowski, K., Kurzynski, M., Wozniak, M., Zolnierek, A. (eds.) Proceedings of the 8th International Conference on Computer Recognition Systems CORES 2013, Advances in Intelligent Systems and Computing, vol. 226, pp. 187–196. Springer International Publishing (2013). http://dx.doi.org/10.1007/978-3-319-00969-8_18
16. Viola, P., Jones, M.: Robust real-time face detection. Int. J. Comput. Vis. **57**(2), 137–154 (2004)
17. Zini, L., Odone, F., Cavallaro, A.: Multiview matching of articulated objects. IEEE Trans. Circuits Syst. Video Technol. **24**(11), 1920–1934 (2014)

Color Space Optimization for Lacunarity Method in Analysis of Papanicolaou Smears

Dorota Oszutowska–Mazurek, Przemysław Mazurek, Kinga Sycz and Grażyna Waker–Wójciuk

Abstract There are numerous cells nuclei analysis algorithms. The lacunarity is the texture analysis algorithm and could be applied for binary or grayscale images of cells nuclei. The cells in Papanicolaou process are stained so numerous conversions to grayscale or binary images are possible. The optimization of RGB color space using weights is proposed for polynomial based analysis using lacunarity and the cell area of binary image. Obtained results show significant differences for best and worst cases for the number of cells of atypical and correct classes with similar cells area.

Keywords Lacunarity · Image analysis · Cytology

1 Introduction

Image analysis of cell nuclei is important because Computer Aided Diagnosis allows rapid discrimination of cells automatically. Such systems are applied for full analysis of specimen or as a tool for the detection of most important cells due to atypia. Image analysis of cells is based on the grayscale images typically and algorithms are defined for such image class. Selected image analysis algorithms are used for binary images only. The conversion from grayscale image to binary using fixed or adaptive threshold

D. Oszutowska–Mazurek (✉)
Higher School of Technology and Economics in Szczecin,
Klonowica 14 St., 71244 Szczecin, Poland
e-mail: adorotta@op.pl

P. Mazurek
Department of Signal Processing and Multimedia Engineering, West–Pomeranian University of Technology, 26. Kwietnia 10 St., 71126 Szczecin, Poland
e-mail: przemyslaw.mazurek@zut.edu.pl

K. Sycz · G. Waker–Wójciuk
Department of Pathomorphology, Independent Public Voivodeship United Hospital, Arkońska 4 St., 71455 Szczecin, Poland
e-mail: grazynka@blue.net.pl

R.S. Choraś (ed.), *Image Processing and Communications Challenges 7*,
Advances in Intelligent Systems and Computing 389,
DOI 10.1007/978-3-319-23814-2_23

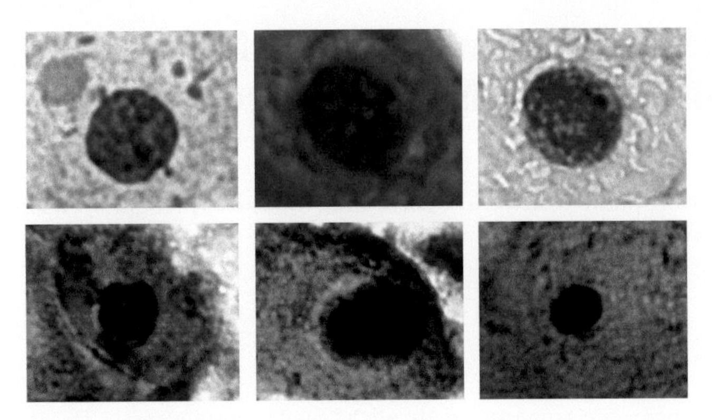

Fig. 1 Example microscopic images of cell nuclei

is possible. Such conversion reduces information, but could be applied if binary input data of the image is accepted. There are numerous image analysis algorithms that could be applied for biological objects. Exemplary like wavelets analysis is not unique approach and many variants are possible due to selection of wavelet, analysis tree and estimators that are applied for specified details or approximations.

The color images of cells are obtained by the selection of proper staining technique (Fig. 1). The Papanicolaou process (Pap) is applied in cervical screening and is a good example of staining process [3, 4, 8]. It is a kind of chemical segmentation that is necessary for cytoscreneers. The color of cells depends on a few factors and could be applied for analysis. This process is related to cytoplasm intentionally, so colors are visible in cytoplasm. The cell nucleus is inside of cell and stained also, but the dark color—black is dominant. The color of cell nucleus is an effect of staining and cytoplasm folding during the preparation of slide. This phenomenon could be applied for better discrimination between atypical and correct cells classes. The optical density of cell nucleus is observed due to specific transparency as well as volumetric light effects for such thin object, also.

1.1 Related Works

Cell nuclei are fractals or multifractals [19] and there are many algorithms for multiresolutional analysis of cell nuclei, like box–counting, triangular prism method [14, 15], variogram [16], lacunarity [17], Slit–Island Method, Area–Perimeter Method [10]. Multiresolutional analysis using fractals or wavelet could be applied for nucleus of different cells [1, 5, 11]. The most sophisticated problem is the segmentation of cells and cell nuclei, and many research are related to this topic [6, 7, 9, 12]. There are numerous texture analysis algorithms [13, 21]. In [2] 'haar' wavelet is proposed for cell nuclei and surrounding area. Multispectral images are analyzed in [20] using 'daub2', 'daub16', 'bior2.2' and Gabor transform.

1.2 Content and Contribution of the Paper

The lacunarity estimator for binary images is considered in Sect. 2. Image preprocessing as well as estimation is considered in Sect. 3. Proposed color space conversion to grayscale image for further binarization, results and discussion are provided in Sect. 4. Final conclusions are provided in Sect. 5.

2 Estimation of Lacunarity

Lacunarity is applied for the analysis of binary images typically [18]. Numerous extensions to grayscale images are available also, based on multiple analysis with different threshold values for example.

Lacunarity, like variogram, is not single value but function, which properties of binary texture depend on scale. Lacunarity is applied for analysis of holes in such texture typically. A few techniques of analysis are available including tiles or sliding window. Sliding window is assumed in this paper. Box window with r–pixels is assumed, so the area of analysis have $R \times R$ resolution. The number of 1's inside window is from the following range:

$$s \in \langle 0, r \rangle. \tag{1}$$

The number of pixels with 1's inside window W_i is calculated:

$$s_i = \sum_i W_i, \tag{2}$$

and i denotes different 2D positions of window W in image.

The frequency table is initialized simply:

$$n(s, r) = 0 \tag{3}$$

and updated using recursive formula:

$$n(s, r) \leftarrow n(s, r) + 1 \tag{4}$$

because for different position i of window s_i is calculated. The size of window r is also modified, so this value is not fixed.

The normalization of the frequency table is necessary after processing of all position and all allowed sizes of window:

$$Q(., r) = \frac{n(., r)}{\sum n(., r)}. \tag{5}$$

The results of this operation is the probability table Q. The obtained probability table is applied for the computation of two moments. The first moment is:

$$Z_1(r) = \sum_i s_i Q(s_i, r) \tag{6}$$

and second moment is:

$$Z_2(r) = \sum_i s_i^2 Q(s_i, r). \tag{7}$$

The lacunarity $\Lambda(r)$ for $r = R \times R$ window size is calculated using the following formula:

$$\Lambda(r) = \frac{Z_2(r)}{[Z_1(r)]^2}. \tag{8}$$

Normalized lacunarity is obtained using scaling:

$$\Lambda_{norm}(r) = \frac{\Lambda(r)}{\Lambda(1)}. \tag{9}$$

The proposed approximation in [17] of lacunarity function uses the following polynomial:

$$\log_{10}(\Lambda(r)) = a_3(\log_{10}(r))^3 + a_2(\log_{10}(r))^2 + a_1(\log_{10}(r)) + a_0, \tag{10}$$

and the most significant coefficient (a_2) is selected for the classification purposes.

Obtained results [17] depends on the selection of color space, and additionally grayscale and channel switching method are applied. The differences show the sensitivity to selection of color space, so optimization of color space could be technique for results improvement.

3 Image Preprocessing

Automatic segmentation is outside of the research scope. Acquired images are segmented manually, so cell nuclei are obtained. Color image of cell nucleus is converted to grayscale (C) using mean of all color channels (C_R, C_G and C_B of RGB color model):

$$C(x, y) = (C_R(x, y) + C_G(x, y) + C_B(x, y))/3. \tag{11}$$

The obtained image is normalized, so fixed contrast is obtained. It is necessary, due to further thresholding and light condition variances. Adaptive threshold does not need normalization. The median filter is applied for the selection of threshold level $MED(C)$:

$$C_{bin}(x,y) = \begin{cases} 1 : C(x,y) > MED(C) \\ 0 : C(x,y) \leq MED(C) \end{cases}, \quad (12)$$

where C is the image of cell nucleus only.

4 Color Space Optimization—Experiments and Discussion

There are numerous color spaces that could be applied for the image analysis. Original images obtained from camera uses RGB color space. The aim of optimization is the selection of weight related to appropriate channels. Such linear transformation changes threshold level as well as binary texture. The classification process uses cell area and a_2 coefficient of polynomial so this coefficient will be different. The cell area is fixed and independent on weights of color channels.

$$C(x,y) = (w_R C_R(x,y) + w_G C_G(x,y) + w_B C_B(x,y))/3 \quad (13)$$

The main discriminant is the area of cell nuclei, but is not sufficient. There is small subclass of atypical cells inside correct cells region. The ellipsoid discriminant of 95 % confidence region for this subclass is depicted in Fig. 2. All weights are set to 1, so the color space is not optimized. There are 12 of correct cells that are examined by cytoscreener as a possible atypical.

The application of weights allows the determination of better color space as a combination of R, G and B channels. The number of correct cells is calculated inside 95 % confidence region related to atypical cells. The results are shown in Fig. 3.

The relation between weights due to adaptive threshold obtained by the application of median occurs, but redundant results are depicted intentionally in Fig. 3. The best

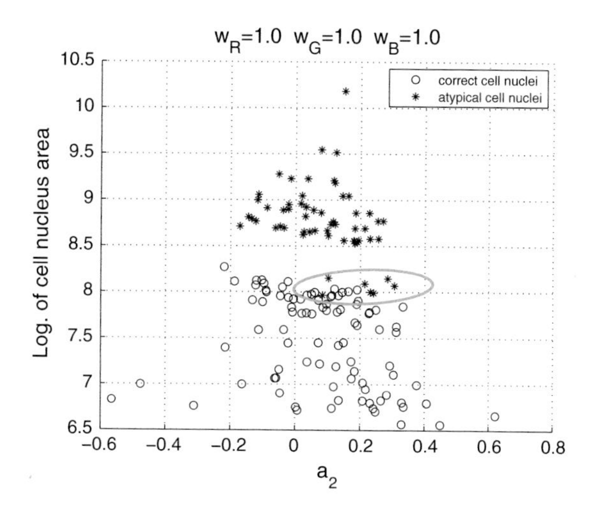

Fig. 2 Distribution of atypical and correct cell nuclei—no weights

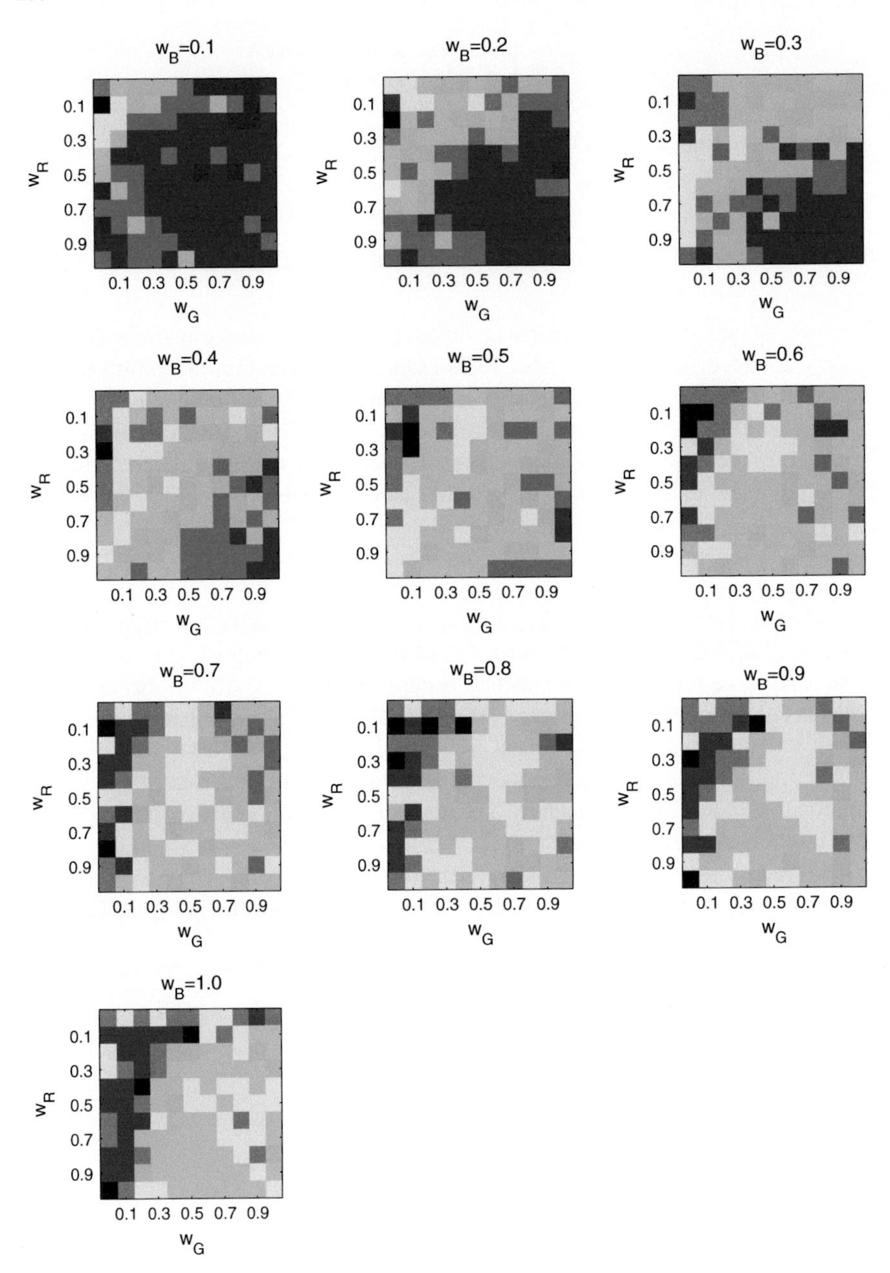

Fig. 3 Number of correct cells inside atypical 95 % confidence region (*dark blue*—7, *red*—17)

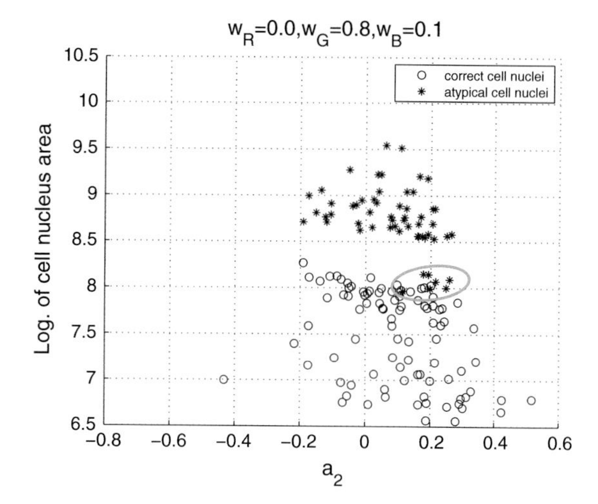

Fig. 4 Distribution of atypical and correct cell nuclei—best results

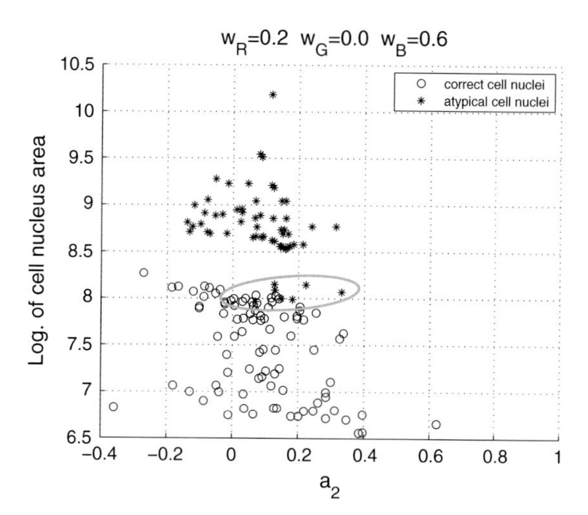

Fig. 5 Distribution of atypical and correct cell nuclei—worst results

significant solution is the green channel with small weight of blue channel. The red channel is rejected ($w_R = 0$). Red and blue cytoplasm cells are obtained in Papanicolaou process. The green channel preserves important information about differences between classes. The best result is 7 correct cells that are examined by cytoscreener as a possible atypical (Fig. 4).

Improper selection of weights gives poor results (Fig. 5). The worst result is 17 correct cells that are examined by cytoscreener as a possible atypical.

5 Conclusions

The proposed approach of color space optimization could be applied for other color space and estimation algorithms. The analysis is based on discrete sampling of color space weights, but gradient optimization could be also applied.

The number of correct cells that are assigned to atypical region is reduced significantly for our database. There most important result is the selection of color channel weights that shows importance of green channel that is not available in Papanicolaou process directly.

The effect of light condition and chemical process related to cell region and are complex and considered for further work.

References

1. Adam, R., Silva, R., Pereira, F., Leite, N., Lorand-Metze, I., Metze, K.: The fractal dimension of nuclear chromatin as a prognostic factor in acute precursor B lymphoblastic leukemia. Cell. Oncol. **28**, 55–59 (2006)
2. Bondarenko, A., Katsuk, A.: Extracting feature vectors of biomedical. In: KORUS'2005, pp. 52–56 (2005)
3. Chosia, M., Domagała, W.: Cytologia szyjki macicy. Fundacja Pro Pharmacia Futura (2010)
4. Cibas, E., Ducatman, B.: Cytology. Diagnostic Principles and Clinical Correlates. Saunders Elsevier (2009)
5. Ferro, D., Falconi, M., Adam, R., Ortega, M., Lima, C., de Souza, C., Lorand-Metze, I., Metze, K.: Fractal characteristics of May-Grünwald-Giemsa stained chromatin are independent prognostic factors for survival in multiple myeloma. PLoS ONE **6**(6), 1–8 (2011)
6. Filipczuk, P., Wojtak, W., Obuchowicz, A.: Automatic nuclei detection on cytological images using the firefly optimization algorithm. In: Lecture Notes in Computer Science (including subseries Lecture Notes in Artificial Intelligence and Lecture Notes in Bioinformatics) 7339 LNBI, pp. 85–92 (2012)
7. Frejlichowski, D.: Detection of erythrocyte cells in microscopy images. Electr. Rev. **88**(10b), 264–267 (2012)
8. Hoda, R., Hoda, S.: Fundamentals of Pap Test Cytology. Humana Press (2007)
9. Hrebień, M., Korbicz, J., Obuchowicz, A.: Hough transform, (1 + 1) search strategy and watershed algorithm in segmentation of cytological images. Adv. Soft Comput. **45**, 550–557 (2007)
10. Mazurek, P., Oszutowska-Mazurek, D.: From slit-island method to Ising model - analysis of grayscale images. Int. J. Appl. Math. Comput. Sci. **24**(1), 49–63 (2014)
11. Metze, K.: Fractal dimension of chromatin and cancer prognosis. Epigenomics **2**(5), 601–604 (2010)
12. Obuchowicz, A., Hrebień, M., Nieczkowski, T., Marciniak, A.: Computational intelligence techniques in image segmentation for cytopathology. Stud. Comput. Intell. **151**, 169–199 (2008)
13. Okarma, K., Frejlichowski, D., Czapiewski, P., Forczmański, P., Hofman, R.: Similarity estimation of textile materials based on image quality assessment methods. Lect. Notes Comput. Sci. **8671**, 478–485 (2014)
14. Oszutowska, D., Purczyński, J.: Estimation of the fractal dimension using tiled triangular prism method for biological non-rectangular objects. Electr. Rev. **R.88**(10b), 261–263 (2012)
15. Oszutowska-Mazurek, D., Mazurek, P., Sycz, K., Wójciuk, G.W.: Estimation of fractal dimension according to optical density of cell nuclei in Papanicolaou smears. In: Information

Technologies in Biomedicine 2012 (ITIB'2012). Lecture Notes in Computer Science, vol. 7339, pp. 456–463 (2012)

16. Oszutowska-Mazurek, D., Mazurek, P., Sycz, K., Wójciuk, G.W.: Advances in Intelligent Systems and Computing, vol. 184, chap. Variogram Based Estimator of Fractal Dimension for the Analysis of Cell Nuclei from the Papanicolaou Smears, pp. 47–54. Springer (2013)
17. Oszutowska-Mazurek, D., Mazurek, P., Sycz, K., Wójciuk, G.W.: Lacunarity based estimator for the analysis of cell nuclei. LNCS, vol. 8671, pp. 486–493. Springer (2014)
18. Plotnick, R., Gardner, R., Hargrove, W., Prestegaard, K., Perlmutter, M.: Lacunarity analysis: a general technique for the analysis of spatial patterns. Phys. Rev. E **53**(5), 5461–5468 (1996)
19. Steven, I.: Linear Richardson plots from non-fractal data sets. Dutch Math. Geol. **25**(6), 737–751 (1993)
20. Zhao, T., Wachman, E., Farkas, D.: A novel scheme for abnormal cell detection in pap smear images. In: Proceedings of SPIE 5318, Advanced Biomedical and Clinical Diagnostic Systems II, 151 (July 1, 2004) (2004)
21. Zieliński, K., Strzelecki, M.: Komputerowa analiza obrazu biomedycznego. Wstęp do morfometrii i patologii ilościowej, Wydawnictwo Naukowe PWN (2002)

Toward Texture-Based 3D Level Set Image Segmentation

Daniel Reska, Cezary Boldak and Marek Kretowski

Abstract This paper presents a three-dimensional level set-based image segmentation method. Instead of the typical image features, like intensity or edge information, the method uses texture feature analysis in order to be more applicable to image sets withs distinctive patterns. The current implementation makes use of a set of Grey Level Co-occurrence Matrix texture features that are generated and selected according to the characteristics of the initial region. The region is then deformed using the level set-based algorithm to cover the desired image area. The generation of the texture features and the level set surface deformation scheme are performed with graphics card hardware acceleration. The preliminary experiments, performed on synthetic data sets, show promising segmentation results.

Keywords Image segmentation · Deformable models · Level set · Texture analysis

1 Introduction

Level set methods are a group of algorithms broadly used in many areas of computer science, particularly in computer vision [14]. Level set-based deformable models [7] are especially useful for the task of image segmentation, i.e. partitioning of the image into distinct regions. The idea of a level set segmentation is based on an implicitly represented shape (a curve for a 2D image or a surface for a 3D volume), defined within the image domain. This shape is then deformed in an evolution process driven by external (image data) and internal (smoothness/curvature) forces. The resulting shape contains the targeted image region of desired characteristics.

D. Reska (✉) · C. Boldak · M. Kretowski
Bialystok University of Technology, Bialystok, Poland
e-mail: d.reska@pb.edu.pl

C. Boldak
e-mail: c.boldak@pb.edu.pl

M. Kretowski
e-mail: m.kretowski@pb.edu.pl

R.S. Choraś (ed.), *Image Processing and Communications Challenges 7*,
Advances in Intelligent Systems and Computing 389,
DOI 10.1007/978-3-319-23814-2_24

Level sets have an important ability to adapt their topology, which is a big advantage over the traditional parametric deformable models [4]. A level set shape can contract or expand, but also naturally divide or merge, therefore can handle changes in the segmented region topology without any additional frameworks [6]. The image data forces can be also modified to incorporate region [2, 13] or texture-based [9, 15] features. A known downside of the level sets is their computational intensity, especially in 3D. Recently, this limitation has been addressed by utilization of the graphics card (GPU) acceleration, which can significantly increase the performance of the 3D methods [5, 12] and open new possibilities for more interactive applications.

In this paper, we propose a texture-based level set method for segmentation of three-dimensional image volumes. The method replaces the traditional intensity-based data term in the level set speed equation with a term based on image texture features. In its current form, the method utilizes features based on Gray-Level Co-occurrence Matrices (GLCM) [3]. The texture feature generation process, as well as the level set evolution algorithm, uses hardware-accelerated implementations on GPU. The preliminary experiments were performed on synthetic images, where the texture-based approach showed an improvement over traditional methods.

2 Texture-Based Level Set Method

The proposed segmentation method incorporates a texture-based image data term into a 3D level set algorithm. In the image domain, the level set defines a surface that can deform according to the characteristics of the initial region: the surface can expand into a region with a uniform texture that is similar to the initial region or contract from an area where the texture does not meet the similarity criteria.

The initial spherical surface is manually initialized inside the segmented region. The texture features are then generated for the part of the volume inside the bounding box of the surface. Next, the algorithm selects the most uniform features for the initial region and uses them in the level set-based deformation process.

2.1 Texture Feature Generation and Selection

The texture feature generation process creates a set of feature maps for the volume inside the initial surface and then selects the features that have a low dispersion inside the initial region. This condition selects the texture features that will distinguish the segmented area from regions with different textures.

Currently, the method uses 2D texture features, generated separately for each of the images in the data stack (that composes the volume). To find the initial texture feature set, the algorithm selects the 2D images intersected by the bounding box of the initial surface and calculates partial texture maps only for the areas of the intersection. A mask of the initial surface region is also generated and used to calculate the characteristics of the features only for the volume points (voxels) inside the start region. For each of the features m_i, the mean of the feature values $\bar{x}_i$, standard deviation σ_i and Relative Standard Deviation $\%RSD_i = \frac{\sigma_i}{\bar{x}_i} \times 100$ are calculated. The texture feature used for the segmentation process must have the $\%RSD$ lower than a user-specified threshold (equal to 65 % by default). After this selection step, all of the selected features are calculated for the entire input data set (again in 2D: each selected feature for every image). Additionally, the selection step can reduce the number of maps by analyzing their similarity for different orientations in groups with the same feature, window size and displacement. The details of this process are described in [10].

The currently used texture features, generated from the Grey-Level Co-occurrence Matrix, are: Entropy, Correlation, Homogeneity, Contrast and Energy. The maps are generated for different sets of GLCM parameters: window size (from 3×3 to 11×11 by default), displacement (from 1 to 3 pixels) and orientation ($0°$, $45°$, $90°$, $135°$ and for all four angles). The algorithm, however, is not limited to the GLCM approach – any method that can generate a feature map of the segmented image can be used.

2.2 Level Set Surface Deformation

The manually initialized surface is deformed using the level set method. In this approach, we can define the surface S in a form of a set of points $p = (x, y, z)$ in the image domain as $S = \{p|\phi(p, t) = 0\}$, where $\phi(p, t) : \Re^3 \mapsto \Re$ and t is the time step of the simulation. During the simulation process, the surface points are moved according to a speed function $F(p, t)$, which allows the surface to expand or contract in order to eventually enclose the segmented region.

In this work, we modify the speed function proposed by Lefohn et al. [5], defined as:

$$F(p, t) = \alpha D(p) + (1 - \alpha)C(p, t), \tag{1}$$

where $D(p)$ is the image data term that drives the deformation, $C(p, t)$ is the surface curvature term and $\alpha \in [0, 1]$ is a user-defined parameter for balancing the influence of the two terms. The original image intensity-based data term was defined as:

$$D_{int}(p) = \epsilon - |I(p) - T|, \tag{2}$$

where $I(p)$ is the intensity value in point p and ϵ and T specify the target intensity range: $I(p)$ between $T - \epsilon$ and $T + \epsilon$ promotes the expansion of the surface, while the intensity out of the range encourages a shrinkage.

Our proposed texture term D_{tex} influences the surface motion by taking all the previously generated texture features into consideration: the surface is encouraged to expand to a point p if a feature similarity condition is fulfilled for all the selected texture features, or to contract otherwise, as defined in:

$$D_{tex}(p) = \begin{cases} v & \text{if } \forall m \in M_{best} : |val_m(p) - \bar{x}_m| \leq \theta \times \sigma_m \\ -v & \text{otherwise,} \end{cases} \tag{3}$$

where m is a texture feature in the selected set M_{best}, $\bar{x}_m$ and σ_m are the feature mean and standard deviation in the initial surface, $val_m(p)$ is the value of the texture feature m in the point p, v is a predefined constant value and θ is a user-defined parameter that denotes the term sensitivity.

The level set equation is solved with a GPU-accelerated implementation of the upwind scheme [8]. The simulation time is currently determined by specifying the desired number of iterations. Due to the numerical stability requirements of the method, during a single iteration each surface point can be moved only by one position.

3 Experimental Results

This section shows the experimental segmentation results of the proposed method. As the algorithm was mostly GPU-bound, the experiments were performed on workstations with three types of Nvidia graphics cards: GeForce GT630 (with 96 CUDA cores), Nvidia Quadro K2000 (384 CUDA cores) and GeForce GTX780 (2304 cores). The total segmentation time for the presented examples is presented in Table 1. In comparison to the intensity-based level set method (see Eq. (2)), the proposed algorithm was about 2 times slower (on average). Each experiment consisted of 1000 iterations of the level set method.

The algorithm was implemented using the MESA system [11]—a platform for designing and evaluation of the deformable model-based segmentation methods. MESA provides a template system for construction of active contours from exchangeable elements (i.e., models, energies and extensions), allowing an easy comparison of the proposed approach with other methods. The GPU-accelerated algorithms were

Table 1 Segmentation time of the proposed method on different GPUs (in seconds)

Data set	GPU type		
	GF GT630	Quadro K2000	GF GTX780
Volume 1	38	12	3.5
Volume 2	120	42	9

implemented in OpenCL [16] and integrated with the existing code using a Java binding library (JOCL from www.jocl.org).

The method was tested on synthetic volumes, created using the Brodatz texture database [1]. The initial surfaces were initialized manually. The experiments usually required an adjustment of the θ and α parameters.

The first example (see Fig. 1) shows a segmentation of a synthetic volume ($256 \times 256 \times 9$ voxels) with an ellipsoidal region of a high contrast texture. The intensity variance of the pattern makes it impossible to segment with default intensity-based speed function (see Fig. 1c). The texture-based speed function, however, managed to correctly drive the surface to the region boundaries (see Fig. 1d).

The second example (see Fig. 2) presents a more challenging case: a volume ($256 \times 256 \times 20$ voxels) containing a region with a complex surface and a texture identical to the background, but rotated by 90°. The default method failed to extract the region (see Fig. 2c). The texture-based method, however, managed to extract a rough outline of the surface due to the strong directional characteristics of the region pattern. The final form of the extracted surface is visible in Fig. 2f.

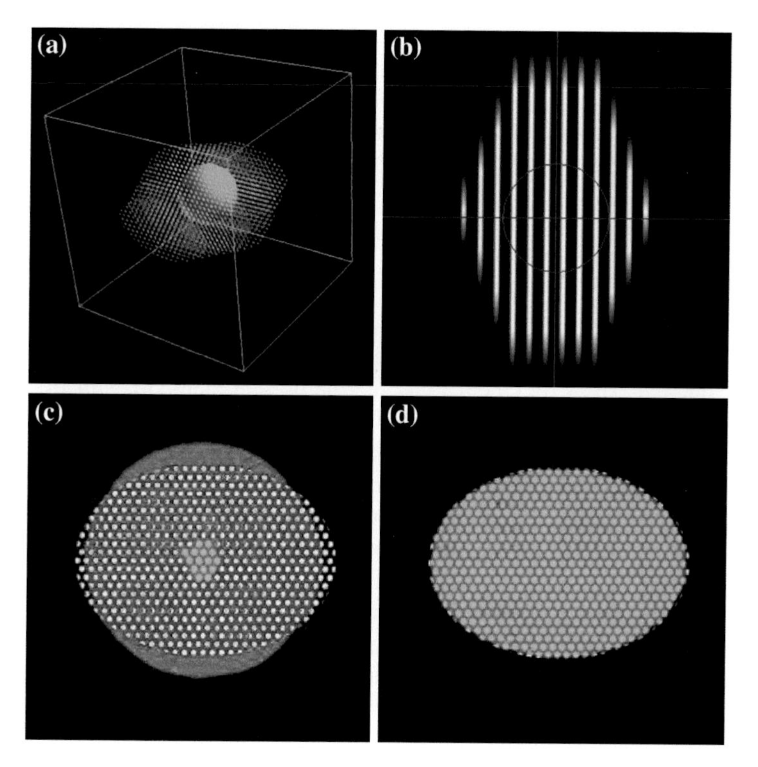

Fig. 1 Segmentation results of the first data sets: **a** 3D visualization of the set and initial surface, **b** initial surface in a 2D cross-section of the set, **c** sample result of the algorithm with the default image term, **d** result with the texture data term

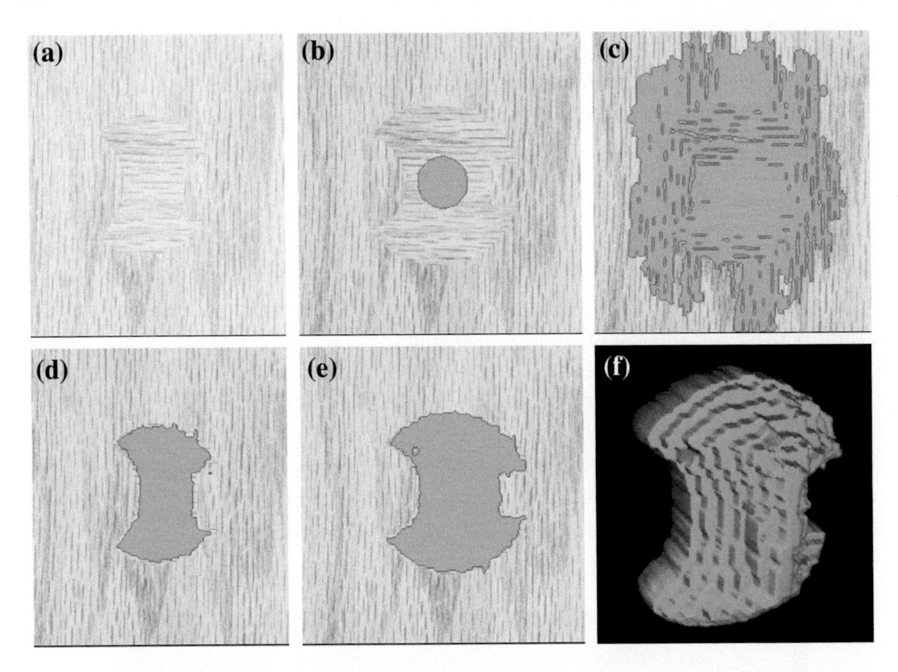

Fig. 2 Segmentation results of the second volume: **a**, **b** 2D slices from the volume with initialization visible on (**b**), **c** example volume slice with a result of the algorithm with the default image term, **d**, **e** results with the texture data term, **f** 3D visualization of the extracted surface

4 Conclusions and Future Work

In this paper a texture-based level set method for segmentation of three-dimensional image volumes is proposed. The method implementation makes a heavy use GPU acceleration, achieved with OpenCL. While the present method is still in an early stage of development, the preliminary results on synthetic images show a potential for other applications, possibly in natural or medical image analysis.

In the future work, an important enhancement would be an utilization of fully 3D texture features [17] instead of the currently used combination of 2D textures. Moreover, the utilized feature set (based on GLCM) can be easily extended by incorporation of other texture feature extraction methods. Furthermore, while the current numerical level set solving method is relatively much faster than a CPU implementation, the nature of the method can suggest an employment of a different, more effective algorithm. As the surface is initialized inside the segmented region and is generally expected only to expand, a less computationally intensive method like fast marching [14] or narrow band [12] could be more suitable here.

Acknowledgments This work was supported by Bialystok University of Technology under Grant W/WI/5/2014 and S/WI/2/2013.

References

1. Brodatz, P.: Textures: A Photographic Album for Artists and Designers. Dover Publications (1966)
2. Chan, T., Vese, L.: Active contours without edges. IEEE Trans. Image Proc. **10**(2), 266–277 (2001)
3. Haralick, R., Shanmugam, K., Dinstein, I.: Textural features for image classification. IEEE Trans. Syst. Man Cybern. **6**, 610–621 (1973)
4. Kass, M., Witkin, A., Terzopoulos, D.: Snakes: active contour models. Int. J. Comput. Vis. **1**(4), 321–331 (1988)
5. Lefohn, A., Cates, J., Whitaker, R.: Interactive, GPU-based level sets for 3D segmentation. In: Medical Image Computing and Computer-Assisted Intervention-MICCAI 2003, pp. 564–572. Springer (2003)
6. Mcinerney, T., Terzopoulos, D.: T-snakes: Topology adaptive snakes. In: Medical Image Analysis, pp. 840–845 (1999)
7. Moore, P., Molloy, D.: A survey of computer-based deformable models. In: International Machine Vision and Image Processing Conference, IMVIP, pp. 55–66 (2007)
8. Osher, S., Sethian, J.: Fronts propagating with curvature-dependent speed: algorithms based on Hamilton-Jacobi formulations. J. Comput. Phys. **79**(1), 12–49 (1988)
9. Paragios, N., Deriche, R.: Geodesic active regions and level set methods for supervised texture segmentation. Int. J. Comput. Vis. **46**(3), 223–247 (2002)
10. Reska, D., Boldak, C., Kretowski, M.: A texture-based energy for active contour image segmentation. Adv. Intell. Syst. Comput. Image Process. Commun. Challenges **6**(313), 187–194 (2015)
11. Reska, D., Jurczuk, K., Boldak, C., Kretowski, M.: MESA: complete approach for design and evaluation of segmentation methods using real and simulated tomographic images. Biocybernetics Biomed. Eng. **34**, 146–158 (2014)
12. Roberts, M., Packer, J., Sousa, M., Mitchell, J.: A work-efficient GPU algorithm for level set segmentation. In: Conference on High Performance Graphics, pp. 123–132 (2010)
13. Ronfard, R.: Region-based strategies for active contour models. Int. J. Comput. Vis. **13**(2), 229–251 (1994)
14. Sethian, J.: Level Set Methods and Fast Marching Methods: Evolving Interfaces in Computational Geometry, Fluid Mechanics, Computer Vision, and Materials Science, vol. 3. Cambridge University Press (1999)
15. Shen, T., Zhang, S., Huang, J., Huang, X., Metaxas, D.: Integrating shape and texture in 3D deformable models: from metamorphs to active volume models. In: Multi Modality State-of-the-Art Medical Image Segmentation and Registration Methodologies, pp. 1–31. Springer (2011)
16. Stone, J., Gohara, D., Shi, G.: OpenCL: a parallel programming standard for heterogeneous computing systems. Comput. Sci. Eng. **12**(3), 66 (2010)
17. Tesař, L., Shimizu, A., Smutek, D., Kobatake, H., Nawano, S.: Medical image analysis of 3D CT images based on extension of Haralick texture features. Comput. Med. Imaging Graph. **32**(6), 513–520 (2008)

The Calibration of the Pan-Tilt Units for the Active Stereo Head

Adam Schmidt, Li Sun, Gerardo Aragon-Camarasa and Jan Paul Siebert

Abstract We present a method for precise calibration of the pan-tilt unit based on the Levenberg-Marquardt optimization. Two different error functions are proposed—the first based on the position of the calibration marker and the second one indirectly incorporating the orientation of the marker. The proposed method was experimentally evaluated and the obtained results confirm its usefulness for the precise stereo-head operation.

Keywords Pan-tilt calibration · ptu · Active vision

1 Introduction

The pan-tilt units (PTUs) have been widely used to increase the effective field of view of cameras and to facilitate objects tracking. The applications of PTUs include, but are not limited to, acquisition of multi-scale images [15], driver assistance [4] and rail detection [12], surgical imaging [9], biometrics [7, 17].

Beside facilitating the operation of single cameras the PTUs constitute a crucial element of the active stereo heads used in robotics such as the Yorick head [3] or the Medusa head [2]. The ability to rotate the cameras is especially important for mimicking the version and vergence movement and designing bio-inspired visual perception and recognition systems [1, 14].

Regardless of the particular application it is important to know the pose of the camera for given pan and tilt angles. However, the precision of the pose estimation

A. Schmidt (✉)
Institute of Control and Information Engineering, Poznan University of Technology, Piotrowo 3a, 60-965 Poznan, Poland
e-mail: adam.schmidt@put.poznan.pl

L. Sun · G. Aragon-Camarasa · J.P. Siebert
School of Computing Science, University of Glasgow, 17 Lilybank Gardens, Glasgow G12 8RZ, UK
e-mail: l.sun.1@research.gla.ac.uk

R.S. Choraś (ed.), *Image Processing and Communications Challenges 7*,
Advances in Intelligent Systems and Computing 389,
DOI 10.1007/978-3-319-23814-2_25

gets even more crucial if the stereo head is used for accurate depth reconstruction as in [16].

Over the years several approaches to the calibration of the PTUs have been proposed. In his work Hartley [8] presented a two-stage algorithm for the self-calibration of the pan-tilt camera. The first non-iterative stage calibrates the camera intrinsics and kinematics parameters. The second, iterative refines the calibration. Neubert and Ferrier [13] proposed a closed-form solution based on the Lie theory approach to kinematic modelling. Davis and Chen developed a calibration method for wide-area surveillance networks [6]. The algorithm uses a single LED target observed simultaneously by several cameras to establish the geometric relationships between the cameras. Sinha and Pollefeys presented a method based on the bundle adjustment. However, the accuracy of their solution suffers from the assumption that the intersection of the pan and tilt axes coincides with the camera's optical center. Huang et al. [10] noticed that calibrating PTUs with telephoto lenses is hindered by near-ortographic projection. Instead of using an object of known size they introduced an external stereo-head to establish a set of 3D points for the PTU calibration. The same problem has been noticed by Kim [11]. Their solution was based on using direct points' observations instead of inter-image homographies to reduce the influence of degenerate configurations.

Here we present an algorithm for calibration of the pan-tilt unit aimed at providing precise estimates of the camera's pose according to the set pan and tilt angles. The Levenberg-Marquardt algorithm is used to find the kinematic parameters of the PTU. The preliminary results show that a very high precision of the pose estimates can be obtained.

2 System Overview

The considered system consists of a PTU with a camera attached to it. The PTU-D46 unit and the Nikon digital SLR camera D5100 were used in the experiments (Fig. 1). Moreover, a static chessboard marker is used for the calibration.

The PTU unit contains two perpendicular rotation axes intersecting in a point considered to be the origin of the PTU's coordinate system. The camera's pan is modeled as a rotation around the z axis and the tilt is modeled as a rotation around the x axis. There is no assumption regarding the coincidence of the camera's optical center and the origin of the PTU coordinate system. Moreover, the camera's axes are not necessarily aligned with the PTU's axes. The transformation between the PTU's mounting point and the camera's optical center is represented with a translation vector $t_c = \left[t_c^x \; t_c^y \; t_c^z \right]^T$ and the rotation represented according to the Rodrigues' formula as a vector $\text{rod}_c = \left[\text{rod}_c^x \; \text{rod}_c^y \; \text{rod}_c^z \right]^T$. Thus, the transformation between the PTU's coordinates and the camera's coordinates is given as:

Fig. 1 The active stereo head used in the experiments

$$T_{PTU}^{C}(\alpha,\beta) = T_{z,\alpha} T_{x,\beta} \begin{bmatrix} R(\mathrm{rod}_c) & t_c \\ 0 & 1 \end{bmatrix} \tag{1}$$

$$T_{z,\alpha} = \begin{bmatrix} \cos(\alpha) & -\sin(\alpha) & 0 & 0 \\ \sin(\alpha) & \cos(\alpha) & 0 & 0 \\ 0 & 0 & 1 & 0 \\ 0 & 0 & 0 & 1 \end{bmatrix} \tag{2}$$

$$T_{x,\beta} = \begin{bmatrix} 1 & 0 & 0 & 0 \\ 0 & \cos(\beta) & -\sin(\beta) & 0 \\ 0 & \sin(\beta) & \cos(\beta) & 0 \\ 0 & 0 & 0 & 1 \end{bmatrix} \tag{3}$$

$$\theta = \sqrt{\mathrm{rod}_c^{x\,2} + \mathrm{rod}_c^{y\,2} + \mathrm{rod}_c^{z\,2}} \tag{4}$$

$$r_c = \begin{bmatrix} r_x & r_y & r_z \end{bmatrix}^T = \mathrm{rod}_c^T \frac{1}{\theta} \tag{5}$$

$$R(\mathrm{rod}_c) = \cos(\theta) I + (1 - \cos(\theta))\, r_c r_c^T + \sin(\theta) \begin{bmatrix} 0 & -r_z & r_y \\ r_z & 0 & -r_x \\ -r_y & r_x & 0 \end{bmatrix} \tag{6}$$

where α and β are the pan and tilt rotation angles respectively.

Similarly, the pose of the marker is represented with the translation vector t_m and rotation vector rod_m. The transformation between the marker's and the PTU's coordinates can be calculated as:

$$T^M_{PTU} = \begin{bmatrix} R(\text{rod}_m) & t_m \\ 0 & 1 \end{bmatrix} \tag{7}$$

3 Pan-Tilt Calibration

The camera was calibrated using the rational lenses model presented in [5]. Once the intrinsic parameters of the camera are known it is easy to calculate the pose of the marker w.r.t. the camera's coordinate points. This pose depends on the current pan and tilt angles of the PTU and its estimate is denoted as:

$$T^M_C(\alpha, \beta) = \begin{bmatrix} R^M_C(\alpha, \beta) & t^M_C(\alpha, \beta) \\ 0 & 1 \end{bmatrix} \tag{8}$$

The proposed calibration procedure requires capturing N images of the marker for different pan and tilt angles α_i and β_i where $i \in \langle 1, N \rangle$. Once the observations are gathered the Levenberg-Marquardt algorithm is used to find the vector of parameters $x = \begin{bmatrix} t_c & \text{rod}_c & t_m & \text{rod}_m \end{bmatrix}$ minimizing the assumed error function.

In the simplest form the error function compares the observed position of the marker w.r.t. cameras coordinates with the coordinates calculated according to the current estimate of the system's parameters:

$$\text{err}_1 = \sum_{i=1}^{N} e_i^T e_i \tag{9}$$

$$e_i = T^C_{PTU}(\alpha_i, \beta_i)^{-1} t_m - t^M_C(\alpha_i, \beta_i) \tag{10}$$

However, such error function does not take into the account neither the marker's orientation w.r.t. PTU's coordinates (rod_m) nor its observed orientation ($R^M_C(\alpha, \beta)$). Unfortunately, due to the different units used, it is impossible to directly incorporate position and rotation errors in the same error function without using arbitrarily selected weights. This issue is alleviated by introduction of the second error function comparing the positions of three corners of the calibration marker:

$$\text{err}_2 = \sum_{i=1}^{N} \sum_{j=1}^{3} e_{i,j}^T e_{i,j} \tag{11}$$

$$e_{i,j} = T^C_{PTU}(\alpha_i, \beta_i)^{-1} p_j - T^M_C(\alpha_i, \beta_i) p_j \tag{12}$$

Table 1 The parameters estimated for both error functions

	$t_c^x\,[m]$	$t_c^y\,[m]$	$t_c^z\,[m]$	rod_c^x	rod_c^y	rod_c^z	$t_m^x\,[m]$	$t_m^y\,[m]$	$t_m^z\,[m]$	rod_m^x	rod_m^y	rod_m^z
err_1	−0.001	0.106	0.082	−1.592	−0.022	0.000	0.049	1.774	−0.124	0.000	0.000	0.000
err_2	0.001	0.106	0.083	−1.595	−0.023	0.001	0.049	1.774	−0.127	1.487	0.029	−0.078

$$p_1 = \begin{bmatrix} 0 & 0 & 0 & 1 \end{bmatrix}^T \tag{13}$$

$$p_2 = \begin{bmatrix} w & 0 & 0 & 1 \end{bmatrix}^T \tag{14}$$

$$p_3 = \begin{bmatrix} 0 & h & 0 & 1 \end{bmatrix}^T \tag{15}$$

where w and h stand for the width and height of the marker.

4 Experiments and Results

1240 images of the calibration marker were taken with the pan angle α ranging from $-20°$ to $19°$ and tilt angle ranging from $-20°$ to $10°$ with the increment of $1°$. Half of the images was used for the calibration while the second half constituted the verification set.

The kinematic parameters obtained using both the proposed error functions are given in Table 1. It is worth noting that for the function err_1 the marker's orientation rod_m was not estimated.

The obtained parameter vectors were compared using three metrics:

- the distance between the estimated and the observed position of the marker:
- the minimum rotation angle between the estimated and the observed orientation of the marker
- the distance between the estimated and the observed image positions of the chessboard corners (the reprojection error).

Table 2 contains the mean values of the metrics for both the error functions used. The histograms of the metrics are presented in Fig. 2. It is clearly visible that both error functions minimize the error of the marker's position estimates. However, the simpler function fails to correctly reduce the rotation and reprojection errors. On the other hand, using the second error function resulted in obtaining precise estimates of the camera's pose in terms of position, rotation and reprojection.

Table 2 The error metrics for the estimated parameters

	err_1	err_2
Distance (m)	0.0014	0.0014
Angle (°)	6.3669	0.3794
Reprojection (pixels)	16.8591	2.9048

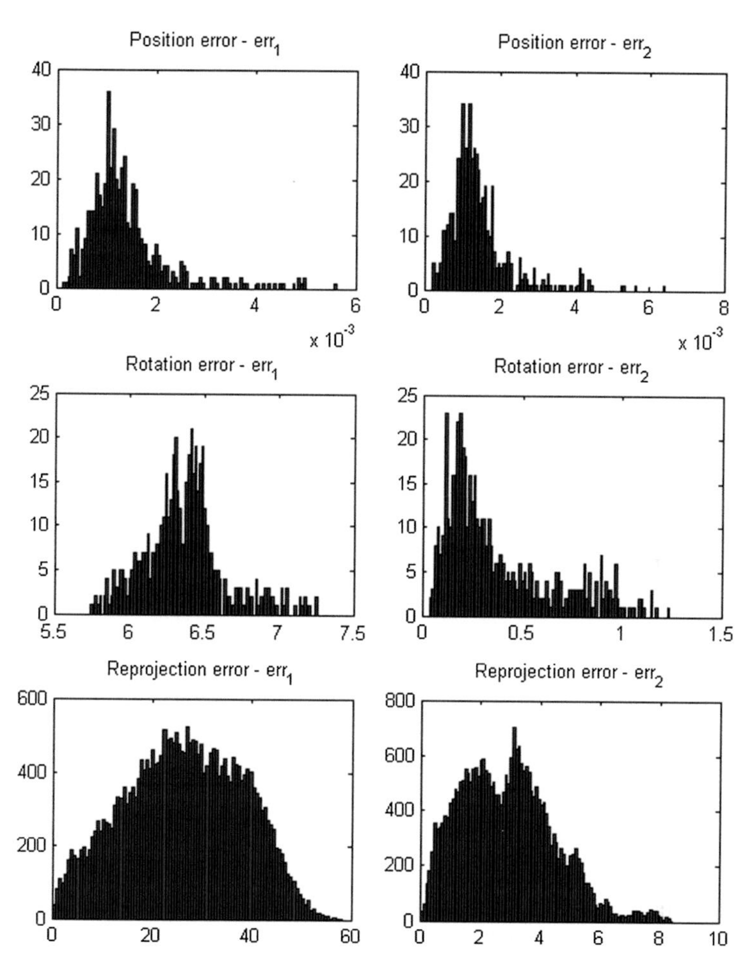

Fig. 2 The histograms of the error metrics

5 Conclusion

We presented an optimization based method for calibration of the PTU used in an active stereo head. Instead of assuming that the camera's optical center and the PTU's rotations' center coincide the full pose of the camera w.r.t. PTU's coordinate system was estimated. Abandoning this simplification resulted in a precise calibration of the PTU's parameters.

The rotation component of the measurements has been indirectly incorporated into the error function significantly improving the precision of the calibration. Such an approach allowed us to avoid the bi-objective optimization and arbitrary weighting of position and rotation errors.

In the future we will extend the method to work with both PTUs of the stereo-head simultaneously and to estimate the pose of the PTU w.r.t. robot's base frame. The results of the calibration will be evaluated in the robot manipulation tasks.

Acknowledgments This article was financially supported within the project "Engineer of the Future. Improving the didactic potential of the Poznan University of Technology"—POKL.04.03.00-00-259/12, implemented within the Human Capital Operational Programme, co-financed by the European Union within the European Social Fund.

References

1. Aragon-Camarasa, G., Fattah, H., Siebert, J.P.: Towards a unified visual framework in a binocular active robot vision system. Robot. Auton. Syst. **58**(3), 276–286 (2010). (Towards Autonomous Robotic Systems 2009: Intelligent, Autonomous Robotics in the UK)
2. Bernardino, A., Santos-Victor, J.: Binocular tracking: integrating perception and control. IEEE Trans. Robot. Autom. **15**(6), 1080–1094 (1999)
3. Björkman, M., Eklundh, J.O.: Attending, foveating and recognizing objects in real world scenes. In: British Machine Vision Conference (BMVC), pp. 227–236. London, UK (2004)
4. Cindy, X., Collange, F., Jurie, F., Martinet, P.: Object tracking with a pan-tilt-zoom camera: application to car driving assistance. In: IEEE International Conference on Robotics and Automation, 2001. Proceedings 2001 ICRA, vol. 2, pp. 1653–1658 (2001)
5. Claus, D., Fitzgibbon, A.W.: A rational function lens distortion model for general cameras. In: Proceedings of the 2005 IEEE Computer Society Conference on Computer Vision and Pattern Recognition (CVPR'05), vol. 1–vol. 01. pp. 213–219. CVPR'05, IEEE Computer Society, Washington, DC, USA (2005)
6. Davis, J., Chen, X.: Calibrating pan-tilt cameras in wide-area surveillance networks. In: Ninth IEEE International Conference on Computer Vision, 2003. Proceedings, vol. 1, pp. 144–149 (Oct 2003)
7. Haque, M., Nasrollahi, K., Moeslund, T.: Real-time acquisition of high quality face sequences from an active pan-tilt-zoom camera. In: 10th IEEE International Conference on Advanced Video and Signal Based Surveillance (AVSS), pp. 443–448 (Aug 2013)
8. Hartley, R.: Self-calibration of stationary cameras. Int. J. Comput. Vis. **22**(1), 5–23 (1997)
9. Hu, T., Allen, P., Nadkarni, T., Hogle, N., Fowler, D.: Insertable stereoscopic 3d surgical imaging device with pan and tilt. In: 2nd IEEE RAS EMBS International Conference on Biomedical Robotics and Biomechatronics, BioRob 2008, pp. 311–316 (Oct 2008)
10. Huang, X., Gao, J., Yang, R.: Calibrating pan-tilt cameras with telephoto lenses. In: Yagi, Y., Kang, S., Kweon, I., Zha, H. (eds.) Computer Vision—ACCV 2007. Lecture Notes in Computer Science, vol. 4843, pp. 127–137. Springer, Berlin (2007)
11. Kim, H.: Practical self-calibration of pan-tilt cameras. IEEE Proc. Vis. Image Signal Process. **148**(6), 349–355 (2001)
12. Nassu, B., Ukai, M.: A vision-based approach for rail extraction and its application in a camera pan and tilt control system. IEEE Trans. Intell. Transp. Syst. **13**(4), 1763–1771 (2012)
13. Neubert, J., Ferrier, N.: Robust active stereo calibration. In: IEEE International Conference on Robotics and Automation, Proceedings. ICRA'02. vol. 3, pp. 2525–2531 (2002)
14. Samarawickrama, J., Sabatini, S.: Version and vergence control of a stereo camera head by fitting the movement into the Hering's law. In: Fourth Canadian Conference on Computer and Robot Vision, CRV'07, pp. 363–370 (May 2007)
15. Senior, A., Hampapur, A., Lu, M.: Acquiring multi-scale images by pan-tilt-zoom control and automatic multi-camera calibration. In: Seventh IEEE Workshops on Application of Computer Vision, WACV/MOTIONS'05, vol. 1, pp. 433–438 (Jan 2005)

16. Sun, L., Aragon-Camarasa, G., Rogers, S., Siebert, J.P.: Accurate garment surface analysis using an active stereo robot head with application to dual-arm flattening. In: 2015 IEEE International Conference on Robotics and Automation (ICRA) (May 2015)
17. Yoon, S., Jung, H.G., Suhr, J.K., Kim, J.: Non-intrusive iris image capturing system using light stripe projection and pan-tilt-zoom camera. In: IEEE Conference on Computer Vision and Pattern Recognition, CVPR'07, pp. 1–7 (June 2007)

Some Remarks on the Optimization-Based Trajectory Reconstruction of an RGB-D Sensor

Adam Schmidt, Marek Kraft, Dominik Belter and Andrzej Kasiński

Abstract In this paper we present an analysis of the optimization-based trajectory reconstruction of an RGB-D sensor. Several approaches varying in the error function formulation as well as the camera's poses and features' positions initialization are considered. Their performance in terms of both the accuracy and the processing time is evaluated within a simulated environment.

Keywords Sparse bundle adjustment · Structure from motion · RGB-D · Trajectory reconstruction

1 Introduction

Spatial relationship between two camera poses at which the images of same scene were taken is a central problem in vision-based navigation. An accurate estimation of a series of such transformations is a central problem in many applications. While the most accurate methods are certainly based on a rich set of correspondences and global, batch optimization, such an approach is in many cases too time-consuming and impractical. Therefore, two alternative approaches to real-time structure from motion (SFM) or simultaneous localization and mapping (SLAM) operating on a sparse set of input data have evolved in the research community.

The first group are the methods based on filtering, aimed at accumulating knowledge from past measurements to update the structure and motion information in the form of possibly accurate probability distribution with minimum scatter. Common approaches to filtering-based SLAM include the use of various forms of Kalman filter [1] and particle filter [10].

A. Schmidt (✉) · M. Kraft · D. Belter · A. Kasiński
Institute of Control and Information Engineering, Poznań University of Technology,
Piotrowo 3A, 60-965 Poznań, Poland
e-mail: adam.schmidt@put.poznan.pl

R.S. Choraś (ed.), *Image Processing and Communications Challenges 7*,
Advances in Intelligent Systems and Computing 389,
DOI 10.1007/978-3-319-23814-2_26

Keyframe-based methods, on the other hand, aim at solving the problem using optimisation approach, but due to the real-time constraints they operate on a limited number of past frames or features. For keyframe-based methods, estimation of the structure of the scene and the motion of the sensor can be performed using several approaches. Out of these approaches, the most commonly used is an iterative technique called sparse bundle adjustment (SBA) [15]. SBA finds the relationships between camera poses and a sparse 3D model of the scene by minimizing the reprojection error [4, 8]. The intrinsic camera parameters can also be estimated in the process. Trajectory estimated by using SBA is often used in conjunction with SLAM to impose constraints on the trajectory of the robot [5].

The two approaches have been recently bridged in graph-based SLAM, where the problem is formulated as a graph optimization task. Every node of the graph corresponds to a pose of the robot and the edges in the graph express the spatial constraints between the nodes established by performing measurements. Once the graph is constructed, a configuration of the nodes that is maximally consistent with the measurements is found by solving an error minimization problem. Interestingly, the graph-based SLAM was first proposed in 1997 [9], but it took several years for it to become popular due to its relatively high complexity. However, recent advancements in sparse linear algebra and the SFM itself sparked a new interest in graph-based SLAM methods, which are currently considered to be the state-of-the-art techniques w.r.t. speed and accuracy [3].

Advancements in the field of robot sensors and perception keep up with the progress in the field of algorithm development for robot navigation. Historically, 3D mapping relied on expensive and bulky laser scanners. With the recent launch of compact, inexpensive RGB-D sensors based on structured light [6] or time of flight cameras [7] provided the robotics community with an attractive and powerful alternative solution. The depth data from such sensors can be easily integrated into a SFM framework [2].

In this paper an analysis of optimization-based trajectory reconstruction of an RGB-D sensor is presented. A range of approaches varying in the choice of objective function and the use of additional data for optimization initialization were tested, both in terms of accuracy, as well as computation speed.

2 State Model

The state model consists of a number of point features and a series of the camera's poses. The point features represent the spatial structure of the environment. Each of them is modeled as a Cartesian position vector:

$$x_f^i = \left[\, p_x^i \; p_y^i \; p_z^i \,\right]^T \tag{1}$$

The camera poses represent the camera's trajectory i.e. the spatio-temporal evolution of the camera's state. The camera's pose at moment j is defined as:

$$x_c^j = \left[t^j \ \mathrm{rod}^j \right]^T = \left[t_x^j \ t_y^j \ t_z^j \ \mathrm{rod}_x^j \ \mathrm{rod}_y^j \ \mathrm{rod}_z^j \right]^T \tag{2}$$

where t^j is the Cartesian position vector and rod^j encodes the orientation using the Rodrigues' notation. This particular representation of orientation was selected as it is compact (only 3 variables) and, unlike rotation matrices or quaternions, does not require any regularization, which facilitates numerical optimization.

The complete state of the system is then defined as:

$$x = \left[x_f^i \ \dots \ x_f^{fN} \ x_c^j \ \dots \ x_c^{cN} \right] \tag{3}$$

where fN is the number of features and cN stands for the number of the camera's poses.

During its movement the camera constantly observes the environment. If a feature is detected, its position in the current image is stored. Moreover, due to the availability of the RGB-D sensor, the measurement of the 3D position of the feature w.r.t. the current camera's coordinates is registered as well. Thus, two observation vectors h_{2D} and h_{3D} are defined. Each of them consists of particular observations of i-th feature at moment j:

$$h_{2D}^{i,j} = \left[u^{i,j} \ v^{i,j} \right] \tag{4}$$

$$h_{3D}^{i,j} = \left[x^{i,j} \ y^{i,j} \ z^{i,j} \right] \tag{5}$$

3 Trajectory Reconstruction

The process of camera's trajectory reconstruction can be expressed as finding the estimate of the state vector x that minimizes the difference between the actual observations and their values predicted using the state vector's estimate. Once properly formulated such a problem can be solved using various optimization methods e.g. the Levenberg-Marquardt algorithm.

In this paper, two variants of the error function are presented. The first one, traditionally used in the bundle adjustment approach, is based on the reprojection error of the observed features:

$$\mathrm{err}_{2D} = \sum_{h_{2D}} {e_{2D}^{i,j}}^T e_{2D}^{i,j} \tag{6}$$

$${e_{2D}}^{i,j} = H\left(R(\mathrm{rod}^j)^T (x_f^i - t^j) \right) - h_{2D}^{i,j} \tag{7}$$

where H stands for the camera's projection function (e.g. the pinhole model) and the rotation matrix $R(\text{rod}^j)$ is calculated as:

$$\theta = \sqrt{\text{rod}_x^{j\,2} + \text{rod}_y^{j\,2} + \text{rod}_y^{j\,2}} \tag{8}$$

$$r_c = \left[r_x \; r_y \; r_z \right]^T = \text{rod}^j \frac{1}{\theta} \tag{9}$$

$$R(\text{rod}^j) = \cos(\theta) I + (1 - \cos(\theta))\, r_{cr_c}^T + \sin(\theta) \begin{bmatrix} 0 & -r_z & r_y \\ r_z & 0 & -r_x \\ -r_y & r_x & 0 \end{bmatrix} \tag{10}$$

The second one uses the 3D positions of the features, which is possible due to the availability of depth measurements:

$$\text{err}_{3D} = \sum_{h_{3D}} e_{3D}^{i,j\,T} e_{3D}^{i,j} \tag{11}$$

$$e_{2D}^{i,j} = R(\text{rod}^j)^T (x_f^i - t^j) - h_{3D}^{i,j} \tag{12}$$

The analytic form of the Jacobian can be easily calculated for both of the error functions. Moreover, as each observation depends only on state of a single feature and a single camera's pose the Jacobians are sparse, which significantly facilitates the operation of the Levenberg-Marquardt algorithm.

The success of the optimization procedure strongly depends on the accuracy of the initial guess of the estimated parameters vector. In the described case the consecutive poses of the camera can be either initialized as coincident or with some arbitrarily chosen values. However, the initial estimate of the camera's displacement can be obtained by the means of a visual odometry algorithm, e.g. the one described in [11]. The initial positions of the features can be calculated using either the depth obtained during the first observation or an arbitrary value of the initial depth.

4 Experiments

4.1 Data

To perform comparison experiments we designed a simulation environment. The simulation environment provides ground truth position of the camera and ground truth for the map (features' positions). To provide measurements for the localization method we create a set of 3D points. In the presented experiment, we create a cubical environment, which represents a room (5.5 × 5.5 × 5.5 m). On each wall 1000 points are randomly generated. The camera is then moved along a reference trajectory (Fig. 1). Then we determine the set of 3D points which are visible from current

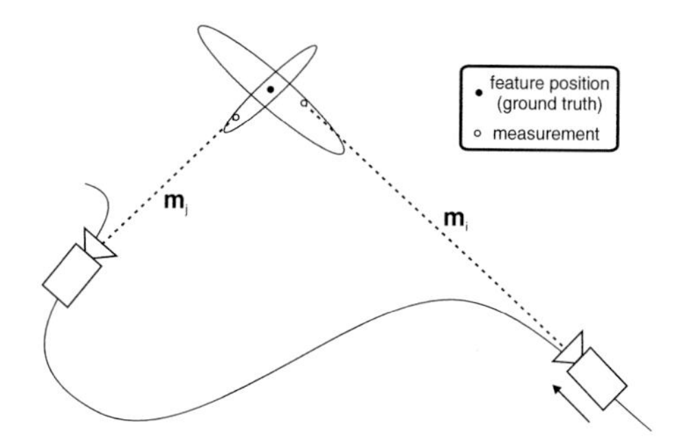

Fig. 1 Example which demonstrates the experiment in the simulation environment. The camera moves along reference trajectory. The measurement $\mathbf{m} = [u, v, d]^T$ is generated according to the uncertainty model of the RGB-D sensor

viewpoint. To this end, we compute the position of each 3D point in the current camera frame. The inverse model of the camera is then applied to compute the projection of the point on the camera image $[u, v, d]^T$:

$$\begin{bmatrix} u \\ v \\ d \end{bmatrix} = \begin{bmatrix} \frac{xf_x}{z} + C_x \\ \frac{yf_y}{z} + C_y \\ z \end{bmatrix}. \tag{13}$$

If the projection of a selected point is located on the camera image we apply the uncertainty model of the camera to generate measurement.

The uncertainty model of the camera is based on the approach proposed by Park et al. [12]. We compute the covariance matrix $\boldsymbol{\Sigma}_{f(3\times3)}$ of each observed feature:

$$\boldsymbol{\Sigma}_f = \mathbf{J}_p \cdot \boldsymbol{\Sigma}_p \cdot \mathbf{J}_p^T, \tag{14}$$

where $\mathbf{J}_{p(3\times3)}$ is Jacobian of the forward model of the camera $[x, y, z]^T = f(u, v, d)$ with respect to u, v and d, while $\boldsymbol{\Sigma}_{p(3\times3)}$ is the covariance matrix of image coordinates u, v, and depth d. The values of pixel uncertainty variances u and v are constant but the variance σ_d of depth d increases with the measured depth (c.f. Fig. 1). The obtained uncertainty matrix is used to randomly select a measurement which is inside the ellipsoid determined by the sensor model (Fig. 1).

The trajectory generated for the experiment consisted of 160 camera poses evenly distributed over a square path. The length of the square's side equaled 3.5 [m].

4.2 Results

Six scenarios differing in the error function and initialization methods used were considered in the experiment. Their summary is given in Table 1. For obvious reasons combinations of the 3D measurements and arbitrary initialization of the features were omitted.

Table 1 The six experiment scenarios

Scenario	Error function	Poses initialization	Feature initialization
s_1	err_{2D}	Coincident	Arbitrary
s_2	err_{2D}	Coincident	Measurement
s_3	err_{2D}	Odometry	Arbitrary
s_4	err_{2D}	Odometry	Measurement
s_5	err_{3D}	Coincident	Measurement
s_6	err_{3D}	Odometry	Measurement

Table 2 The results for the six experiment scenarios

Scenario	RMS ATE	Processing time (s)
s_1	14.2633	60.5
s_2	1.9699	162.47
s_3	0.2138	4395.19
s_4	0.0036	1028.89
s_5	0.0168	3264.57
s_6	0.0168	40.28

The obtained trajectories were compared with the ground truth data using the absolute trajectory error (ATE) metric proposed by Sturm et al. [14]. The comparison is based on aligning the two trajectories in the terms of least-squares to compensate for the effects of the scale ambiguity and the initial pose estimate. The processing time on a computer with the Intel Core i5 2.6 GHz and 12 GB RAM for each scenario was registered as well. Table 2 presents the obtained results and Fig. 2 shows the trajectories obtained in the successful experiments.

The trajectory reconstruction failed when no initial data regarding the camera poses was provided and err_{2D} was used even if the initial positions of the features were used.

If the initial estimate of the camera poses was given the err_{2D} was sufficient to obtain satisfactory results. Moreover, if the initial positions of the features were provided as well, the system achieved the best accuracy. However, it is worth noting that the processing time required to reconstruct the trajectory makes it impossible to use such an approach in even close-to-real-time situations.

Using the err_{3D} error function resulted in almost 5 times bigger RMS of the ATE. However, the accuracy is still sufficient for many applications. It is worth noting that the error was exactly the same regardless of features' positions initialization. Providing the initial estimates of the features' positions tremendously reduced the processing time of the system. It is especially important considering the possible applications in the visual robot navigation.

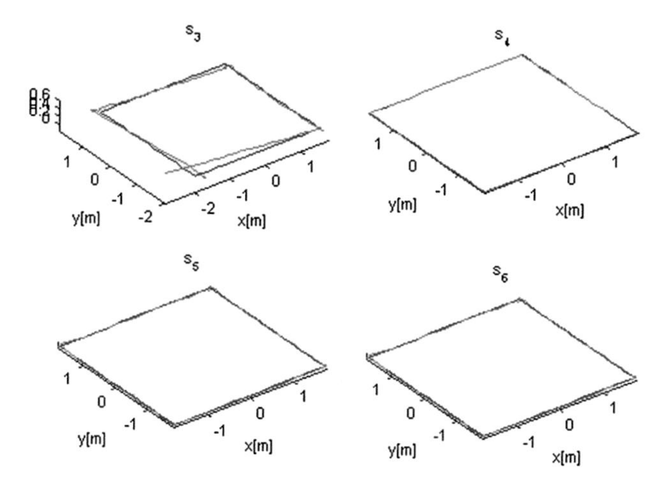

Fig. 2 Trajectories obtained for the successful scenarios. *Blue*—GT, *green*—aligned

5 Conclusions

In this paper we presented a comparison of different approaches to the reconstruction of the RGB-D camera's trajectory. The proposed method was based on the Levenberg-Marquardt optimization algorithm and minimized either the reprojection error of the point features or their 3D position error. The validating experiments were performed in a simulated environment to ensure maximal control over the environment parameters.

Surprisingly, the best results in the terms of accuracy were obtained through minimizing the reprojection error and using the depth measurements only for initialization. Such behaviour can be explained by indirect incorporation of the features' depth into the measurements—the more distant the feature is the less is the influence of its position estimate on the reprojection error.

Minimizing the 3D position error gave worse (yet still acceptable) accuracy. However, a significant improvement in the terms of processing time was observed.

The future work will include testing the presented approach with more complicated trajectories with both the simulated data and trajectories obtained with a physical motion registration system (e.g. the PUT RGB-D dataset [13]).

Acknowledgments This research was financed by the Polish National Science Centre grant funded according to the decision DEC-2013/09/B/ST7/01583, which is gratefully acknowledged.

References

1. Davison, A., Reid, I., Molton, N., Stasse, O.: MonoSLAM: real-time single camera SLAM. IEEE Trans. Pattern Anal. Mach. Intell. **29**(6), 1052–1067 (2007)
2. Endres, F., Hess, J., Sturm, J., Cremers, D., Burgard, W.: 3D mapping with an RGB-D camera. IEEE Trans. Robot. (T-RO) **30**(1), 177–187 (2013)
3. Grisetti, G., Kümmerle, R., Stachniss, C., Burgard, W.: A tutorial on graph-based SLAM. IEEE Intell. Trans. Syst. Mag. **2**(4), 31–43 (2010)
4. Hartley, R., Zisserman, A.: Multiple View Geometry in Computer Vision. Cambridge University Press (2003)
5. Kaess, M., Dellaert, F.: Probabilistic structure matching for visual SLAM with a multi-camera rig. Comput. Vis. Image Underst. **114**(2), 286–296 (2010), special issue on Omnidirectional Vision, Camera Networks and Non-conventional Cameras
6. Khoshelham, K., Elberink, S.O.: Accuracy and resolution of Kinect depth data for indoor mapping applications. Sensors **12**(2), 1437–1454 (2012)
7. Lachat, E., Macher, H., Mittet, M.A., Landes, T., Grussenmeyer, P.: First experiences with kinect v2 sensor for close range 3D modelling. ISPRS—International Archives of the Photogrammetry, Remote Sensing and Spatial Information Sciences (2015)
8. Lourakis, M.I., Argyros, A.A.: SBA: a software package for generic sparse bundle adjustment. ACM Trans. Math. Softw. (TOMS) **36**(1), 2 (2009)
9. Lu, F., Milios, E.: Globally consistent range scan alignment for environment mapping. Auton. Robots **4**(4), 333–349 (1997)
10. Montemerlo, M., Thrun, S., Koller, D., Wegbreit, B.: FastSLAM: A factored solution to the simultaneous localization and mapping problem. In: Eighteenth National Conference on Artificial Intelligence. pp. 593–598. American Association for Artificial Intelligence, Menlo Park, CA, USA (2002)
11. Nowicki, M., Skrzypczyński, P.: Combining photometric and depth data for lightweight and robust visual odometry. In: 2013 European Conference on Mobile Robots (ECMR), pp. 125–130 (Sept 2013)
12. Park, J.H., Shin, Y.D., Bae, J.H., Baeg, M.H.: Spatial uncertainty model for visual features using a Kinect^{TM} sensor. Sensors **12**(7), 8640–8662 (2012)
13. Schmidt, A., Fularz, M., Kraft, M., Kasiński, A., Nowicki, M.: An indoor RGB-D dataset for the evaluation of robot navigation algorithms. In: Blanc-Talon, J., Kasinski, A., Philips, W., Popescu, D., Scheunders, P. (eds.) Advanced Concepts for Intelligent Vision Systems, Lecture Notes in Computer Science, vol. 8192, pp. 321–329. Springer International Publishing (2013)
14. Sturm, J., Engelhard, N., Endres, F., Burgard, W., Cremers, D.: A benchmark for the evaluation of RGB-D slam systems. In: 2012 IEEE/RSJ International Conference on Intelligent Robots and Systems (IROS), pp. 573–580. IEEE (2012)
15. Triggs, B., McLauchlan, P.F., Hartley, R.I., Fitzgibbon, A.W.: Bundle adjustment—a modern synthesis. In: Vision algorithms: theory and practice, pp. 298–372. Springer (2000)

Combining Multiple Nearest-Neighbor Searches for Multiscale Feature Point Matching

Michał Swiercz, Marcin Iwanowski, Grzegorz Sarwas and Arkadiusz Cacko

Abstract Multiscale feature point descriptors are used to describe the surrounding of image feature points, taking into account neighboring image details at various levels. The feature vectors are divided into several parts, each of which describes surrounding at increasing distance from the feature point. In this paper a method for matching such descriptors is proposed. It is based on multiple nearest-neighbors searches that are applied to match parts of the descriptor, followed by combining the partial matching results into final indication of the closest descriptor.

Keywords Feature points · Matching · Nearest-neighbor · Feature extraction

1 Introduction

In this paper a method for matching of multiscale feature point descriptors is proposed. Feature points, well established concept in modern image processing, allows to find correspondence between points detected on various images, and consequently—to reason about the similarity between images or their parts. In this paper we propose the method for matching feature descriptors based on multiple nearest-neighbors searched. In order to apply this method, the feature descriptor must consist of separable levels of features detected on multiple scales (multiscale descriptor). Such

M. Swiercz (✉) · M. Iwanowski · A. Cacko
Institute of Control and Industrial Electronics, University of Technology,
Ul.Koszykowa 75, 00-662 Warszawa, Poland
e-mail: kubaswiercz@gmail.com

M. Iwanowski
e-mail: iwanowski@ee.pw.edu.pl

A. Cacko
e-mail: arkadiusz.cacko@ee.pw.edu.pl

M. Swiercz · G. Sarwas
Lingaro, Ul. Puławska 99a, 02-595 Warszawa, Poland
e-mail: grzegorz.sarwas@lingaro.com

R.S. Choraś (ed.), *Image Processing and Communications Challenges 7*,
Advances in Intelligent Systems and Computing 389,
DOI 10.1007/978-3-319-23814-2_27

a descriptor is also presented in this paper. It consists of statistical measures of groups of pixels—fields located around the feature point. These fields are positioned at increasing distances from the feature point—in the case presented in the paper, two distance-levels are investigated. Features are computed for fields at each distance separately. Owing to that, in fact, feature point is described by multiple (in our case—two) feature vectors. In the matching phase each of them is considered independently when using the nearest-neighbor matchings. Every search finds k closest feature vectors in the reference set. Next, the results of all searches, consisting of set of closest vectors and distances to them, are combined together using distance sorting. Finally the list of closest points is based of combined distances from different searches.

Feature points became one of the hottest topics in image processing over the last decade [1, 2, 4, 7, 8, 11, 12, 15]. They allow to describe the content of an image by extracting meaningful features of the visual scene in a manner invariant to various image transformations. Most of the feature point approaches are based on two-step principle consisting of feature point detector followed by extractor of the description its neighborhood. One of the principal issues concerning feature point descriptors is their matching [3, 5, 9, 10, 13, 14, 16]. As most feature point methods, formulating a feature-based description of the image is a two-step process. A necessary first step is finding points in the image that represent the area containing some important information about the objects in the image. This should be a well-defined, area with a high gradient, and an obvious choice for such an area is the corner of an object in the image. The corner detection used in this paper is a modified Harris corner algorithm [6], but our descriptor method can be used in conjunction with any corner detection method displaying sufficient stability and the ability to consistently detect the same corner under different parameters of the visual scene. Analysis of the feature point detection step is beyond the scope of this paper, we just assume it meets the above criteria of stability.

The paper is organized as follows. Section 1 describes a multicales feature point descriptor that is used in the current study. In the Sect. 3, the proposed approach to feature vector matching is presented. Section 4 shows the results of tests and, finally, Sect. 5 concludes the paper.

2 Multiscale Feature Point Descriptor

To provide strong descriptive properties and invariance to rotations, the proposed method recognizes the surroundings of a detected feature point as a series of circular *rings* around the feature point. Each ring is associated with a certain, fixed radius R around the feature point. The neighborhood of the feature point is probed by placing a *probing window* in evenly distributed *probing points* along the circumference of each ring, and extracting *probing sets* of pixels from the feature point's neighborhood, that lie underneath the probing mask.

The probing masks used in the described method are defined by Eq. 1 for the inner ring, and by Eq. 2 for the outer ring. For each ring, the position of the probing mask is defined by a function, which genetares the offsets for position of the mask's center, rounded to the nearest integer (3). These offsets, combined with the coordinates of the feature point, mark the final position of the probing mask, to indicate which pixels should be extracted as part of a given probing set.

$$P_1 = \begin{bmatrix} 1 & 1 & 1 \\ 1 & 1 & 1 \\ 1 & 1 & 1 \end{bmatrix} \quad (1)$$

$$P_2 = \begin{bmatrix} 0 & 1 & 1 & 1 & 0 \\ 1 & 1 & 1 & 1 & 1 \\ 1 & 1 & 1 & 1 & 1 \\ 1 & 1 & 1 & 1 & 1 \\ 0 & 1 & 1 & 1 & 0 \end{bmatrix} \quad (2)$$

$$F_n = \{(\alpha, r) : \alpha = \{0, \frac{1}{k} * 360, \frac{2}{k} * 360, \ldots\}, r = R_n\} \quad (3)$$

where α and r are the polar coordinates of the probing mask's center, relative to the position of the feature point, R_n is the radius of the n-th ring, and k is the number of probing points along the ring's circumference. These polar coordinates are then transformed into Cartesian coordinates and rounded to the nearest integer to obtain final offsets of the probing mask's position. The pixels laying under the probing mask superimposed on the image constitute a probing set.

In the presented method, each ring n is associated with a separate feature vector, constructed from several sub-vectors (*probing set feature vectors*), each associated with a certain probing set of pixels along the ring. Elements of the probing set feature vectors are the median values of red, green and blue channels of probing set pixels' colours, as defined by Eq. 4

$$m_{i_n} = [med(R_{i_n}), med(G_{i_n}), med(B_{i_n})] \quad (4)$$

where med is the median operator, R_{i_n}, G_{i_n}, B_{i_n} are, appropriately, the red, green and blue components of the RGB color of the pixels of the i-th probing set on the n-th ring.

These sub-vectors are subsequently combined into the complete feature vector D_n for the n-th ring (*ring feature vector*), describing a certain circular area (which can be intuitively imagined as a "donut") around the feature point, as specified by Eq. 5

$$D_n = [m_{1_n}, m_{2_n}, \ldots, m_{k_n}] \quad (5)$$

where m_{i_j} is the i-th feature sub-vectors of the n-th ring (as described by Eq. 4), and k is the number of probing sets in the n-th ring.

In the presented method, the descriptor (the describing dataset of a specific feature point) is a structure, consisting of feature vectors for all the rings. These feature vectors are not combined into a single vector for the purposes of descriptor-space searches and comparisons, but rather function as separate entities, that can be independently compared—i.e., a ring feature vector can be compared with any other ring feature vector provided that their length is equal.

3 Feature Vectors Matching

A simple nearest-neighbour search provides a matching method that is not sufficient for such a descriptor structure. Therefore, a new method was developed, which chooses the best matches not on the basis of a simple distance in the descriptor space, but on the basis of the smallest sum of distances for the corresponding best matches across all of the rings. Also, since there is no priority direction in the Cartesian coordinates space of the image, we cannot be sure that the order of the probing sets is the same between two compared images, as rotation of the tested image versus the reference image will change that order.

The problem of scale in the context of the proposed method can be partially solved by introducing a set of predefined scales of comparison. Each feature point in the reference image can have a set of descriptors extracted with different ratio of ring radiuses and probing window sizes, such that, for example, if we enlarge the image by the factor of 2, and simultaneously increase the radiuses of all the rings and the size of the probing window by the same factor, we will obtain the characteristics of a similar portion of the image and in a similar way, but on a different scale. By producing such dummy descriptors with different scale factors for every reference feature point, we increase the chance of a correct match, even if the scales of the test and reference images vary. The matching process is multi-stage and is performed as follows:

1. For a tested descriptor, for each ring and its feature vector D_n k-nearest-neighbour search is utilized to find the k nearest neighbors for this vector among the n-th Ring feature vectors Q_n in the reference set. Each of these k nearest neighbours must be within a threshold distance th_n of the tested point's feature vector, set separately for each ring n, otherwise it is eliminated from further matchings. However, due to the difficulties with test image rotation discussed above, we utilize a multi-comparison scheme, where we also perform a series of circular shifts of the feature vector D_n by the multiples of the length of the feature sub-vector m (so, with a probing set sub-vector consisting of 3 features—medians—these shifts would be in multiples of 3). Each of these shifted vectors is also compared to the reference set, and for each ring, 5 best matches are taken into account.

$$\mathrm{dist}(D_n, Q_n) = \{\mathrm{dist}\,(\mathrm{sh}_{i*l}(D_n), Q_n)\}\,, \tag{6}$$

Table 1 Best matches and their normalized distances (in brackets), in rings

Ring	1-st best	2-nd best	3-rd best	4-th best	5-th best
1	1(10)	5(13)	2(20)	7(33)	4(50)
2	5(5)	1(22)	7(25)	8(30)	9(60)

Table 2 Best matches and their total distances

1-st best	2-nd best	3-rd best	4-th best	5-th best	6-th best	7-th best
5(18)	1(32)	7(58)	2(–)	4(–)	9(–)	8(–)

where l is the length of the probing set sub-vector, *sh* is the circular shift vector operator, m_n is the number of probing points in n-th ring and $i : \{i = 1, 2, \ldots, m_n\}$

2. Once step 1 is complete, we select the final best match by normalizing the distances in each ring by the number of probing points in that ring and summing the distances in the feature space for each match, across all the rings, taking the smallest sum. A match needs to present in all of the rings to be taken into account. Table 1 presents an example of this process. In ring 1, the best match to our tested feature point is feature point 1 from the reference set (with distance of 10 in the feature space), the second best match is point 5 in the reference set, etc. After summing the distances for each match, across all the rings we obtain results as presented in Table 2. It is then determined that the best match is point 5 from the reference set. Points 2, 4, 8 and 9 are disregarded, as they do not appear in every ring matching.

4 Results

In order to assess the performance of the proposed ring descriptor, we have tested it on a set of images with varied characteristics, as presented in Fig. 1. Furthermore, additional test images were artificially generated by subjecting the test set to a number of affine transformations and other distortions (full 360° rotations, blur, Gaussian noise, gamma distortions, intensity changes, scale changes). This image base was tested against the reference images and the match correctness was evaluated. Matching feature points were considered to be the closest reference and detected test points in the Cartesian coordinate space of the image, within the distance of 4 pixels from the reference points. The test were performed under the configuration of $n = 2$ rings with radiuses, respectively, 1.5 and 3.5 pixels, the probing window specified in Eq. 1, 8 probing points in ring 1 and 16 probing points in ring 2. The aggregated results are presented in Table 3 (Fig. 2).

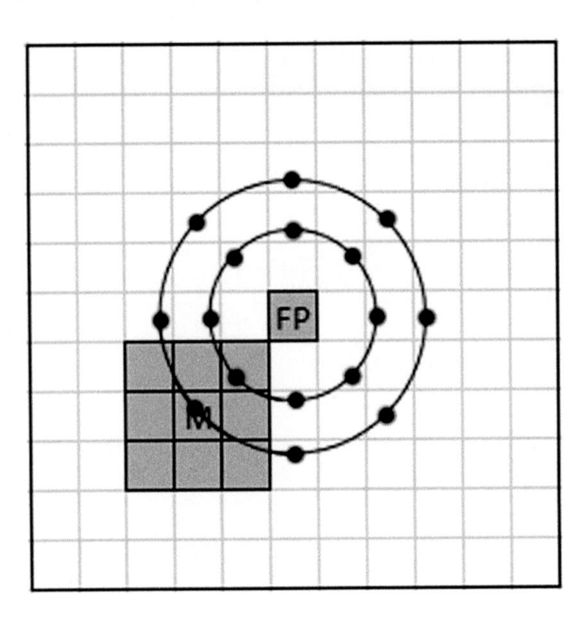

Fig. 1 Example: *Rings* around the feature point (FP) with marked probing points along the rings and position of an example probing mask (M)

Table 3 Test results for different families of distortions

Types of distortion	Average accuracy (%)	Worst accuracy (%)
Rotation, blur, scale, noise	76.2	55.1
Rotation	71.6	51.2
Blur, noise, intensity	85.4	68.2

Fig. 2 Test images

5 Conclusions

The problem of constructing a feature point based description of an image is complex and requires one to develop a robust algorithm of extracting meaningful characteristics of each feature point's surroundings. In this paper, a robust method for building feature point descriptors is presented. We have shown that the method is resistant to various image transformations and distortions, and that it can be utilized to partially solve the problem of scale in image comparison. Furthermore, it is relatively light in terms of computation effort and can be easily parallelised by dividing the processing of feature extraction and feature point matching into multiple independent workloads. The results presented in the paper are preliminary one, but since the results are promising, the research will be continued in two directions: investigation of other variants of multiscale descriptors, as well as an application of other schemes of multilevel matching.

Acknowledgments This work was co-financed by the European Union within the European Regional Development Fund.

References

1. Alahi, A., Ortiz, R., Vandergheynst, P.: FREAK: fast retina keypoint. In: 2012 IEEE Conference on Computer Vision and Pattern Recognition (CVPR), pp. 510–517 (2012)
2. Bay, H., Ess, A., Tuytelaars, T., Van Gool, L.: Speeded-up robust features (SURF). Comput. Vis. Image Underst. **110**(3), 346–359 (2008) (similarity Matching in Computer Vision and Multimedia)
3. Broder, A.Z., Charikar, M., Frieze, A.M., Mitzenmacher, M.: Min-wise independent permutations. J. Comput. Syst. Sci. **60**, 327–336 (1998)
4. Calonder, M., Lepetit, V., Strecha, C., Fua, P.: Brief: Binary robust independent elementary features. In: Daniilidis, K., Maragos, P., Paragios, N. (eds.) Computer Vision—ECCV 2010. Lecture Notes in Computer Science, vol. 6314, pp. 778–792. Springer, Berlin (2010)
5. Chum, O., Matas, J.: Fast computation of min-hash signatures for image collections. In: 2012 IEEE Conference on Computer Vision and Pattern Recognition (CVPR), pp. 3077–3084 (2012)
6. Harris, C., Stephens, M.: A combined corner and edge detector. In: Proceedings of Fourth Alvey Vision Conference, pp. 147–151 (1988)
7. Leutenegger, S., Chli, M., Siegwart, R.: BRISK: binary robust invariant scalable keypoints. In: 2011 IEEE International Conference on Computer Vision (ICCV), pp. 2548–2555 (2011)
8. Lowe, D.: Object recognition from local scale-invariant features. In: The Proceedings of the Seventh IEEE International Conference on Computer Vision, vol. 2, pp. 1150–1157 (1999)
9. Muja, M., Lowe, D.G.: Fast approximate nearest neighbors with automatic algorithm configuration. In: International Conference on Computer Vision Theory and Application (VISSAPP'09), pp. 331–340. INSTICC Press (2009)
10. Muja, M., Lowe, D.G.: Fast matching of binary features. In: Proceedings of the 2012 Ninth Conference on Computer and Robot Vision. pp. 404–410. CRV '12, IEEE Computer Society, Washington, DC, USA (2012)
11. Rosten, E., Porter, R., Drummond, T.: FASTER and better: a machine learning approach to corner detection. IEEE Trans. Pattern Anal. Mach. Intell. **32**, 105–119 (2010)

12. Rublee, E., Rabaud, V., Konolige, K., Bradski, G.: ORB: an efficient alternative to SIFT or SURF. In: 2011 IEEE International Conference on Computer Vision (ICCV), pp. 2564–2571 (2011)
13. Salakhutdinov, R., Hinton, G.: Semantic hashing. Int. J. Approximate Reasonig **50**(7), 969–978 (2009)
14. Slaney, M., Casey, M.: Locality-sensitive hashing for finding nearest neighbors. Sig. Process. Mag. IEEE **25**(2), 128–131 (2008)
15. Smith, S., Brady, J.: SUSAN, a new approach to low level image processing. Int. J. Comput. Vis. **23**(1), 45–78 (1997)
16. Tang, F., Tao, H.: Fast multi-scale template matching using binary features. In: IEEE Workshop on Applications of Computer Vision, WACV '07, pp. 36–36 (2007)

Detection of Face Position and Orientation Using Depth Data

Mariusz Szwoch and Paweł Pieniążek

Abstract In this paper an original approach is presented for real-time detection of user's face position and orientation based only on depth channel from a Microsoft Kinect sensor which can be used in facial analysis on scenes with poor lighting conditions where traditional algorithms based on optical channel may have failed. Thus the proposed approach can support, or even replace, algorithms based on optical channel or based on skeleton or face tracking information. The accuracy of proposed algorithms is 91 % and was verified on Facial Expressions and Emotions Database using 169 recordings of 25 persons. As the processing time is below 20 ms per frame on a standard PC, the proposed algorithms can be used in real-life applications. The presented algorithms were validated in a prototype application for user emotion recognition based on depth channel information only.

Keywords Face detection · Depth image processing · Kinect sensor

1 Introduction

In many situations, emotions may play an important role in the human-computer interaction as they have great impact on the way of using software, learning or training results, and also overall experience received from using computers. Affective computing is one of the emerging research areas on human behavior that develops emotion recognition, interpretation, and processing methods to create affective and affect-aware software to better adapt its behavior to user's needs. Such applications can have a significant impact in many fields such as healthcare, education, entertainment, software engineering, e-learning, etc. [13].

M. Szwoch (✉) · P. Pieniążek
Gdansk University of Technology, Gdansk, Poland
e-mail: szwoch@pg.gda.pl

P. Pieniążek
e-mail: pawpieni@pg.gda.pl

R.S. Choraś (ed.), *Image Processing and Communications Challenges 7*,
Advances in Intelligent Systems and Computing 389,
DOI 10.1007/978-3-319-23814-2_28

Affect recognition methods can use different information channels, such as video [7], audio [17], standard input devices [9], physiological signals [16], depth information [3], and others. As single channel approach can be limited by different factors, combination of multiple types of inputs from different modalities, or different features over the same modality, can significantly improve the system's classification abilities [7]. Unfortunately, in many cases only few input channels are available, which limits the range of possible solutions, as well as recognition accuracy.

The most popular source of non-invasive and non-intrusive information is a video camera as it allows for recognition of facial expressions, gestures, and body movement. Some algorithms concentrate only on facial expression recognition (FER) as they are typically used by humans to analyze the affect of other people [5]. Unfortunately, algorithms using video camera are very sensitive to face illumination conditions, causing great problems with a dark or unevenly illuminated scene. One of the possible solutions is to use additional channel with scene depth information. As depth sensors use additional light source (e.g. infrared) the information is generally insensitive to different ambient light conditions. Rapid development and increasing availability of relatively cheap RGB-D consumer sensors allows for creation of real-time systems for recognition of facial expressions and, going further, human emotions and moods.

In this paper, we focus on real-time processing of depth channel information in order to obtain a reliable detection of the face location and its orientation regardless of the scene illumination conditions. In the next section the important problems of depth image processing and face detection are presented. In Sect. 3 we propose an original approach based on some low-pass filtering and shape analysis methods. Next, we present some experiments and preliminary results obtained using prototype emotion recognition software. Finally, we present some conclusions and future works.

2 Related Work

Many approaches for affect recognition have been proposed during the last years and most of them focus on facial analysis, as facial expressions are one of the most natural ways for humans to express their emotions. Although, in general, multimodal analysis gives better results, many PC systems are equipped with video cameras and standard input devices only. As visual channel carries usually most information in human interaction it is the most common input device used in human affect recognition. Unfortunately, optical image analysis is often difficult or even impossible due to insufficient or uneven illumination of the scene. In such cases analysis of depth channel information may be used to provide lacking information about face localization and its orientation or even for further facial expression recognition. Analysis of depth data usually consists of several stages, such as depth image preprocessing and segmentation followed by detection of a face position and its orientation.

2.1 Preprocessing of Depth Images

Preprocessing is an essential stage in analysis of depth images obtained from consumer-grade range sensors, such as Microsoft Kinect. This is caused by several factors including: relatively low depth measurement accuracy, its temporal inconsistencies due to random temporal noise, and existence of regions (*holes*) where depth measurement is impossible due to object's occlusion or other reasons [8]. Additionally, the precision of depth images highly depends on the properties of materials covering the subjects, as well as their distance and even orientation relative to the camera axis. Depth image preprocessing consists usually of several steps, such as image denoising, background removal, and data reconstruction.

Denoising methods depend on the assumed noise model of the acquisition process. Using median filters can efficiently remove single salt and pepper noise. Unfortunately, the holes tend to group in larger formations in depth images. Moreover, the size of holes regions increases for scenes containing rapid objects movements. In [4] estimation of special noise covariance matrix is proposed for noise removal in depth images. Another possible method is temporal averaging of subsequent frames, possibly with greater weights assigned to the latest ones [12]. The problem is that averaging too many frames causes significant delay in frames processing and excessive blur of moving objects due to low-pass filtering.

Background removal, which is a part of segmentation stage, is a rather simple operation for depth images. In the case of using Kinect sensor or another device of comparable class and accuracy, the useful depth for face recognition is in the range from 0.6 m to about 2 m determining a kind of natural background boundary. After identifying of near located, interesting objects (possibly human heads) other, more distant pixels can be marked as background and removed. The only problem appears when two or more people are close to the depth camera or one to another.

The characteristic features of depth images are jagged edges of objects and holes on their surface (Fig. 1a), which are caused by problems with depth measurements. Some of this degradation elements can be eliminated during denoising phase. Remaining artifacts can be removed by other methods, such as morphological filters, shape analysis or temporal filtering. Removal of small holes can be achieved by morphological *closing*. Greater holes can be closed by using slightly increased values of their edges, assuming that they are entirely located on the same surface (Fig. 1).

In general the reconstruction results, especially those using information from subsequent frames, gives better results for images with a static and slow-changing content. In the case of rapid changes, the reconstruction results are much worse and can even lead to depth image degradation filling it with relatively large regions containing improper depth values. Fortunately, facial expressions are usually slow enough to correctly preprocess each frame.

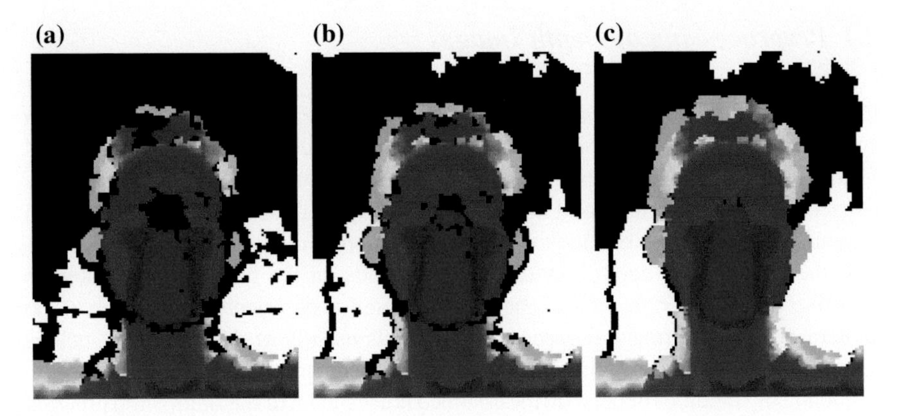

Fig. 1 Sample results of preprocessing and reconstruction of a depth image: an original depth image (**a**), an averaged depth image for $n = 3$ (**b**), and the reconstructed depth image (**c**)

2.2 Face Detection

Face detection in depth images is, generally, an easier task comparing to analysis of visual images. The main advantage of depth images is the possibility of easy and quick segmentation of consistent regions e.g. areas with similar depth location and roughly continuous depth gradient. Moreover, earlier background elimination allows to significantly reduce the analyzed area.

There are several approaches possible to detect a face in depth images. One possible approach is to treat depth images as grey level images and use them as the input to classical algorithms used for face detection, such as Viola-Jones and others. In the case of using algorithms depending on sliding windows it is also possible to change the window size according to the typical face size, depth camera resolution, and the region distance from a camera [10].

Another approach is to use a shape matching algorithm. Classical Active Appearance Models [11] or 3D Morphable Objects [1] can be supported by additional constraints, such as typical pose of human who sits in front of the camera, as well as possibility of determining approximate real size of objects in a depth scene. Using the information about the typical size of an average human face [14] and its distance from a depth sensor allows to create a list of candidate objects of a proper size. Such selection can significantly increase face detection efficiency. However, it should be taken into account that a face can be rotated to some extent that would influence its effective size.

2.3 Detection of Face Orientation

Detection of face orientation in a certain coordination system, usually relative to the camera axis, is another problem in the facial analysis systems. In most cases user's face is not perpendicular to the camera axis due to user's head movement and camera or sensor location below or above the monitor frame. Using any shape matching algorithm allows for automatic detection of a face orientation. In other cases, at least three nonlinear, characteristic points should be found that define a face plane in 3D space. Such algorithms usually start from nose detection as one of the most characteristic landmarks. In [2] the maximum direction vector approach is proposed, while in [6] the algorithm of effective energy is proposed.

3 Detection of Face Position and Orientation Using Depth Channel

The proposed approach consists of several steps including: depth image preprocessing and its segmentation, detection of a face position, and its orientation. These steps will be described in the next subsections. One of the main assumption of proposed algorithms is to work in real time on a modern PC. This speed criterion forces the application to use rather simple but fast algorithms which often use various heuristics but still gives satisfying results.

3.1 Preprocessing of Depth Images

In the proposed approach two averaging stages are used that are performed in time and space domain, respectively. The preprocessing in time domain is based on weighted averaging of n subsequent frames. The exact value of n is a compromise between the wish to increase the averaging accuracy on the one hand and the possibility of introducing too much delay in the measurement, and loss of a facial movement information on the other hand. The temporal averaging formula is as follows:

$$d'_{(x,y)}[i] = \frac{1}{n} \sum_{j=0}^{n} w[j] \cdot d_{(x,y)}[i-j] \tag{1}$$

where $d_{(x,y)}[i]$ is a depth value of a pixel P at (x, y) coordinates in the depth frame $D_F[i]$, and $w[j]$ is a weight of the frame $D_F[i-j]$. Experiments proved that in the case of analysis of human face the value of $n = 3$ combined with the weight vector $w^T = [9, 4, 1]$ gave satisfactory results. Additional experiments proved that elimination of uncertain measurements from averaging formula (1) can further improve the quality of a preprocessed image. In that case, all pixels $P_{(x,y)}$ which are marked as too near, too far or unrecognized, are not taken into account in the formula (1), e.g. their values are temporary reset together with decreasing the denominator n. A sample result of the averaging process is presented in Fig. 1b.

Unfortunately, averaging of a small number of frames ($n = 3$) cannot eliminate all uncertain pixels in the depth image. This problem occurs mainly for surfaces which lay in the shadow and cannot be exposed to the structural infrared pattern. The problem can also be observed for many specular materials, very detailed surfaces (e.g. hair) and for surfaces that are almost perpendicular to the viewing plane. The good example of such artifacts in depth images containing a face are areas around the nose which is usually separated from the rest of the face by a few-pixel-wide frame (black pixels in Fig. 1a).

In the proposed approach an additional spatial filtering step is performed to eliminate small areas of such pixels. For each uncertain pixel its new depth value is taken from the nearest neighboring pixel of a proper value in the 7×7 N^4-neighborhood. The final effect is somewhat similar to the result of the morphological dilation used three times. Using the simple N^4-neighborhood speeds up the preprocessing time while still giving satisfying results. Using the full N^8-neighborhood does not significantly improve results while increasing the complexity from linear to square one. The neighborhood size is limited to 7×7 which is sufficient for images containing faces in the small distance from the depth sensor. For further located faces the considered neighborhood can be reduced. Such simple filtering does not allow for exact image reconstruction but it quickly eliminates small holes in the face area. The sample results of additional filtering are presented in Fig. 1c.

3.2 Image Segmentation and Detection of Face Position

The face of a person sitting in front of the monitor is a very specific and characteristic element of a depth image (Fig. 1) but there are two potential problems with its detection. The first concerns other segments of the depth image of a similar size and shape (e.g. other faces) or segments located near the face (e.g. user's body or a hand) that can merge or occlude the face area. The second problem is a face rotation that can significantly change its properties in the depth image. In the proposed approach four criteria are used to extract consistent and homogeneous candidate segments S from the preprocessed depth image D_F and to classify them as the face area F. These criteria are based on anthropometric features of the face and some assumed constraints of user's location and orientation relative to the depth sensor.

The first criterion is *depth continuity* which specifies the maximum distance between neighboring pixels within the face area F. The threshold value was estimated as $\epsilon_1 = 1\,\text{cm}$. Such relatively high value results mostly from possible depth discontinuities in the region around the nose. Using any region growing algorithm, such as using breadth-first search (BFS) or other, allows for easy creation of a set of candidate segments S.

For each segment S the second criterion of *depth homogeneity* is applied which specifies the maximum allowed depth distance between pixels within the face area. For the faces directed towards the depth sensor this distance can be estimated as $\epsilon_2' = 5\,\text{cm}$ which results from the distance between the nose top and the end of cheeks. Unfortunately, for faces rotated around the horizontal axis X the maximum depth distance within the face area is often much higher, and can be estimated as $\epsilon_2'' = Hsin(\alpha)$, where $H \approx 22\,\text{cm}$ is the typical height of the human face [10]. For the angle $\alpha = 45°$ it gives $\epsilon_2'' \approx 15\,\text{cm}$. Choosing the greater of these two values $\epsilon_2 = max(\epsilon_2', \epsilon_2'')$ for the threshold of the second criteria allows for a rough elimination of these candidate segments that are too stretched along the Z axis.

The third criterion of *background separability* defines the minimum depth distance between the face and its surrounding pixels. The threshold value $\epsilon_3 = 5\,\text{cm}$ appeared to be sufficient for effective face separation.

The forth criterion of the *face size* finally verifies the candidate segments S for the proper dimensions of the analyzed region regardless of their distance from a depth sensor. This criterion assumes that for a typical face its height $H \approx 22\,\text{cm}$ and width $W \approx 15\,\text{cm}$ [10]. For further calculations an average face distance f_d from a depth sensor can be estimated by (2):

$$f_d = \frac{1}{\bar{\bar{S}}} \sum_{j=0}^{\bar{\bar{S}}} d_{(x,y)} : P_{(x,y)} \in S \tag{2}$$

where $\bar{\bar{S}}$ is the number of pixels P in the analyzed region S. In that case the expected face's dimensions h_f (height) and w_f (width) in pixels can be evaluated using [8]:

$$h_f = \frac{size_h \cdot H}{\tan(\frac{FOV_h}{2}) \cdot 2 \cdot f_d}, \; w_f = \frac{size_w \cdot W}{\tan(\frac{FOV_w}{2}) \cdot 2 \cdot f_d} \tag{3}$$

where $size_h$ and $size_w$ are the image dimensions and FOV is the depth sensor's field of view.

Evaluation of expected h_f and w_f values allows for their easy comparison with the actual height h_S and width w_S of the analyzed region S using (4):

$$|h_f - h_F| < \epsilon_4 \wedge |w_f - w_F| < \epsilon_5 \tag{4}$$

where h_S is the distance between the highest and lowest points of region S and w_S is calculated similarly as the distance between the leftmost and rightmost edge points. Unfortunately, it has occurred during validation tests that evaluation of effective threshold values ϵ_4 and ϵ_5 is impossible in practice due to their high volatility caused by rotation of the user's head and its frequent merging with his/her neck and shoulders. This problem results mainly from the sensor location below the monitor.

In the case of lower sensor location another approach is proposed that is based on the analysis of line widths in the potential face area S which should not be too narrow nor too wide comparing with the expected face width w_f. This *width* criterion can be formulated as (5):

$$w_f \cdot \epsilon_6 < w^{'}[y] < w_f \cdot \epsilon_7 \quad (5)$$

where $w^{'}[y]$ is the width of line y in the potential face area S.

The number $h^{'}$ of lines meeting the formula (5) should not exceed certain part of the total lines in the area that is expressed by the last formula (6):

$$\epsilon_8 \cdot h^{'} < h_f \quad (6)$$

Analysis of training data allowed to empirically determine the best values of these three thresholds as $\epsilon_6 = 0.4$, $\epsilon_7 = 1.45$, and $\epsilon_8 = 0.41$.

The first three criteria and additional formulas (5), (6) allow to efficiently detect a face area F in the depth image D_F assuming that only one person is sitting in front of the sensor.

3.3 Detection of Face Orientation

The next stage is detection of face rotation angles around each axis. In the proposed approach we start with determining the rotation γ around the Z axis, which can be approximated by a relative shift of the gravity centers $P_1(x_1, y_1)$ and $P_2(x_2, y_2)$ of the upper and lower part of the face (light grey areas in Fig. 2a):

$$\gamma = arctan(\frac{x_1 - x_2}{y_1 - y_2}) \quad (7)$$

Further processing aims at determining the face plane by approximation of its rotation angles α and β around the X and Y axes, respectively (Fig. 2b, c). Evaluation of these angles is performed on partially straightened up face area which can be received by its rotation by $-\gamma$ angle around the Z axis. For the rotated face area the gravity centers are calculated: $P_3(x_3, y_3, d_3)$, $P_4(x_4, y_4, d_4)$, $P_5(x_5, y_5, d_5)$, and

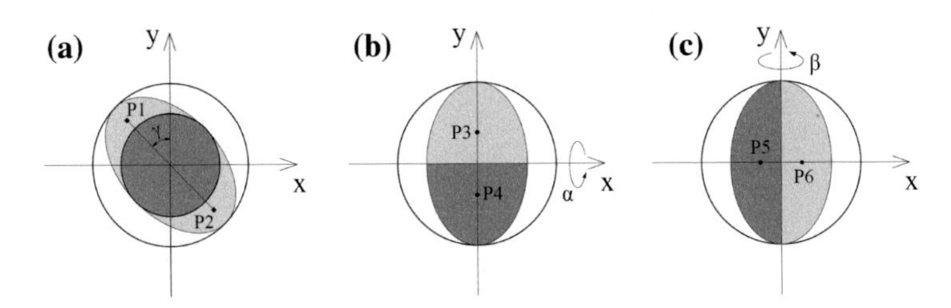

Fig. 2 Centers of gravity of: *upper* and *lower* facial areas (**a**), *upper* and *lower* halves (**b**), *left* and *right* halves (**c**)

$P_6(x_6, y_6, d_6)$ of the proper half of the face according to Fig. 2b, c). In order to estimate the α and β rotation angles the following formulas can be used:

$$\alpha = arctan(\frac{d_3 - d_4}{y_3 - y_4}), \beta = arctan(\frac{d_5 - d_6}{x_5 - x_6}) \quad (8)$$

where $d_i \equiv d(x_i, y_i)$ is an average depth value in the corresponding region.

3.4 Face Image Normalization

The final stage of a face image processing in the proposed approach is its normalization which should allow for effective facial expression recognition regardless of its position, distance and rotation relatively to the sensor. Information estimated during the previous stages allows to estimate values for these affine transformations of the face in the camera's field of view. The normalization is performed by transforming real pixel coordinates $P_{(x,y,d(x,y))}$ into uniform analysis space using three dimensional transformation matrix $\mathcal{M}$ which is a superposition of the scaling $\mathcal{M}^S$, rotation $\mathcal{M}^R$, and translation $\mathcal{M}^T$ matrices:

$$P'_{(x',y',d'(x',y'))} = P_{(x,y,d(x,y))}\mathcal{M}^S\mathcal{M}^R\mathcal{M}^T \quad (9)$$

The $\mathcal{M}_x^S$ and $\mathcal{M}_y^S$ scaling factors are calculated basing on an average face distance f_d from a sensor as $\mathcal{M}_x^S = \mathcal{M}_y^S = f_d/d_{min}$, where $d_{min} = 400$ is the minimum object distance that is recognizable by the Kinect sensor. Depth value is not scaled thus $\mathcal{M}_z^S = 1$. The rotation matrix $\mathcal{M}^R$ is determined by the three angles α, β, γ, while the translation matrix $\mathcal{M}^T$ is based on a face gravity center

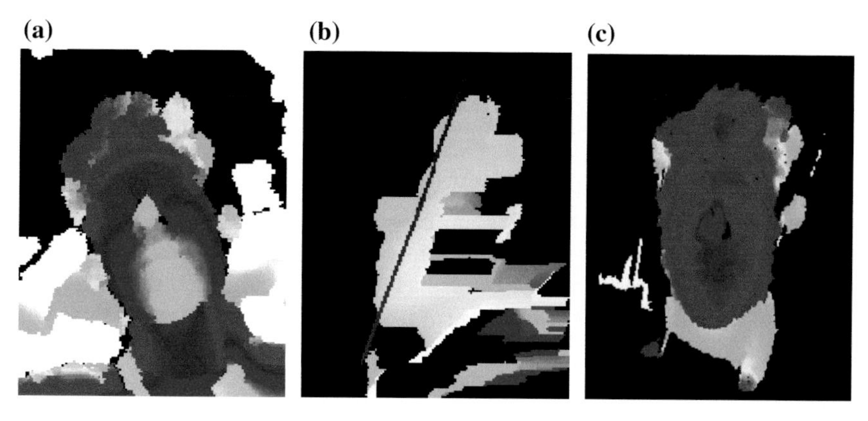

Fig. 3 Sample of the face normalization: detected face (**a**), face profile with a facial plane (**b**), normalized face in a uniform analysis space (**c**)

$P_{(x_{avg}, y_{avg}, d(x_{avg}, y_{avg}))}$ as $\mathcal{M}_x^T = x_{avg}$, $\mathcal{M}_y^T = y_{avg}$, and $\mathcal{M}_z^T = (x_{avg}, y_{avg}) - d_{min}$. Superposition of the above three matrices defines a uniform transformation matrix $\mathcal{M}$ that allows for affine transformation of the face area into the analysis space. As the resulting $F^{'}$ area may contain undefined values for some coordinates $P^{'}_{(x^{'}, y^{'}, d^{'}(x^{'}, y^{'}))}$ additional linear approximation is performed for such points. In this way any potential hole in the face area is filled by average values of its horizontal neighbors. Examples of normalized faces in a uniform analysis space are given in Fig. 3.

4 Experiments and Preliminary Results

The algorithms described in the previous section were implemented as the initial depth image processing and segmentation stages in $EmotiDET$ prototype application that recognizes user's emotions basing on depth stream from the Kinect sensor. Several facial landmarks are determined in the recognized face area F in the next stage. Then, specific movement patterns of the landmark areas are recognized and classified as one of 9 emotions, namely *neutral*, *joy*, *surprise neutral* and *positive*, *euphoria*, *fear*, *fright*, *anger*, and *scorn*.

EmotiDET application is able to manage depth stream directly from Kinect sensor, as well as from video files. In order to verify the recognition efficiency of proposed approach FEEDB database was used as a testing dataset [15]. FEEDB contains 1650 recordings of 33 emotions and facial expressions expressed by 50 young students, who seem to be quite representative in some applications such as video games. All recordings were acquired using Microsoft Kinect sensor and contain both RGB color, as well as depth streams from the sensor. The recordings were done in standard scenery of an IT laboratory with mixed natural and fluorescent illumination. No special control on illumination or other people in the background were done. All the images were taken against an artificially lighted non-homogeneous background (walls and ceiling). Participants sat in an upright, frontal position at different places allowing for differentiated background and lighting conditions. Participants were instructed to begin and end each expression with a neutral pose, and to estimate a difficulty in expressing of particular emotions. Kinect sensor was placed just below the monitor frame. All participants were asked to keep the optimal distance to the sensor that is about 60–80 cm.

Unfortunately, analysis of FEEDB recordings showed that many of them do not suit the experiment requirements. In many cases students were unable to express the requested emotion or were overriding their faces with hands. Some recordings were expressed carelessly, artificially long or did not start or end with neutral expression. Finally, 169 recordings were chosen which best satisfied the above conditions and were accepted by a human expert.

In the first experiment the face detection efficiency of the face position and rotation was estimated as 91 %. The main problem for 15 recordings was identified as too low location of the Kinect sensor that caused a frog perspective in the depth images in

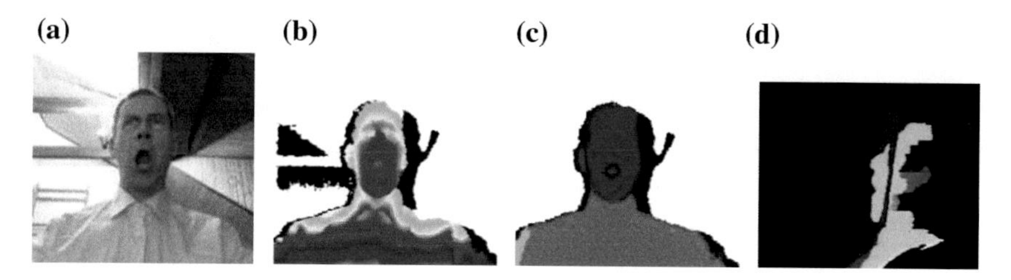

Fig. 4 Examples of face detection stages: original RGB (**a**), and depth image (**b**), detected face area (**c**), and face plane (**d**)

which user's face often merges with his/her neck and shoulders. The frame processing time is about 20 ms on a standard PC that seems enough for further analysis steps. Sample results of detecting user's face and its orientation are presented in Fig. 4a–d.

In the second experiment the results of the previous stage were used for detection of some facial landmarks and their movement patterns in order to recognize user's emotions. Although the average recognition efficiency for nine emotions was only slightly above 50 % one should take into account that these results were obtained basing on analysis of depth channel only. Fusing this approach with recognition results obtained from other information channels would certainly increase the final efficiency of emotion recognition.

5 Conclusions and Future Works

In this paper an original approach was presented for detection of the face position and rotation using only depth channel information. The initial requirements of processing in a real-time enforced some compromises and usage of simple heuristic algorithms for depth image preprocessing, detection of user's face position and its orientation. Depending on the processing power and availability of other information channels it is possible to use more sophisticated algorithms at any of the processing pipeline stages, as well as fusing the results at any stage with results from other channels such as RGB video, user's skeleton, or physiological signals.

The carried out experiments proved that depth channel contains sufficient information for efficient and fast detection of the face location and its orientation. Its advantage over processing a video channel is independence of lighting conditions in the analyzed scene. Time instability and quite high noise and measurement errors of depth images can be efficiently corrected using low pass filtering in time and spatial domains. Unfortunately, the usual location of the Kinect sensor below the monitor may cause merging of user's face and body area, which is hard to analyze and may cause face detection failure. It is very probable that using smaller RGB-D sensors, such as Intel RealSense, built-in the monitor or located just above its upper frame would significantly improve the detection rate.

Further research proved that using only depth information for recognition of facial expressions and user's emotion leads to, generally, unsatisfactory results. Nevertheless, depth information seems to be very useful in the case of poor scene illumination. Thus, our further research will focus on the fusion of RGB and depth channels in order to reliably and efficiently recognize facial expressions from depth sensors in real time.

Acknowledgments The research leading to these results has received partial funding from the Polish-Norwegian Research Programme operated by the National Centre for Research and Development under the Norwegian Financial Mechanism 2009–2014 in the frame of Project Contract No Pol-Nor/210629/51/2013.

References

1. Blanz, V., Vetter, T.: Face recognition based on fitting a 3D morphable model. IEEE Trans. Pattern Anal. Mach. Intell. **25**(9), 1063–1074 (2003)
2. Breitenstein, M., Kuettel, D., Weise, T., Van Gool, L., Pfister, H.: Real-time face pose estimation from single range images. In: IEEE Conference on Computer Vision and Pattern Recognition. CVPR 2008, pp. 1–8. IEEE (2008)
3. Burgin, W., Pantofaru, C., Smart, W.D.: Using depth information to improve face detection. In: Proceedings of the 6th International Conference on Human-robot Interaction, pp. 119–120.HRI'11 (2011)
4. Cai, Q., Gallup, D., Zhang, C., Zhang, Z.: 3D deformable face tracking with a commodity depth camera. In: Proceedings of the 11th European Conference on Computer Vision Conference on Computer Vision: Part III, pp. 229–242. ECCV'10 (2010)
5. Castrillón-Santana, M., Déniz-Suárez, O., Antón-Canalís, L., Lorenzo-Navarro, J.: Face and facial feature detection evaluation performance evaluation of public domain haar detectors for face and facial feature detection. In: Third International Conference on Computer Vision Theory and Applications, VISAPP08 (2008)
6. Chew, W.J., Seng, K.P., Ang, L.M.: Nose tip detection on a three-dimensional face range image invariant to head pose (2009)
7. Gunes, H., Piccardi, M.: Affect recognition from face and body: early fusion vs. late fusion. In: 2005 IEEE International Conference on Systems, Man and Cybernetics, vol. 4, pp. 3437–3443 (2005)
8. Khoshelham, K.: Accuracy analysis of Kinect depth data. ISPRS—International Archives of the Photogrammetry, Remote Sensing and Spatial Information Sciences, vol. XXXVIII-5/W12, pp. 133–138 (2011)
9. Kolakowska, A.: A review of emotion recognition methods based on keystroke dynamics and mouse movements. In: 2013 The 6th International Conference on Human System Interaction (HSI), pp. 548–555 (2013)
10. Kolb, A., Koch, R. (eds.): Dynamic 3D Imaging. Lecture Notes in Computer Science, vol. 5742 (2009)
11. Martin, C., Werner, U., Gross, H.M.: A real-time facial expression recognition system based on active appearance models using gray images and edge images. In: 8th IEEE International Conference on Automatic Face Gesture Recognition, FG'08, pp. 1–6 (2008)
12. Matyunin, S., Vatolin, D., Berdnikov, Y., Smirnov, M.: Temporal filtering for depth maps generated by kinect depth camera. In: 3DTV Conference: The True Vision—Capture, Transmission and Display of 3D Video (3DTV-CON), pp. 1–4 (2011)
13. Picard, R.W.: Affective computing: from laughter to IEEE. IEEE Trans. Affect. Comput. **1**(1), 11–17 (2010)

14. States., U.: Human engineering design data digest [microform]: human factors standardization systems/human factors standardization SubTAG. The Group Washington, D.C (2000)
15. Szwoch, M.: FEEDB: a multimodal database of facial expressions and emotions. In: 2013 The 6th International Conference on Human System Interaction (HSI), pp. 524–531 (2013)
16. Szwoch, W.: Using physiological signals for emotion recognition. In: 2013 The 6th International Conference on Human System Interaction (HSI), pp. 556–561 (2013)
17. Zeng, Z., Pantic, M., Roisman, G., Huang, T.: A survey of affect recognition methods: audio, visual, and spontaneous expressions. IEEE Trans. Pattern Anal. Mach. Intell. **31**(1), 39–58 (2009)

Part II
Communications

A Comparative Study of Statistical Models with Long and Short-Memory Dependence for Network Anomaly Detection

Tomasz Andrysiak, Łukasz Saganowski and Adam Marchewka

Abstract Protection of systems and computer networks against novel, unknown attacks is currently an intensively examined and developed domain. One of possible solutions to the problem is detection and classification of abnormal behaviors reflected in the analyzed network traffic. In the presented article we attempt to resolve the problem by anomaly detection in the analyzed network traffic described with the use of five different statistical models. We tested two groups of models which differed in autocorrelation dependences. The first group was composed of AR, MR and ARMA models which are characterized by short memory dependences. The second group, on the other hand, included statistical attempts described with ARFIMA and FIGARCH models which are characterized by long memory dependences. In order to detect anomalies in the network traffic we used differences between real network traffic and its estimated model. Obtained results of the performed experiments show purposefulness of the conducted comparative study of exploited statistical models.

Keywords Anomaly detection · Statistical models · Network traffic prediction

1 Introduction

For many years, there have been used safety systems based on formerly isolated and classified patterns of threats, named signatures. Anti-virus software, systems for detection and breaking-in counteraction and protection against information leaks are just examples from a long and diversified list of application of those techniques.

T. Andrysiak (✉) · Ł. Saganowski · A. Marchewka
UTP University of Science and Technology Institute of Telecommunications,
ul. Kaliskiego 7, 85-789 Bydgoszcz, Poland
e-mail: andrys@utp.edu.pl

Ł. Saganowski
e-mail: luksag@utp.edu.pl

A. Marchewka
e-mail: adimar@utp.edu.pl

R.S. Choraś (ed.), *Image Processing and Communications Challenges 7*,
Advances in Intelligent Systems and Computing 389,
DOI 10.1007/978-3-319-23814-2_29

Nevertheless, there is one aspect in common, namely, they are able to protect systems and computer networks from known attacks described by the mentioned patterns. However, does lack of traffic matching known signatures mean there is no threat?

A means to defend from novel, unknown attacks is a rather radical change in operation concept. Instead of searching for attack signatures in network traffic it is necessary to browse for abnormal behavior which is a deviation from the normal traffic characteristic. The strength of such an approach is visible in solutions which are not based on knowledge a priori of attack signatures but on what does not respond particular norms, profiles of the analyzed network traffic. The techniques based on the above mentioned assumptions should be able to detect both: simple attacks of DoS type (Denial of Service) or DDoS (Distributed Denial of Service), and intelligent network worms up to hybrid attacks which are a combination of numerous different destruction methods. The consequence of such kind of attacks is inception of network anomalies, which creates a possibility to detect them, or even prevent from unwanted actions. The hardest challenge, however, is differentiation between dangerous behavior and normal movement in its initial stage in order to limit the usage of network resources. Anomalies are abnormalities, variations from the adopted rule. Anomalies in network traffic can signify device damage, an error in software or attack on resources and network systems. The essence of anomaly disclosure in computer networks is therefore detecting abnormal behaviors or actions which in particular can constitute a source of a potential attack [6]. One of possible solutions to the presented problem is implementation of Anomaly Detection Systems. They are currently used as one of the main mechanisms of safety supervision in computer networks. Their action consists in monitoring and detecting attacks directed onto information system resources on the basis of abnormal behaviors reflected in parameters of network traffic. Anomaly detection methods have been a topic of numerous surveys and review articles. In works describing the methods there were used techniques consisting in machine learning, neural networks, clustering techniques and expert systems. At present, anomaly detection methods that are particularly intensively developed are those based on statistical models describing the analyzed network traffic [6]. The most often used models are autoregressive ARMA or ARIMA, and Conditional Heteroscedastic Models ARCH and GARCH, which allow to estimate profiles of a normal network traffic [18]. In the present article we propose using estimation of statistical models AR, MR, ARMA, ARFIMA and FIGARCH for defined behavior profiles of a given network traffic. The process of anomaly detection (a network attack) is realized by comparison of parameters of a normal behavior (predicted on the basis of the tested statistical models) and parameters of real network traffic. This paper is organized as follows. After the introduction, in Sect. 2 we present the definition of long and short memory dependence. In Sect. 3 the different statistical models for date traffic prediction are described in details. Then, in Sect. 4 the Anomaly Detection System based on AR, MR, ARMA, ARFIMA and FIGARH model estimation is shown. Experimental results and conclusion are given thereafter.

2 Definition of Long and Short-Memory Dependence

Long memory dependences manifest themselves in the existence of autocorrelations of elements creating the given time series. In most cases it is high-order autocorrelation. This means that in the examined series there is a dependence between the observations—even those distant in time. This phenomenon is called long memory and was discovered by a British hydrologist Hurst [13]. In case of long memory existence the autocorrelation function is slowly falling at hyperbolic pace. The time series with long memory feature has in the spectral domain distribution of low frequency. Short memory time series, however, show essential autocorrelation of low frequency only. This means that observations that are separated even by a relatively short time period are no longer correlated. Short memory series are easy to recognize due to the fact that in the time domain the autocorrelation function disappears quickly, and in the spectral domain there are distributions of high frequency. It is said that the stochastic process has long memory with parameter d if its spectral density function $f(\lambda)$ meets the condition

$$f(\lambda) \sim c\lambda^{-2d},\ when \quad \lambda \to 0^{+}, \tag{1}$$

where c is constant, and symbol $\sim$ means that the relation of left and right side is heading to one. When the process meets that condition and when $d > 0$ then its autocorrelation function is disappearing in hyperbolic manner [3, 4, 18] i.e.

$$\rho_k \sim c_\rho k^{2d-1},\ when \quad k \to \infty. \tag{2}$$

Parameter d describes the memory of the process. When $d > 0$, the spectra density function in unlimited in surrounding 0. It is then said that the process has a long memory. When $d = 0$, the spectral density is limited in 0, and the process is described as having short memory. When $d < 0$, then the spectral density equals 0 and the process shows negative memory and is named anti persistent [10, 12].

3 Statistical Models for Network Traffic Prediction

The tested network traffic is represented by means of time series describing variance of parameters characterizing the number of received and sent TCP, UDP and ICMP packages within a time unit. A natural way of describing such series are statistical models which are based on autoregression and moving average in relation to differently realized data variances and autocorrelation of elements creating the given time series.

3.1 Short-Memory Models

In order to describe the properties of short memory time series (essential autocorrelations of low order only) the approach that is often applied is the use of solutions known as autoregression model AR, moving average MR and mixed models ARMA. They can be used for modeling stationary series, i.e. series where there are only random fluctuations around the average, or non-stationary reducible to a stationary form. Their composition is based on autocorrelation phenomenon, i.e. on correlation of the predicted variable value with values of the same variable but delayed in time [5].

Autoregressive Model Numerous time series are composed of interdependent observations which means that it is possible to estimate the models coefficients which describe the following elements of the series on the basis of the delayed in time previous elements of the series $\left(Y_{t-1}, Y_{t-2}, \ldots, Y_{t-p}\right)$, and random component ε_t in current period t. The above can be presented with the use of equation of autoregression of the order (p) as $AR(p)$

$$Y_t = c_0 + \phi_1 Y_{t-1} + \phi_2 Y_{t-2} + \cdots + \phi_p Y_{t-p} + \varepsilon_t, \quad (3)$$

where $\phi_1, \phi_2, \ldots, \phi_p$ describe the models parameters, c_0 is invariable and $\varepsilon_t \left(0, \sigma^2\right)$ is the white noise process with zero mean and variance σ^2. AR is a process with memory of previous realizations of the series. Such a process is called stationary, when there is a condition $p > 1$ and all roots of the polynomial $W(z) = 1 - \phi_1 z - \phi_2 z^2 + \cdots - \phi_p z^p$ of each module are greater than one. For such a model the prediction is built step by step by recurrent substitution of successive values. With stationary processes $AR(p)$ such a prediction is heading for the average value of the process, and the error variance of the forecast aims at the variance of the process.

Moving Average Model It is a linear model in which the realization of Y_t in the current period depends on realization of the random component ε_t in the current period and q in subsequent previous periods. It can be presented by means of an equation of moving average of order q as $MA(q)$

$$Y_t = \varepsilon_t + \theta_1 \varepsilon_{t-1} + \theta_2 \varepsilon_{t-2} + \cdots + \theta_q \varepsilon_{t-q}, \quad (4)$$

where $\theta_1, \theta_2, ..., \theta_p$ describe the models parameters, and $\varepsilon_t \sim \left(0, \sigma^2\right)$ is the white noise process with zero mean and variance σ^2. MA is a process with memory of previous values of the random component. Every MA process which can be reduced to a stationary autoregressive process is called invertible. In general case this condition is fulfilled when the roots of the polynomial $W(z) = 1 + \theta_1 z + \theta_2 z^2 + \cdots + \theta_p z^q$ lie outside the unit circle. The prediction made with the use of $MA(q)$ model is obtained in the recurrent way, as it seeks the average value.

Autoregressive Moving Average Model For a stationary series, instead of applying separate models of AR and MR classes, in order to describe the connections between observations from the subsequent periods we use autoregressive models of moving average [3], i.e. $ARMA(p, q)$ models with delay order (p, q) written as

$$Y_t = \phi_1 Y_{t-1} + \phi_2 Y_{t-2} + \cdots + \phi_p Y_{t-p} + \varepsilon_t + \theta_1 \varepsilon_{t-1} + \theta_2 \varepsilon_{t-2} + \cdots + \theta_q \varepsilon_{t-q} \quad (5)$$

where $\phi_1, \phi_2, \ldots, \phi_p$, and $\theta_1, \theta_2, \ldots, \theta_p$ describe the models parameters, and $\varepsilon_t \sim (0, \sigma^2)$ is the white noise process with zero mean and variance σ^2. As a result, by means of a lower number of AR and MR components than separately for AR model and MR model, any linear process can be described, which is beneficial from the perspective of the models estimation possibility and its use in predicting. $ARMA$ process contains properties of both AR and MR which is most easily visible in decomposition of the ACF function. $ARMA$ model generates stationary process if its components are: stationary AR and reversible MA. The prediction made by means of $ARMA(p, q)$ model is obtained in a recurrent way.

3.2 Long-Memory Models

An interesting approach towards describing the features of long memory time series was the use of solutions with movable autoregressive averaging in the process of fractional differentiation. In a result $ARFIMA$ (Fractional Differenced Noise and Autoregressive Moving Average) model was created [10], which is a generalization of $ARMA$ and $ARIMA$ models. Another approach towards time series description was including conditional variance dependence of the process from its previous values using the $ARCH$ model (Autoregressive Conditional Heteroskedastic Model) introduced by Engel [6]. Generalization of this approach was the $FIGARCH$ model (Fractionally Integrated $GARCH$), which autocorrelation function of squared residuals of the model decreases in a hyperbolic way. Such a behavior of an autocorrelation function enables naming $FIGARCH$ a model of long memory in the context of the autocorrelation function of squared residuals of the model.

3.3 Introduction to ARFIMA Model

The Autoregressive Fractional Integrated Moving Average model called $AR-FIMA$ (p, d, q) is a combination Fractional Differenced Noise and Auto Regressive Moving Average which is proposed by Grange, Joyeux and Hosking, in order to analysis the Long-Memory property [10, 12].

The $ARFIMA(p, d, q)$ model for time series Y_t is written as:

$$\Phi(L)(1 - L)^d y_t = \Theta(L)\varepsilon_t, \quad t = 1, 2, \ldots\Omega, \quad (6)$$

where y_t is the time series, $\varepsilon_t \sim (0, \sigma^2)$ is the white noise process with zero mean and variances σ^2, $\Phi(L) = 1 - \phi_1 L - \phi_2 L^2 - \cdots - \phi_p L^p$ is the autoregressive polynomial and $\Theta(L) = 1 + \theta_1 L + \theta_2 L^2 + \cdots + \theta_p L^q$ is the moving average polynomial, L is the backward shift operator and $(1 - L)^d$ is the fractional differencing operator given by the following binomial expansion:

$$(1 - L)^d = \sum_{k=0}^{\infty} \binom{d}{k} (-1)^k L^k \tag{7}$$

and

$$\binom{d}{k} (-1)^k = \frac{\Gamma(d+1)(-1)^k}{\Gamma(d-k+1)\Gamma(k+1)} = \frac{\Gamma(-d+k)}{\Gamma(-d)\Gamma(k+1)}, \tag{8}$$

where $\Gamma(*)$ denotes the gamma function and d is the number of differences required to give a stationary series and $(1 - L)^d$ is the dth power of the differencing operator. When $d \in (-0.5, 0.5)$, the $ARFIMA(p, d, q)$ process is stationary, and if $d \in (0, 0.5)$ the process presents long-memory behavior.

Forecasting $ARFIMA$ processes is usually carried out by using an infinite autoregressive representation of (1), written as $\prod(L) y_t = \varepsilon_t$, or

$$y_t = \sum_{i=1}^{\infty} \pi_i y_{t-i} + \varepsilon_t, \tag{9}$$

where $\prod(L) = 1 - \pi_1 L - \pi_2 L^2 - \cdots = \Phi(L)(1 - L)^d \, \Theta(L)^{-1}$. In terms of practical implementation, this form needs truncation after k lags, but there is no obvious way of doing it. This truncation problem will also be related to the forecast horizon considered in predictions (see [12]). From (9) it is clear that the forecasting rule will pick up the influence of distant lags, thus capturing their persistent influence. However, if a shift in the process occurs, this means that pre-shift lags will also have some weight on the prediction, which may cause some biases for post-shift horizons [8].

3.4 FIGARH Model

The model enabling description of long-memory in variance series is $FIGARCH(p, d, q)$ (Fractionally Integrated $GARCH$) introduced by Baillie et al. [1]. The $FIGARCH(p, d, q)$ model for time series y_t can be written as:

$$y_t = \mu + \varepsilon_t, \quad t = 1, 2, \ldots \Omega, \tag{10}$$

$$\varepsilon_t = z_t \sqrt{h_t}, \quad \varepsilon_t | \Theta_{t-1} \sim N(0, h_t), \tag{11}$$

$$h_t = \alpha_0 + \beta(L) h_t + \left[1 - \beta(L) - [1 - \phi(L)](1 - L)^d\right] \varepsilon_t^2, \tag{12}$$

where z_t is a zero-mean and unit variance process, h_t is a positive time dependent conditional variance defined as $h_t = E\left(\varepsilon_t^2 | \Theta_{t-1}\right)$ and Θ_{t-1} is the information set up to time $t-1$.

The $FIGARCH(p, d, q)$ model of the conditional variance can be motivated as $ARFIMA$ model applied to the squared innovations

$$\varphi(L)(1 - L)^d \varepsilon_t^2 = \alpha_0 + (1 - \beta(L)) \vartheta_t, \quad \vartheta_t = \varepsilon_t^2 - h_t, \tag{13}$$

where $\varphi(L) = \varphi_1 L - \varphi_2 L^2 - \cdots - \varphi_p L^p$ and $\beta(L) = \beta_1 L + \beta_2 L^2 + \cdots + \beta_q L^q$ and $(1 - \beta(L))$ have all their roots outside the unit circle, L is the lag operator and $0 < d < 1$ is the fractional integration parameter. If $d = 0$, then FIGARCH model is reduced to GARCH; for $d = 1$ though, it becomes IGARCH model. However, FIGARCH model does not always reduce to GARCH model. If GARCH process is stationary in broader sense, then the influence of current variance on its forecasting values decreases to zero in exponential pace. In IGARCH case the current variance has indefinite influence on the forecast of conditional variance. For FIGARCH process the mentioned influence decreases to zero far more slowly than in GARCH process, i.e. according to the hyperbolic function [1, 18]. In practical implementation of prediction FIGARH model see [18].

4 Parameters Estimation and Choice of Model

The aim of searching for a useful forecasting model is not utilization of the greatest number of parameters which will most accurately describe variance of the analyzed time series. It is due to the fact that too big matching may embrace the description not only of the part of the process called signal but also of random noise, for which in finished trials one can discern as random regularity. The objective of the research is rather discovery of such a model which will describe the most important properties of the analyzed time series by means of a finite number of statistically essential parameters [7]. The most often used method of parameter estimation of autoregressive models is the Maximum Likelihood Estimation (MLE). The basic problem appearing while using this method is the necessity to define the whole model and consequently sensitivity of the obtained estimator to the presumptive errors in the specification of polynomials AR and MA, which are responsible for the process dynamics [9]. There is no universal criterion for the choice of the model. Usually, the more complex the model, the bigger is its likelihood function. Therefore, there is a searching for a compromise between the number of parameters occurring in the model and the value of the likelihood function. The choice of a sparing form of the model is performed on the basis of information criteria such as Akaike (AIC) or Schwarz (SIC). In our article, for parameter estimation and choice of the model, we utilized the Maximum

Likelihood Method. It was due to its relative simplicity and computational efficiency. In order to estimate the order of the AR and MA models we used the autocorrelation function ACF and $PACF$. For $ARMA$ model, however, we used Box-Jenkins procedure [2]. For $ARFIMA$ model we applied HR estimator (described in the Haslett and Rafterys work [11]) and automatic models order selection algorithm based on information criteria (see Hyndman and Khandakar [14]). For estimation of $FIGARCH$ model we used the methodology described in the article [18].

5 Experimental Results

Experimental results are based on traffic features set taken from SNORT [17] based preprocessor which we proposed in [16]. We have used 26 traffic features presented in Table 1. For algorithms evaluation we used Kali Linux [15] tools for simulating real

Table 1 Network traffic features used for experiments

f_1	number of TCP pockets
f_2	in TCP pockets
f_3	out TCP pockets
f_4	number of TCP pockets in LAN
f_5	number of UDP datagrams
f_6	in UDP datagrams
f_7	out UDP datagrams
f_8	number of UDP datagrams in LAN
f_9	number of ICMP pockets
f_{10}	out ICMP pockets
f_{11}	in ICMP pockets
f_{12}	number of ICMP pockets in LAN
f_{13}	number of TCP pockets with SYN and ACK flags
f_{14}	out TCP pockets (port 80)
f_{15}	in TCP pockets (port 80)
f_{16}	out UDP datagrams (port 53)
f_{17}	in UDP datagrams (port 53)
f_{18}	out IP traffic [kB/s]
f_{19}	in IP traffic [kB/s]
f_{20}	out TCP traffic (port 80) [kB/s]
f_{21}	in TCP traffic (port 80) [kB/s]
f_{22}	out UDP traffic [kB/s]
f_{23}	in UDP traffic [kB/s]
f_{24}	out UDP traffic (port 53) [kB/s]
f_{25}	in UDP traffic (port 53) [kB/s]
f_{26}	in TCP traffic (port 4444)

world attacks in controlled network environment (for example: Application specific DDos, various port scanning, DoS, DDoS, Syn Flooding, pocket fragmentation, spoofing and others).

For anomaly detection we used statistical algorithms with short and long memory dependence: $ARMA$, $FIGARH$ and $ARFIMA$. In Tables 2 and 3 there are results of DR [%] and FP [%] for mentioned three algorithms. Most promising results in terms of DR and FP were achieved for $ARFIMA$ long memory statistics (with FP less then 10 %).

Table 2 Detection Rate DR [%] for a given network traffic features

Feature	ARMA	FIGARH	ARFIMA
f_1	4.24	5.40	6.26
f_2	9.22	10.20	12.24
f_3	9.22	10.20	12.24
f_4	9.22	10.20	12.24
f_5	9.22	10.20	12.24
f_6	0.00	0.00	0.00
f_7	0.00	0.00	0.00
f_8	30.52	32.20	35.64
f_9	88.68	90.42	96.52
f_{10}	87.23	90.24	95.45
f_{11}	0.00	0.00	0.00
f_{12}	78.82	80.24	82.24
f_{13}	9.22	10.20	12.24
f_{14}	9.22	10.20	12.24
f_{15}	9.22	10.20	12.24
f_{16}	0.00	0.00	0.00
f_{17}	4.42	5.40	6.26
f_{18}	9.22	10.20	12.24
f_{19}	9.22	10.20	12.24
f_{20}	4.42	5.40	6.26
f_{21}	9.22	10.20	12.24
f_{22}	0.00	0.00	0.00
f_{23}	0.00	0.00	0.00
f_{24}	0.00	0.00	0.00
f_{25}	0.00	0.00	0.00
f_{26}	75.24	78.00	80.00

Table 3 False Positive FP [%] for a given network traffic features

Feature	ARMA	FIGARH	ARFIMA
f_1	6.01	5.24	4.22
f_2	5.85	5.45	4.12
f_3	5.92	5.24	4.15
f_4	5.72	5.22	4.11
f_5	5.24	4.28	3.54
f_6	4.42	3.34	2.23
f_7	7.24	6.75	5.98
f_8	6.15	5.24	4.15
f_9	6.85	6.22	5.05
f_{10}	2.53	1.46	0.48
f_{11}	4.24	3.52	2.56
f_{12}	2.18	1.04	0.05
f_{13}	6.21	5.46	4.14
f_{14}	5.12	4.45	3.24
f_{15}	5.08	4.38	3.32
f_{16}	2.11	1.24	0.02
f_{17}	2.25	1.82	0.39
f_{18}	5.12	4.55	3.82
f_{19}	5.14	4.62	3.26
f_{20}	6.22	5.34	4.55
f_{21}	5.32	4.22	3.11
f_{22}	3.17	2.46	1.60
f_{23}	5.77	4.44	3.42
f_{24}	0.00	0.00	0.00
f_{25}	1.12	0.45	0.02
f_{26}	1.12	0.45	0.02

6 Conclusion

Ensuring a sufficient level of safety to resources and information systems is a question that is currently intensively surveyed and developed by many research centers in the world. A growing number of novel attacks, their global reach and level of complexity enforce dynamic development of network safety systems. Most often implemented mechanism aiming to ensure security are methods of detection and classification of abnormal behaviors reflected in the analyzed traffic. In the present article, we compare properties of predicated analyzed statistical models in terms of their effectiveness to detect anomalies in network traffic. The analyzed models were those of a long and short memory reflected in the autocorrelation strength of elements composing a given time series. Parameter estimation and identification of the range of the model were realized as a compromise between the models coherence and size of its estimation error. While realizing implementation processes of the described models there were achieved diverse statistical estimations for the analyzed signals of the network traffic. In order to detect anomalies in the network traffic we used differences between the real network traffic and its estimated model for the analyzed parameters characterizing number of received or sent TCP, UDP and ICMP packages within a time unit. The results obtained after the performed experiments show advantage of predictive models ARFIMA and FIGARCH in the network traffic anomaly detection.

References

1. Baillie, R., Bollerslev, T., Mikkelsen, H.: Fractionally integrated generalized autoregressive conditional heteroskedasticity. J. Econom. **74**, 3–30 (1996)
2. Beran, J.A.: Statistics for Long-Memory Processes. Chapman and Hall, New York (1994)
3. Box, G., Jenkins, G., Reinsel, G.: Time Series Analysis. Holden-day, San Francisco (1970)
4. Box, G.E., Jenkins, M.G.: Time Series Analysis Forecasting and Control, 2nd edn. Holden-Day, San Francisco (1976)
5. Brockwell, P., Davis, R.: Introduction to Time Series and Forecasting. Springer, New York (2002)
6. Chondola, V., Banerjee, A., Kumar, V.: Anomaly detection: a survey. ACM Comput. Surv. **41**(3), 1–72 (2009)
7. Crato, N., Ray, B.K.: Model selection and forecasting for long-range dependent pro-cesses. J. Forecast. **15**, 107–125 (1996)
8. Gabriel, V.J., Martins, L.F.: On the forecasting ability of ARFIMA models when infre-quent breaks occur. Econom. J. **7**, 455–475 (2004)
9. Geweke, J., Porter-Hudak, S.: The estimation and application of long memory time series models. J. Time Ser. Anal. (4), 221–238 (1983)
10. Granger, C.W.J., Joyeux, R.: An introduction to long-memory time series models and fractional differencing. J. Time Ser. Anal. **1**, 15–29 (1980)
11. Haslett, J., Raftery, A.E.: Space-time modelling with long-memory dependence: assessing Ireland's wind power resource (with discussion). Appl. Stat. **38**(1), 1–50 (1989)
12. Hosking, J.R.M.: Fractional differencing. Biometrika **68**, 165–176 (1981)
13. Hurst, H.R.: Long-term storage capacity of reservoirs. Transactions of the American Society of Civil Engineers **1**, 519–543 (1951)
14. Hyndman, R.J., Khandakar, Y.: Automatic time series forecasting: the forecast package for R. J. Stat. Softw. **27**(3), 1–22 (2008)
15. Kali Linux: https://www.kali.org (2015)
16. Saganowski, Ł., Goncerzewicz, M., Andrysiak, T.: Anomaly Detection Preprocessor for SNORT IDS System, Image Processing and Communications Challenges 4. Advances in Intelligent Systems and Computing **184**, 225–232 (2013)
17. SNORT IDS: http://www.snort.org (2014)
18. Tayefi, M., Ramanathan, T.V.: An overview of FIGARCH and related time series models. Aust. J. Stat. **41**(3), 175–196 (2012)

Local Statistic Embedding for Malware Behaviour Modelling

Rafał Kozik and Michał Choraś

Abstract In this paper we have presented the preliminary results of the methods of malware detection on the basis of the analysis of network volume properties. The main contribution of our research is the new approach to enrich aggregated features, collected from network flow analysis, with so called local statistics that capture the properties of vectors located in the nearest neighbourhood. Our analyses are conveyed on real-life network samples.

Keywords Anomaly detection · Botnet detection · Machine learning

1 Introduction

Currently, we are facing the problem of reasonable usage of Big Data in the cyber-security domain. Many high-level representatives of corporations and institutions claim that typical signature-based approaches to malware detection is ineffective [1]. Therefore, there is the need to adapt novel techniques such as big data analysis, machine-learning and data mining in the cyber domain. Also our research is focused on those aspects.

Usually, malware samples are carefully crafted pieces of software that aim at staying dormant while performing detailed surveillance of infected infrastructures and assets. Typically, the infection process relies on complicated attack vectors (e.g. social engineering) that are hard to detect. An example could be a Havex malware [9]. Therefore, it is important to detect the infection on time and prevent the malware from spreading.

This paper is structured as follows. First, we provide the analysis of related work in malware detection (Sect. 2). Then, we provide and an overview of the proposed method (Sect. 3). The experiments, results and final conclusions are given after.

R. Kozik (✉) · M. Choraś
Institute of Telecommunications and Computer Science,
UTP University of Science and Technology in Bydgoszcz, Bydgoszcz, Poland
e-mail: rkozik@utp.edu.pl

R.S. Choraś (ed.), *Image Processing and Communications Challenges 7*,
Advances in Intelligent Systems and Computing 389,
DOI 10.1007/978-3-319-23814-2_30

2 Related Work

Commonly the signatures (in form of reactive rules) of an attack for a software like Snort [10] are provided by experts form a cyber community. Typically, for deterministic attacks it is fairly easy to develop patterns that will clearly identify particular attack. It often happens when given malicious software (e.g. worm) uses the same protocol and algorithm to communicate trough network with command and control centre or other instance of such software. However, the task of developing new signatures becomes more complicated when it comes to polymorphic worms or viruses. Such software commonly modifies and obfuscates its code (without changing the internal algorithms) in order to be less predictive and hard to detect.

The development of efficient and scalable method for malware detection is currently challenging also due to the general unavailability of raw network data. Therefore, this aspect while being related to users privacy and administrative and legal reasons, causes additional difficulties for research and development.

Currently, the common alternative is so called NetFlow [2] data that is often captured by ISPs for auditing and performance monitoring purposes. Since NetFlow samples do not contain any sensitive data they are widely available. However, the fact that this kind of samples are lacking raw content of network packets is the disadvantage. In the literature there are different approaches focusing on analysis of NetFlow data. In [3, 4] authors focused on computational paradigms (e.g. MapReduce) for NetFlow data analysis and malware detection. On the other hand, in [7, 8] author proposed statistical techniques for feature extraction from groups of network flows.

In literature there are also different approaches such as SA (Semantic Aware [6]) that are designed to generate signatures of malicious software on a basis of network traffic they generate. Such solutions like [6] can even properly identify malicious behaviour when the traffic is noise-like.

3 Proposed Method

The proposed method uses information data generated by NetFlow [2] tool. The data can be provided by routers or other network equipment and usually contains aggregated information related to the connection (e.g. TCP or UDP) established between network nodes. Example of the captured NetFlow data is shown in Fig. 1.

This kind of aggregated data is (to some extent) antagonized and thus can be used by third party network elements (like IDS or ADS systems) to provide additional level of cyber security of the monitored network. In the proposed approach we use that kind of data to identify malicious behaviour of network elements that communicate with each other. The main contribution of this research is an efficient LSE (Local Statistic Embedding) technique that, as proven in our experiments, can be competitive to classical k-Means approach that is commonly used for similar NetFlows clustering.

StartTime,Dur,Proto,SrcAddr,Sport, Dir, DstAddr,Dport,State,sTos,dTos,TotPkts,TotByt es,SrcBytes,Label	Header
2011/08/10 09:46:59.607825,1.026539,tcp,94.44.127.113,1 577, -> ,147.32.84.59,6881,S_RA,0,0,4,276,156,flow=B ackground-Established-cmpgw-CVUT	Flow #1
2011/08/10 09:47:00.634364,1.009595,tcp,94.44.127.113,1 577, -> ,147.32.84.59,6881,S_RA,0,0,4,276,156,flow=B ackground-Established-cmpgw-CVUT	Flow #2

Fig. 1 Example of data (connection flow) obtained from the NetFlow system. For example such data contains information about IP addresses of sender and receiver, amount of data transferred, protocol etc.

Our algorithm operates as follows:

- Firstly, NetFlow context is established. In order to do that we aggregate flows in 5 min long time widows and compute statistical properties of features. The procedure is explained in Sect. 3.1. This process results in the feature vector instances.
- Secondly, we apply LSE (Local Statistic Embedding) technique to capture statistical properties of vector instances located in a close vicinity. After this step we obtain new feature vectors that are used in the next step. More details are given in Sect. 3.2.
- Finally, we use classifier in order to indicate given instance describing NetFlow context as normal or anomalous. It must be noticed that we do not intend to label every single connection flow, but behaviour of particular network node within the time window.

3.1 Context Extraction

For the context extraction we have used the popular [7, 8] approach. First, we split the analysis of the NetFlow into so called time windows. We used windows that are 5 min long. We have chosen this value during experiments to allow us to have reasonable amount of data to process within a single window and to capture enough of network flows to model malware behaviour.

Within the single time window we group the network flows by source address and calculate statistical properties. The following statistics are used to build the feature vector:

1. Entropy of source port values that is measured using Eq. 1, where S indicates the vector of the source ports captured for a given source address within the considered time window.

$$H(S) = -\sum_{i=0}^{n} P(s_i) log_2(P(s_i)) \quad (1)$$

2. Entropy of destination address values.
3. Entropy of destination port values.
4. Average duration of connection.
5. Maximal duration of connection.
6. Average total bytes transmitted within single connection.
7. Max total bytes transited within single connection.

3.2 Local Statistic Embedding

I our research the LSE (Local Statistics Embedding) captures (for a given feature vector) local statistical properties that are calculated for vectors that are in close vicinity in relation to the analysed vector.

Definition 1 For a given dataset and similarity measure S we define LSE—Local Statistics Embedding as:

$$LSE_k(x) = \left[meanS(Y), mean(Y(\cdot, 4)), mean(Y(\cdot, 6)), max(y_L)\right] \quad (2)$$

where $Y = y_i : y_i \in N(x)$ is set of a k feature vectors that are in a neighbourhood N of vector x. The notation $Y(\cdot, 4)$ indicates that fourth component of all feature vectors are retrieved for computation and y_L indicates the label (1 indicates anomaly, while 0 indicates genuine vector) of feature vector y

The technique adapting LSE encoding allows us to additionally reduce the number of dimensions.

3.3 Classification

The last stage of our algorithm is the classification procedure. In this research we have adapted cost-sensitive linear classifier, due to the fact that we have to deal with strongly imbalanced data. In our experimental data set described in Sect. 4, we have over 2.5 millions of network flows indicating background traffic and only 40 thousands of flows related to malware. Our classifier is defined by Eq. 3.

$$y(x, w, b) = sign(b + x \cdot w) \quad (3)$$

In order to deal with the data imbalance problem, we assign weights to learning samples in the way that majority class (background traffic) is assigned lower value in contrast to minority class (anomalous traffic).

4 Experiments and Results

In order to evaluate our malware detection methodology we have used network flows captured for Neris malware that creates IRC-based botnet in order to send spam, connects to an HTTP CC, and uses HTTP to do ClickFraud. Traffic samples come from CTU-13 dataset [5]. According to description provided by authors of the dataset, the traffic samples were recorded for healthy and infected machines. For over 6 h 2.5 millions of network flows were collected. The whole dataset is fully labelled, meaning that each flow that was recorded, has the assigned label.

Our experiments are divided in two phases, namely learning and testing. During the learning phase we use only the subset of samples to train the classifier. However, before splitting the dataset we first calculate the statistics within the time windows. That way we avoid the situation where particular network flows are pulled out of the context. Therefore, we do not want to classify each separate flow (it may overwhelm network administrator), but identify whenever given sender (IP address) should be considered as infected. In other words we evaluate behaviour of particular IP address within a time window (particular instance of feature vector described in Sect. 3.1).

In our experiments we have measured the following performance indicators to evaluate the proposed method:

- False Positives Ratio (FPR) defined as the number of IP addresses identified wrongly as infected (False Positives—FP) divided by the number of all unique IP addresses within the time window that are not infected (True Negatives + False Positives).

$$FPR = \frac{FP}{TN + FP} \quad (4)$$

- True Positives Ratio (TPR) defined as the number of IP addresses identified correctly as infected (True Positives—TP) divided by the number of all unique IP addresses within the time window that are infected (True Positives + False Negatives).

$$TPR = \frac{TP}{TP + FN} \quad (5)$$

Table 1 Effectiveness of the proposed approach

Experiment	FPR [%]	TPR [%]	TNR [%]	Accuracy [%]
Perception	0.031	75.0	99.968	99.968

- True Negative Ratio (TNR) defined as the number of IP addresses identified correctly as normal (True Negative—TN) divided by the number of all unique IP addresses within the time window that are not infected (True Negatives + False Positives).

$$TNR = \frac{TN}{TN + FP} \tag{6}$$

- Accuracy defined as the sum of True Positives (TP) and True Negatives (TN) divided by the number of all unique IP addresses within the time window.

$$Accuracy = \frac{TN + TP}{TN + TP + FP + FN} \tag{7}$$

The results of the statistics described above are provided in Table 1.

5 Conclusions

In this paper we have proposed preliminary results of the malware detection method. Our approach relies on analysis of malware network activity that is captured by means of NetFlow tool. In our algorithm we propose Local Statistics encoding to capture additional features of the malware behaviour. We also have presented promising results in a real-life scenario, achieving low ratio of false positives and high accuracy. Future work will be dedicated to evaluation of scalability o proposed methods and further improvements towards online machine learning.

References

1. Antivirus is dead, says maker of Norton Antivirus http://www.pcworld.com/article/2150743/antivirus-is-dead-says-maker-of-norton-antivirus.html
2. Claise, B.: Cisco Systems NetFlow Services Export Version 9. RFC 3954 (Informational) (2004)
3. Dean, J., Ghemawat, S.: MapReduce: simplified data processing on large clusters. In: Symposium on Opearting Systems Design and Implementation (OSDI). USENIX Association (2004)
4. Francis, J., Wang, S., State, R., Engel, T.: Bottrack: tracking botnets using netflow and pagerank. In: Proceedings of IFIP/TC6 Networking (2011)
5. Garcia, S., Grill, M., Stiborek, H., Zunino, A.: An empirical comparison of botnet detection methods. Comput. Secur. J. Elsevier **45**, 100–123 (2014)

6. Kong, D., Jhi, Y.C., Gong, T., Zhu, S., Liu, P., Xi, H.: SAS: semantics aware signature generation for polymorphic worm detection. Int. J. Inf. Secur. **10**(5), 269–283 (2011)
7. Lakhina, A., Crovella, M., Diot, C.: Diagnosing network-wide traffic anomalies. ACM SIGCOMM Comput. Commun. Rev. **34**, 357–374 (2004)
8. Lakhina, A., Crovella, M., Diot, C.: Mining anomalies using traffic feature distributions. ACM SIGCOMM Comput. Commun. Rev. **35**, 217–228 (2005)
9. New Havex malware variants target industrial control system and SCADA users. http://www.pcworld.com/article/2367240/new-havex-malware-variants-target-industrial-control-system-and-scada-users.html
10. SNORT. Project homepage. http://www.snort.org/

Analysis and Real Implementation of a Cloud Infraestructure for Computing Laboratories Virtualization

Jose Enrique Munoz-Exposito, Rocio Perez de Prado, Sebastian Garcia-Galan, Rafael Rodriguez-Reche and Adam Marchewka

Abstract In this work, a cloud infrastructure is analysed and deployed to virtualize computing laboratories. Virtualization of computing resources allows the user to face a single computing interface whose capabilities are really offered by a cloud system that makes use of its processing capability on demand. The use of this proposed infrastructure reduces the cost of equipment investments as well as maintenance of real laboratories and further, it makes a more efficient use of available resources with a transparent user experience.

Keywords Computing virtualization · Cloud computing · OpenNebula · Distributed and parallel processing

1 Introduction

The goal of virtualization of resources is to offer a hardware infrastructure that simulates the real hardware demanded by the user [2, 6]. With this aim virtual machines are generated where the operating systems and software required for the labs can be installed. These virtual machines can be created with specific software to create virtual machines such as VMware or Virtualbox. Virtualization presents several advantages [11]:

J.E. Munoz-Exposito · R.P. de Prado (✉) · S. Garcia-Galan · R. Rodriguez-Reche
Telecommunication Engineering Department, University of Jaén,
Alfonso X El Sabio 28, Linares, Spain
e-mail: rperez@ujaen.es

A. Marchewka
Institute of Telecommunications and Computer Science,
UTP University of Science and Technology in Bydgoszcz, Bydgoszcz, Poland
e-mail: adimar@utp.edu.pl

R.S. Choraś (ed.), *Image Processing and Communications Challenges 7*,
Advances in Intelligent Systems and Computing 389,
DOI 10.1007/978-3-319-23814-2_31

- If an installed application in a virtual machine is vulnerable, the attack will take place only in a restricted area, affecting the associated resources of that machine.
- The system can be progammed to back up the entire virtual machine and this is enough to re-start the machine.
- Ease to manage migrations.
- Testing applications installed in the virtual machine without instability problems, since save points can be used within the machine to undo the last performed installation.

Cloud computing is a interesting scenario for virtualization systems [3, 5]. It is a current trend that involves the supply of computing resources as services. There are three types of services offered [1]:

- Iaas, Infrastructure as a Service.
- PaaS, Platform as a Service.
- SaaS, Software as a Service.

The IaaS is based on the outsourcing of data processing machines and storage model. With this model a separation between the perceived infrastructure by users and the real systems where operations are performed is obtained. In the market there are different pay platforms like VMware or free and open platforms. In this project, it is proposed to use the so-called OpenNebula open platform [7, 10]. The full open version of this system allows a complete interoperability with the existing components in the infrastructure.

This paper is organized as follows. Section 2 establishes a background. Section 3 formally introduces the proposed infraestructure. In Sect. 4 some results are shown and finally in Sect. 5 we conclude with final remarks and conclusions.

2 Background

The current economic situation has prompted the virtualization of servers because of two main reasons [9, 11]. Firstly, the need to modify the financial management, leading from models based on investment to models based on cost. Secondly, the present instability in the markets has made the models to adopt services on flexible platforms that allow them to rapidly adapt to changing conditions. The key nowadays is the pay per use of services. This makes it possible for companies to grow without big investments and with the possibility of increasing its computing capabilities, both computational and storing needs, on demand. Furthermore, beyond cost optimization, virtualization offers many benefits to companies, such as the minimization of provision times, since it allows them to create infrastructures for their applications adapted to their business needs and also, the simpler management that involves. In fact, responsible managers of information technologies emphasize the advantage of having a replicable infrastructure that adapts to the needs at every moment, with independence of the providers and hardware obsolescence.

In regard of cloud computing, it must be said that virtualization is one of the necessary tools to offer the services included under this denomination. Their evolution and growth come together. If virtualization allows the optimization of information technologies services, the following step consists on deploying infrastructures that offer services in a real on demand system in a simply manner and that allow end users to keep the maximum control over the system in every moment [6].

In short, virtualization provides servers with the same hardware and cloud computing provides resources on demand only for the real used computing and storage capabilities. This way, virtualization and cloud computing complements to solve the problem to maximize the use of available resources, but the terms are not interchangeable.

3 Proposed Infraestructure

To implement the cloud infrastructure for virtualizing laboratory computers we have performed different tasks. A computer network has been formed using various hardware components that support virtualization infrastructure. We have used two servers Dell Power Edge R710 and R810, a Gigabit network switch and software that allows the distribution of virtual machines between different computers (without the use of a server images). Secondly, OpenNebula has been installed to accommodate the different virtual machines on a cloud system, and thirdly, we have created different virtual machines for the different subjects. All these tasks have been documented in an appropriate way due to the use of manuals and practical guides, which have been developed to allow the students the access to virtual machines from different locations. OpenNebula is a cloud computing platform for managing heterogeneous distributed data center infrastructures. The OpenNebula platform manages a data center virtual infrastructure to build private, public and hybrid implementations of infrastructure as a service. In addition, OpenNebula is free and open-source software. OpenNebula orchestrates storage, network, virtualization, monitoring, and security technologies to deploy multi-tier services as virtual machines on distributed infrastructures, combining both data center resources and remote cloud resources, according to allocation policies (Fig. 1). The toolkit includes features for integration, management, scalability, security and accounting. It also claims standardization, interoperability and portability, providing cloud users and administrators with a choice of several cloud interfaces (Amazon EC2 Query, OGF Open Cloud Computing Interface and vCloud) and hypervisors (Xen, KVM and VMware), and it can accommodate multiple hardware and software combinations in a data center.

It is important to point out that OpenNebula is usually used by hosting providers, services providers, supercomputing centers, research labs, and international research projects.

In order to access virtual machines a web browser is used by students as a frontend, which could be carried out in a remote way. The students use A Raspberry Pi low cost computer to access these virtual machines. We need to install an operating

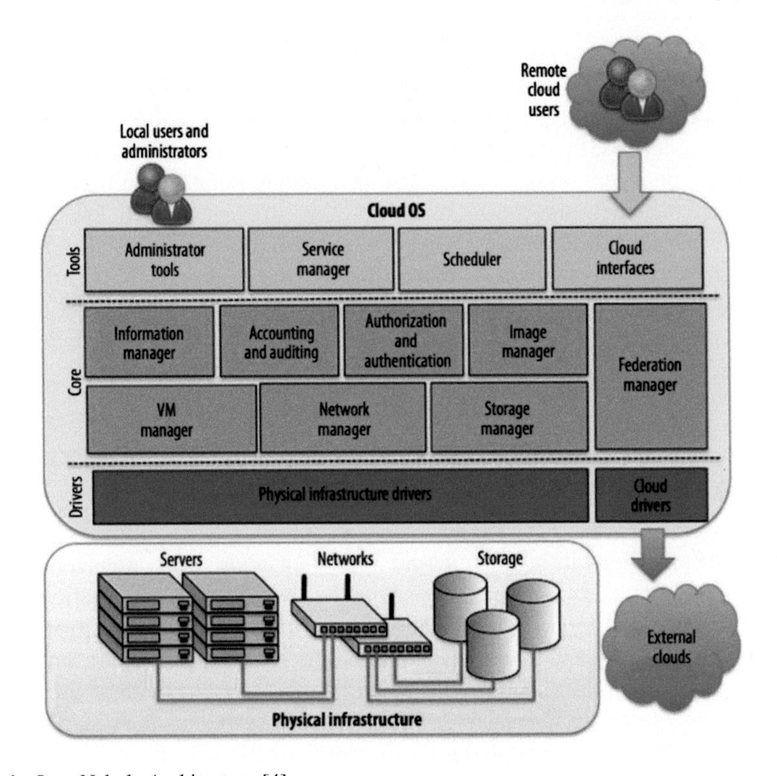

Fig. 1 OpenNebula Architecture [4]

system to control the system and our election is Raspbian. It is a free operating system based on Debian optimized for the Raspberry Pi hardware [8]. An operating system is the set of basic programs and utilities that make your Raspberry Pi run. However, Raspbian provides more than a pure OS: it comes with over 35,000 packages, pre-compiled software bundled in a nice format for easy installation on your Raspberry Pi. Every student must have one Raspberry connected and they run an Epiphany web browser. Epiphany is a free web browser for the GNOME desktop environment. Epiphany originally used the Gecko layout engine and brings a host of neat features to Raspberry Pi including:

- Much-improved HTML5 support.
- A JavaScript JIT.
- Hardware-accelerated video decoding.
- Faster scrolling.
- Better interactivity during page loading.

The use of the browser is neccesary because OpenNebula uses a VNC client for communication between a web browser and the virtual machine running on servers, and it supports a JavaScript JIT and secure protocols [2]. Figure 2 shows a screenshot taken on Raspberry where a user has been connected using VNC and runs a virtual machine in a remote way.

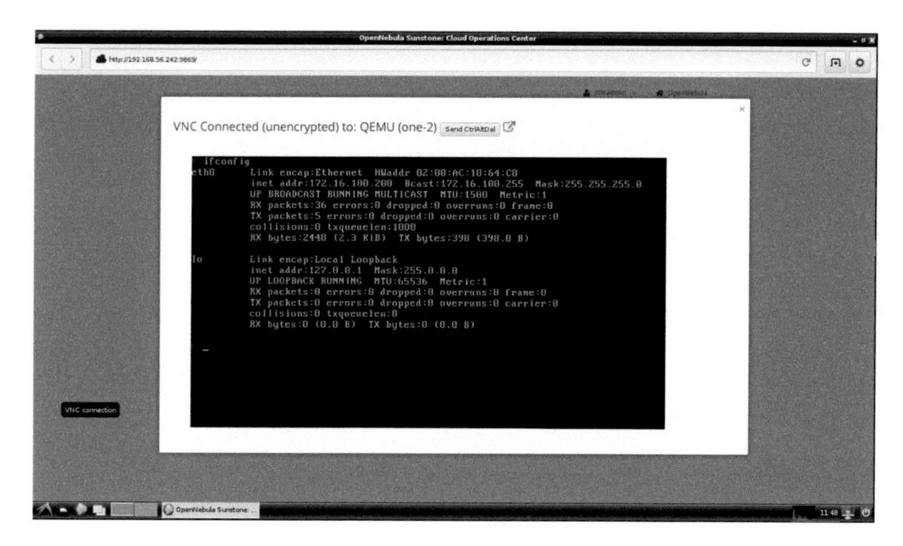

Fig. 2 VNC client using Epiphany web browse

4 Results

By the use of this virtual application, authors intend to provide students a tool to complement the theoretical and practical classes coursed in different subjects. The main advantage of the proposed virtual application is related to the lower equipment maintenance, which has been observed during its use. Consequently, its utilization could be considered as an important way to reduce costs for the academic institution. The gathered experience has been evaluated according to the student participation and their collaboration in discussion forums. To be precise, we have considered the following items:

- The activity assessment done by the students at the end of the course.
- The students participation level in the activity.

Students could enjoy several practical laboratory sessions during the implementation of this project, which involved several elements to complement the theoretical and practical classes in a remote way. This allowed the improvement of the learning outcomes and the acquisition of skills related to the subject. Furthermore, the proposed topic was considered very attractive to students and it has encouraged the independent learning what has given risen to new contributions complementing the basic task initially suggested. Additionally, the virtualized laboratory and virtual machines offer a major flexibility in the preparation of laboratory experimental sessions and practices avoiding high-cost equipment that are typically found in traditional approaches.

5 Conclusions

The virtualized laboratory and virtual machines offer a major flexibility in the preparation of laboratory experimental sessions and practices avoiding high-cost equipments that are typically found in traditional approaches. Specifically, the proposed student station represents a significant saving compared to a traditional station (around 800 euros). Also, the maintenance of this simpler equipment is lower, requiring fewer technical hours and therefore, lower costs for the academic institution. Moreover, students can access the laboratory from everywhere and every time. From the point of view of an economic university management, using virtual machines and instruments over the Internet allows a better use of resources: it increases the number of students who could access to the laboratory. As a consequence of the above-mentioned points, it could be stated that the usage of resources has been optimized and a larger group of students can work with a reduced number of laboratories.

Acknowledgments This work has been financially supported by the Educational Innovation Project (PID17 201416) of the University of Jaen.

References

1. Amazon web services: What is cloud computing? https://aws.amazon.com/what-is-cloud-computing/?nc1=f_cc (2015)
2. GEOTEK: The VNC family of remote control applications. http://ipinfo.info/html/vnc_remote_control.php (2015)
3. Hamdaqa, M., Tahvildari, L.: Cloud computing uncovered: a research landscape. Adv. Comput. **86**, 41–85 (2012)
4. Llorente, I.M.: Publication of the opennebula cloud os architecture in IEEE computer, 9 Jan 2013. http://opennebula.org/publication-of-the-opennebula-cloud-os-architecture-in-ieee-computer (2015)
5. Mell, P., Grance, T.: The NIST definition of cloud computing (2011)
6. Nieh, J., Yang, S., Novik, N.: A comparison of thin-client computing architectures 2000. In: Network Computing Laboratory, Columbia University (2014)
7. OpenNebula Project: Detailed features and functionality. http://archives.opennebula.org/documentation:features (2014)
8. Raspberry Pi Foundation: Documentation. http://www.raspberrypi.org (2014)
9. Richardson, T., Stafford-Fraser, Q., Wood, K.R., Hopper, A.: Virtual network computing. IEEE Internet Comput. **2**(1), 33–38 (1998)
10. Sunstone: Documentation. http://docs.opennebula.org (2015)
11. Zhou, M.: Information and Communication Technology for Education (2 Volume Set), vol. 58. WIT Press (2014)

AAU-Star and AAU Honeyjar: Malware Analysis Platforms Developed by Students

Jens Myrup Pedersen and Matija Stevanovic

Abstract In this paper we will demonstrate how systems for malware testing can be designed, implemented and used by master degree students. In this way, we have established two strong platforms for malware testing, while at the same time provided the students with a strong theoretical and practical understanding of how to execute, analyse, and classify malware based on their network and host activities.

1 Introduction

Malware is becoming an increasing problem in computers and computer systems, with severe economic consequences for individuals, businesses, and societies [8]. On the individual basis, computers can be infected with e.g. spyware that is used for stealing personal information or ransomware, which destroy all date on the infected machine unless a ransom is paid. For businesses, infected machines are vulnerable to the same kind of attacks, but often with more severe consequences such as stealing business secrets or destruction of data that is crucial to the operation of the company. Another particular kind of malware, botnets, makes a strong tool for cyber criminals [6, 9]. Computers infected with bot malware (also known as zombies) can be remotely controlled and used for a number of malicious activities, often targeting business or infrastructure that is critical for society. These include Distributed Denial of Service attacks, where a service is left unavailable due to a huge number of requests coming from different machines. Recent years have seen attacks on e.g. Sony and Mastercard. In Denmark, the personal identification system used to access all public and many private services, was recently successfully attacked. Also other malicious activities such as Click Fraud, Pay Per Install and sending of SPAM emails are supported by botnets.

J.M. Pedersen (✉) · M. Stevanovic
Department of Electronic Systems, Aalborg University, Aalborg, Denmark
e-mail: jens@es.aau.dk

M. Stevanovic
e-mail: mst@es.aau.dk

R.S. Choraś (ed.), *Image Processing and Communications Challenges 7*,
Advances in Intelligent Systems and Computing 389,
DOI 10.1007/978-3-319-23814-2_32

Malware can be fought at different levels: One is of course protection at device level, using anti-virus systems. This is generally efficient, but suffers from zero day attacks (e.g. new malware which is not yet known by the systems) and the risk of users overruling the systems if manipulated through e.g. social engineering. Another approach, which is a research topic at Aalborg University, is to detect malware through the analysis of network traffic. The hypothesis here is that malicious traffic can be distinguished from benign traffic using e.g. machine learning algorithms. Several authors have proposed detection algorithms based on the these assumptions, such as Bilge et al. [2], Gu et al. [4], Strayer et al. [10], Zhao et al. [12]. However, building, improving, analyzing and testing such tools and algorithms is a data intensive task with quite a few challenges. First, we need to collect a large amount of data from many different malware samples, and it needs to be correctly labeled (as either malicious/non-malicious or with a division of the malware into types of malware). Second, it must be done without the risk of harming others, meaning that any test environment should have only limited access to the Internet (with no harmful traffic let through). Third, the computers infected should ideally "look like" normal non-monitored computers, used as a normal user would do, in order to avoid that the malware becomes suspicious of the environment which could trigger a different behavior.

This paper demonstrates how student projects can be used as a mean of building platforms and systems for malware testing, which fulfills these criteria to a large extend.

2 Background

Aalborg University has a long tradition for Problem Based Learning [7], where students spend approximately half of their time during each semester on problem-based project work. This is usually done in groups, in the master programme with 2–4 students per group. There is also some tradition for mega projects, where student groups from different educations and/or semesters develop sub-projects which are part of a larger project, which is often developed over several years. Examples of these are AAU Unicorn racer car [11] and the AAU Student satellites [1].

The first step towards AAU HoneyJar, which is also such a mega project, was taken in 2013, and has been followed up by additional projects in the following years. They are carried out within the master programme of Networks and Distributed Systems. The main motivation for developing the malware testing platforms is to provide an environment for reliable and secure testing of malicious software with the goal of obtaining both network and client-level behavioral forensics. Having a "quality" data sets is one of the most important prerequisite for developing malware detection approaches. Here under quality we mean a substantial amount of data that successfully captures malware activity.

Within the series of student projects that were realized in connection with AAU HoneyJar many of them have relied on machine learning algorithms (MLAs) for identifying the patterns of network traffic produced by malware and the behavior of malware at client machines. As MLAs represent a class of data driven approaches they are dependent on the quality of the data sets used for both training and evaluation. The student projects were using data captured by the HoneyJar in order to develop malware detection approaches based on both supervised and unsupervised MLAs. The developed detection methods targeted malware on both client and host levels. At network level students used supervised machine learning to classify benign from botnet network traffic. At client level supervised MLAs were used to classify software as malicious and then to classify the malicious software to an appropriate malware family. In addition, one project analyzed if malware behavior can be clustered using unsupervised MLAs and if the clusters extracted this way correspond to the malware types defined by anti-virus providers.

3 The Systems Developed at AAU

This section describes the two main projects, which are run and developed by students at Aalborg University: AAU HoneyJar, which is an automatized and secure malware testing environment, and Star-AV, which is a virtualized malware testing development.

3.1 AAU HoneyJar: An Automatized and Secure Malware Testing Environment

AAU HoneyJar is the oldest of the two projects, and has been developed and used since 2013. The basic architecture is shown in Fig. 1. It consists of the following elements, which also makes it a natural division between different student projects:

- The Test Environment consists of a number of computers (inmates), which can be used to run malware samples. It should be automatized as much as possible, and in particular, it should be fast and easy to upload and run malware, to clean/swipe the computer after running a malware sample, and to configure each inmate as needed for the testing purposes. Moreover, it should be possible to have the inmates "behave" as if real users are interacting with them, e.g. by developing scripts to open webpages, login to email/Facebook accounts, perform Google searchers etc. The inmates can be virtual or physical machines, or a combination hereof. Another important part of the Test Environment is an emulated Internet environment, which makes the inmates believe that they are connected to the Internet. By having basic services, such as Windows Time and DNS, available from the emulated Internet, access to the real Internet can be limited. This automatized test setup was developed through a number of student projects during the previous years.

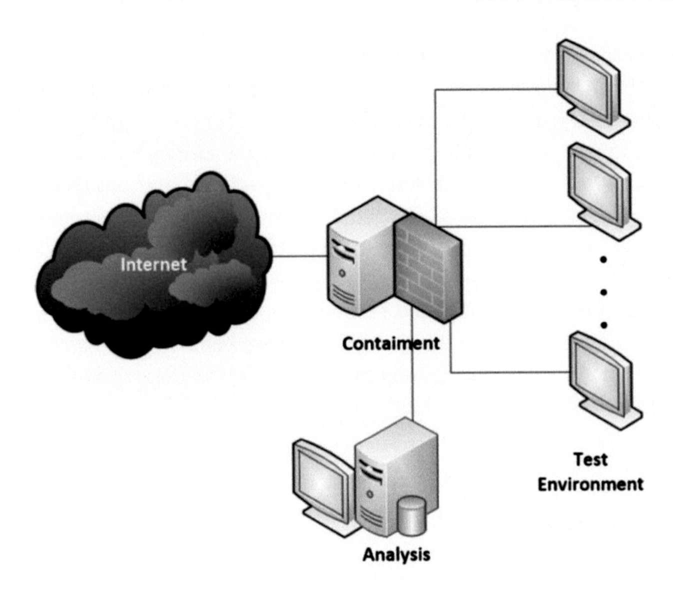

Fig. 1 Architecture of AAU Honey Jar

- The Containment part filters what traffic can enter the real Internet. Giving the infected inmates unfiltered and unlimited Internet access carries a large risk of being active part in DDoS attacks or other malicious activities, including the infection of other user's computers and stealing of personal information. As a university, this risk is unacceptable. On the other hand, with no access to the Internet it is not possible to study the communication between inmates and e.g. command and control servers, which cannot be emulated through the "emulated Internet". Our philosophy is to block all harmful traffic, but to allow traffic through that is considered harmless. This is not easy to define (and even more difficult to decide in real-time). Different strategies has been used in two different student projects, one being to initially block all traffic, but as traffic is observed and deemed harmless to whitelist what is considered harmless traffic. This was partially done in an automatized manner, by using TCP injection to let the inmate reveal the content of the first packet with payload.
- The analysis part is then used for actually analyzing the data and using them for e.g. training and testing of machine learning algorithms. This has been done not only in student projects, but also as a part of PhD research projects, demonstrating the value of AAU HoneyJar.

3.2 Star-AV: Virtualized Malware Testing Environment

The Star-AV project was initiated in 2014. While it is currently a separate project, it might over time be integrated into the "Test Environment" of AAU HoneyJar. The idea behind the Star-AV project is to create a scalable system, that is able to test

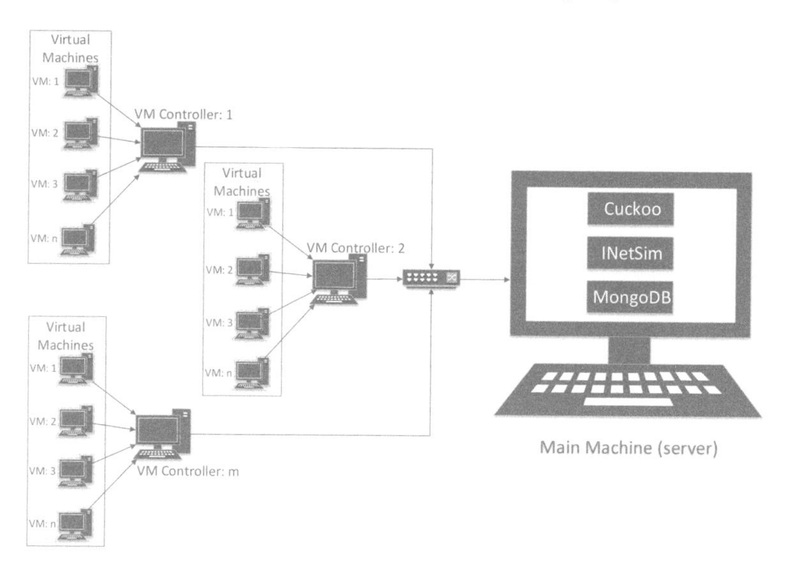

Fig. 2 Architecture of Star-AV

large amounts of malware in a highly automatized fashion. The students has been working on the project during two semesters (Fall 2014, and Spring 2015), and the fall semester part already resulted in a conference paper [5]. Moreover, the students received the Danish Telecommunication Price for Students 2015 based on this work.

Based on a modified version of the Cuckoo Sandbox [3], the system consists of a "master computers" along with a number of computers, where each computer is hosting a number of virtual machines. The basic architecture is shown in Fig. 2. Each machine runs the Microsoft Windows 7 Operating System, along with a number of installed programs including Skype and Adobe Reader. Moreover, each computer runs a script to emulate web activity, to decrease the probability that the malware will detect it is in a contained environment.

The reason for using Cuckoo Sandbox is that it allows for monitoring of the malware behavior on the host side, as opposed to study only network activity. This makes it possible to monitor e.g. accessed files, mutexes, registry keys and windows APIs in addition to network activity.

During the most recent project (a master thesis), the setup was used to test more than 300,000 malware samples, providing valuable data for training algorithms to distinguish between different malware families.

4 Future Work

While both platforms are already fully functional, and useful for both research and education, they also both have a good potential for further development.

For AAU HoneyJar, all of the three subprojects have the potential to be developed even further.

For the Test Setup, it could be particularly interesting to improve the emulation of human behavior as well as the emulated Internet. We believe that this could lead to more malware becoming active when we execute the malware. Moreover, increasing the number of virtual/physical machines would make it possible to run each malware sample for a longer duration of time, also increasing the probability that it will become active.

For the Containment, more sophisticated TCP injection could be an interesting project. In particular, it would be interesting if a conversation could be intercepted and "held in the air" until it was approved. This would require translation of ACK and SEQ numbers, but would be more difficult to detect than in the current implementation (once a connection is whitelisted, it must be reset before a handshake with the server on the Internet is established).

For the Analysis part, it could be interesting to explore the connection between what is measured on the computer using Star-AV, with network activities including analysis of both data and DNS traffic.

For the Star-AV, it could be considered if monitoring of cleanware could be improved. In particular, it would be interesting to see if not only execution, but also installation could be monitored. This is not possible today, due to the large amounts of data involved. Since we can today monitor both installation and execution of malware, we believe that this could improve the training data and thus the accuracy of the resulting classification algorithms.

5 Conclusion

In this paper, we have showed how master student projects can contribute to designing and implementing systems for malware testing. Two strong platforms have been presented, the AAU HoneyJar and Star-AV. These platforms make it possible to execute and monitor behavior of large amounts of malware in a secure setting, thus providing valuable data for education and research purposes. In particular, the projects have made it possible to train and test machine learning algorithms for malware detection. The projects reflect the tradition of Problem Based Learning at Aalborg University, and serves as good examples of mega projects. It is expected that both platforms will be further developed through future student projects.

References

1. AAU StudentSpace: Aalborg University Denmark. http://space.aau.dk (2015)
2. Bilge, L., Balzarotti, D., Robertson, W., Kirda, E., Kruegel, C.: Disclosure: detecting botnet command and control servers through large-scale netflow analysis. In: Proceedings of the 28th Annual Computer Security Applications Conference, ACSAC'12, pp. 129–138. ACM, New York (2012)
3. Cuckoo sandbox. http://cuckoosandbox.org (2015)

4. Gu, G., Perdisci, R., Zhang, J., Lee, W., et al.: Botminer: clustering analysis of network traffic for protocol-and structure-independent botnet detection. USENIX Secur. Symp. **5**, 139–154 (2008)
5. Hansen, S.S., Larsen, T.M.T., Pirscoveanu, R.S., Czech, A., Stevanovic, M., Pedersen, J.M.: Analysis of malware behavior: systemtype classification using machine learning. In: Cyber Situational Awareness, Data Analytics and Assessment (CyberSA 2015). IEEE, C-MRiC (2015)
6. Hogben, G., Plohmann, D., Gerhards-Padilla, E., Leder, F.: Botnets: Detection, measurement, disinfection and defence. In: European Network and Information Security Agency (2011)
7. Kolmos, A., Krogh, L., Fink, F.K.: The Aalborg PBL model: progress, diversity and challenges. Aalborg University Press, Aalborg (2004)
8. Net losses: Estimating the global cost of cybercrime. McAfee, Centre for Strategic & International Studies (2014)
9. Silva, S.S., Silva, R.M., Pinto, R.C., Salles, R.M.: Botnets: A survey. Computer Networks **57**(2), 378–403 (2013)
10. Strayer, W.T., Lapsely, D., Walsh, R., Livadas, C.: Botnet detection based on network behaviour. In: W. Lee, C. Wang, D. Dagon (eds.) Botnet Detection, Advances in Information Security, 36, 1–24. Springer (2008)
11. Unicorn race engineering: Aalborg university's formula SAE team http://unicornraceengineering.dk/ (2015)
12. Zhao, D., Traore, I., Sayed, B., Lu, W., Saad, S., Ghorbani, A., Garant, D.: Botnet detection based on traffic behavior analysis and flow intervals. Comput. Secur. **39**, 2–16 (2013)

System of Distance Measurement and Localization of Wireless Devices Communicating Based on Bluetooth Technologies

Artur Sierszeń, Sławomir Przyłucki, Rafał Wojciechowski and Łukasz Sturgulewski

Abstract Personal wireless networks are very popular nowadays. Wireless networks administrators have access to a great deal of information about users. They may use selected information in order to try to locate a devices who is active in a network. A mechanism has been implemented which enables the user to locate devices communicating based on bluetooth technologies, using coordinates or references to characteristic points.

1 Introduction

Bluetooth technology is an open standard enabling the construction of wireless communication systems aimed at voice and data transmission at short distances. Its properties include no license fees, low costs of use, maximum range up to 100 m, and low energy consumption up to 100 mW. It enables quick connection between several devices, which create a mobile personal network [3, 5].

Although the first Bluetooth devices appeared in 2000, no application has been introduced to the market so far which would enable users to determine a distance between two devices using Bluetooth communication correctly and explicitly. There are also no solutions enabling users to determine a precise position of a Bluetooth device in space. Existing applications operate with a great accuracy errors; in practice,

A. Sierszeń (✉) · R. Wojciechowski · Ł. Sturgulewski
Institute of Applied Computer Science, Lodz University of Technology,
ul. Stefanowskiego 18/22, 90-924 Łódź, Poland
e-mail: artur.sierszen@p.lodz.pl

R. Wojciechowski
e-mail: rafal.wojciechowski@p.lodz.pl

Ł. Sturgulewski
e-mail: lukasz.sturgulewski@p.lodz.pl

S. Przyłucki
Lublin University of Technology, 38A Nadbystrzycka Str, 20-618 Lublin, Poland
e-mail: s.przylucki@pollub.pl

R.S. Choraś (ed.), *Image Processing and Communications Challenges 7*,
Advances in Intelligent Systems and Computing 389,
DOI 10.1007/978-3-319-23814-2_33

their operation comes down to the possibility of determining whether the devices in question are located within or outside their joint range. Locating solutions are also limited to determining an area where the device in question is located. The size of this area, therefore the accuracy, equals a radius of the base station range. A system which would determine distances or locations in space accurately would be a valuable tool, introducing new functions and enabling the use of Bluetooth devices where they are useless now. Such a locating system could be a tool for navigation inside buildings where Global Positioning System (GPS) does not operate. Data collected by the system could be presented to users in a clear way, e.g. on a map of a determined area. This would make it easier for users to move in large and complicated spaces (e.g. hospitals) and to find desired places or items.

2 Project Assumptions

The purpose of this work is to present results of research which enable authors to answer the question of to what extent it is possible to construct an efficient location system for wireless devices communicating using Bluetooth technology and measure distance between them. A great limitation to this project was the assumption of using widely available devices only. The following devices were used for testing solutions: notebook Acer Aspire One 722 (with Broadcom Bluetooth 3.0 + HS module, 1st class, HCI version 5.603, LMP version 5.16899), laptop with external Bluetooth 2.0 (+EDR adapter based on CSR chipset, 1st class, HCI version 3.3164, LMP version 3.3164), laptop with external Bluetooth 2.0 adapter (based on ISSC chipset, 1 class, HCI version 2.500, LMP version 2.500), mobile phone Nokia 6310i with Bluetooth 2.0 (+EDR module, 3rd class, HCI version 2.500, LMP version 2.500), two wireless Nokia BH-104 Bluetooth 2.1 (+EDR module, 3rd class handsets).

Distances from a superior device to all of these devices were estimated. All the devices (apart from wireless handsets) were also used as base units in the procedure of determining position. Based on well-known theories of locating wireless devices [1, 4, 6, 7] it was decided that the authors would use a method of determining distance between two devices based on measurement of Time of Arrival (ToA) and using of Received Signal Strength Indication (RSSI). In the event when the procedure involves only two devices, it is possible to change a location of any of them. When a larger number of devices is used in a measurement, only the device to be located can be mobile; the remaining units have to stay in permanent and clearly defined positions.

3 Measurement of Time of Arrival (ToA)

The first solution is based on measuring the time of arrival of a signal, which equals the time of a radio signal travelling the distance from the transmitter to the receiver. This can be implemented using two methods:

- the first one requires that timers on both stations be synchronized,
- the second one uses compensation of time differences on timers.

Synchronization consists in establishing a connection between devices in order to tune the timers, what provides a joint reference point which enables the calculation of radio signal sending time. In the event of Bluetooth, synchronization of timers is automatic during the establishing of a connection. This also means automatic synchronization of the transmitter and receiver jump frequency. The former is required because of slot transmission. The latter—because of changeable frequency of the active channel. When the logical channel has already been synchronized, both devices should start calculating impulses. Then, the higher unit should send a package to the other one and stop the counter after the transmission. The lower unit should stop the counter after receiving the package. Both devices have to recalculate the received result based on their own clock frequency per time unit according to the formula

$$i = \frac{t}{f} \tag{1}$$

where t—time of radio wave journey in seconds, i—number of calculated impulses, f—clock frequency.

In addition, the lower unit should send the calculated value in response. During the last step, the higher unit subtracts its own value from the received number, receiving time of the radio wave journey as a result.

Time differences between clocks are compensated using Two Way Ranging (TWR), which is based on the measurement of the distance between the two co-operating radio devices when synchronisation is impossible. This eliminates the necessity of establishing a connection between the units. Phase differences between oscillators are compensated by measuring the distance from the transmitter to the receiver and back (Fig. 1). Signal propagation time is calculated according to the formula

$$t_p = \frac{t_c - t_o}{2} \tag{2}$$

where t_c —total time, t_o—time between receipt and response, t_p—one-way propagation time

For calculating time, a counter is also used which will be started in the unit performing measurements upon the end of transmission and stopped upon the receipt of the response. Its value, after recalculation according to the formula (1), will give total time t_c.

In order to obtain time between receiving the package in the lower unit and returning a response, a mechanism should be provided which would start the counter upon the receipt of the pack-age and stop it after the end of transmission. The reading of this counter recalculated according to the formula (2) has to be sent to the measuring unit. This, unfortunately, requires a connection, what eliminates the main advantage of the method. Its drawbacks include also no synchronization of jump frequency,

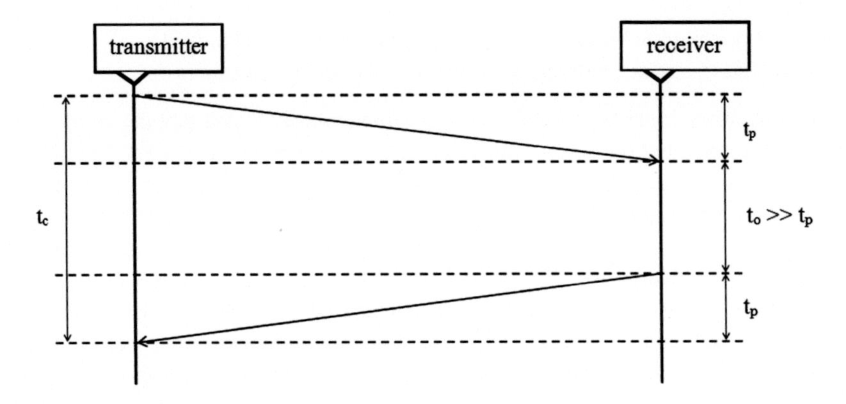

Fig. 1 Two way ranging operation diagram

what involves additional random delays essential to select the same channel. The possibility of calculating time needed for an electromagnetic wave to travel from one unit to another enables measurement of the distance. In order to facilitate calculations, the authors omitted the fact that the electromagnetic wave speed changes in different media. The value of the electromagnetic wave speed was adopted as a speed of light in a vacuum, what is sufficient for initial estimations. Transforming the basic formula for calculating speed, a formula for calculating the path between the points can be obtained.

$$s = t \times \nu \tag{3}$$

where ν—path, s—travelled path, t—time needed for travelling the s path.

Since the devices used for tests provide a range up to 10 m, a reasonable resolution unit is a meter. By transforming the formula (3), assuming the radio wave speed as equal to speed of light in a vacuum and the travelled path as a meter, the following formula (4) was obtained

$$t = \frac{1\,[\text{m}]}{2 \times 10^8\,[\text{m/s}]} = 3.5\,[\text{ns}] \tag{4}$$

This value is also a required minimum value which a clock of the measuring device is able to calculate and which enables determination of the one-metre distance. In the used programming environment (Microsoft .NET), only the System. Diagnostics class with the Stopwatch function (the precision of which is limited with the device clock timing) enables the calculating of time of this interval for devices with Bluetooth class 3 [2].

Unfortunately, devices used in the tests do not have a processor clock quick enough. The value of 285714285 (285.72 MHz) is the quartz clock timing required to provide the interval of 3.5 ns. Although such clocks exist in specialist solutions, it should not be expected that they will be installed in widely used devices in the nearest future. Summing up the first method, in can be concluded that, at present, it is useless due to limited technical possibilities and hardware. However, it is possible that in future this concept may be effective, after the development of technology enabling more precise time measurements in standard solutions.

4 Using Received Signal Strength Indication (RSSI)

The second idea is based on Received Signal Strength Indication RSSI and the relationship indicating that the received signal strength is inversely proportional to the square of the distance between the transmitter and the receiver. In the event of Bluetooth, RSSI is not a direct value of signal strength but information on by how many dB the range of optimum received signal strength was exceeded. Unfortunately, this range is not a precisely defined value and is different for different modules. Therefore, a few tests should be performed before using it in order to determine extreme RSSIs for the target device. It is also important to determine places where the value of 0 is reached and lost. Having obtained these four values, it is possible to calculate rates which help in relating the distance to RSSI. For a positive RSSI, this rate should be calculated using the formula (5); for a negative one—the formula (6).

$$n_1 = \frac{s_1}{R_{max}} \tag{5}$$

where n_1—proportionality rate for positive RSSIs, R_{max}—maximum RSSI for 0 m, s_1—distance for which RSSI changes from a positive value to 0.

$$n_2 = \frac{10 - s_2}{R_{min}} \tag{6}$$

where n_2—proportionality rate for negative RSSIs, R_{min}—maximum RSSI for 10 m, s_2—distance for which RSSI changes from a negative value to 0.

Such a division determines the adopted scheme of RSSI value changes (Fig. 2). The next step consists in relating the rate to a distance using the formula (7). It takes into account the gap for which the rate does not change and assigns it the average of the middle of the range.

$$s = \begin{cases} s_1 - R \times n_1 : R > 0 \\ (s_1 - s_2)/2 \ : R = 0 \\ s_2 + R \times n_2 : R < 0 \end{cases} \tag{7}$$

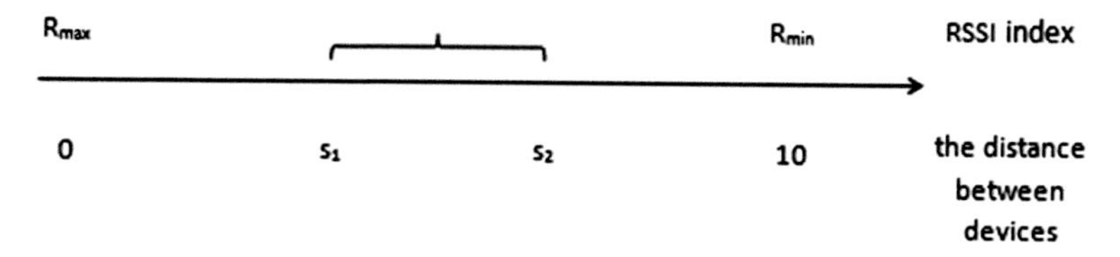

Fig. 2 Relationship between Received Signal Strength Indication (RSSI) and distance

Table 1 Measurements for the Distance of 8 m in an Open Space (A—Acer Aspire One 722 with no active connection; B—Acer Aspire One 722 with an active connection; C—Nokia BH-104 wireless handset; D—Mobile phone Nokia 6310i)

No.	A		B		C		D	
	RSSI*	Time*	RSSI*	Time*	RSSI*	Time*	RSSI*	Time*
1	–16	2495	–20	143	–16	2495	–20	143
2	–15	1065	–17	158	–15	1065	–17	158
3	–15	1079	–21	83	–15	1079	–21	83
4	–16	1066	–20	67	–16	1066	–20	67
5	–14	1071	–20	88	–14	1071	–20	88
6	–16	1074	–20	93	–16	1074	–20	93
7	–15	1068	–18	89	–15	1068	–18	89
8	–16	1125	–18	89	–16	1125	–18	89
9	–16	1069	–17	119	–16	1069	–17	119
10	–17	1071	–20	68	–17	1071	–20	68

RSSI*—RSSI reading
Time*—time of measurement [ms]

The method gives results of low accuracy. In order to assess usefulness of the second method, RSSI values with different distances have to be measured. To this purpose, the authors developed an application operating offline. RSSI values are taken from the package which is a response to an enquiry concerning the support of the service by a remote device. The values were measured in an open space as well as in a closed room. For each distance, the values were measured ten times to eliminate temporary disturbances and random errors. The first table (Table 1) presents the measurements only for the distance of 8 m for three devices. The next table (Table 2) contains only mean values from 10 measurements for each distance.

Based on the results, it can be observed that different values of time required to obtain a response with a RSSI value were measured for the Acer Aspire One 722 notebook. For offline mode this time exceeded a second, what is a satisfactory value for an application trying to locate a device in a time close to real time. When there was an active connection between a smartphone and a notebook, this time was over ten times shorter. The connection was established via an in-built control panel and was not directly related to the operating application. The time difference is only a result of the Bluetooth pile installed in the computer; for other devices, no differences between the active online and offline modes were observed. Therefore, only measurements for offline mode were taken on the wireless handset and on the phone. Another conclusion which can be drawn based on the presented results concerns a longer time required to perform the first data exchange. This is a result of the fact that the tested device does not know the remote device jump frequency and needs additional time to find the right channel. The following enquiries are already sent on the right channel; therefore, the response time is shorter. This has a direct influence on the time of first measurements, which is several times longer than that of the following

Table 2 Mean values of measurements for various distances in an open space (A—Acer Aspire One 722 with no active connection; B—Acer Aspire One 722 with an active connection; C—Nokia BH-104 wireless handset; D—Mobile phone Nokia 6310i)

No.	A		B		C		D	
	RSSI*	Time*	RSSI*	Time*	RSSI*	Time*	RSSI*	Time*
1	5	1267	13	93	21	307	23	186
2	0	1172	0	96	0	432	0	163
3	0	1115	0	82	0	259	–4	144
4	0	1224	–9	86	0	280	–11	262
5	0	1151	–13	77	0	251	–3	163
6	–9	1192	–9	90	–6	226	–12	143
7	–2	1147	–16	91	–4	257	–6	172
8	–8	1148	–23	185	–4	260	–13	258
9	–16	1218	–19	100	–7	207	–13	276
10	–11	1158	–18	84	–13	155	–15	151
11	–6	1144	–20	99	–9	217	–14	159
12	–7	1199	–18	156	–9	154	–16	199
13	–8	1199	–21	99	–8	205	–18	657
14	–9	1175	–21	98	–6	289	–16	273
15	–14	1198	–21	97	–3	276	–17	328
16	–18	1236	–21	105	–7	260	–14	291
17	–23	1302	–22	111	–4	151	–15	271
18	–21	1164	–21	97	–9	186	–14	217
19	–23	1303	–21	96	–7	181	–16	169
20	–22	2287	–21	110	–10	227	–18	211
21	–20	6901	no	no	–14	191	–17	393
22	no	no	no	no	–19	186	–18	720
23	no	no	no	no	–17	164	–19	1589

RSSI*—RSSI reading
Time*—time of measurement [ms]

ones. Another conclusion concerns different RSSI values for separate devices for the same distance. This makes it more difficult to relate RSSI with a distance for different remote devices with a universal formula. An additional difficulty here is the mechanism of adjusting the transmitted signal power, the results of which can be seen in the event of the wire-less handset. Both difficulties were expected due to a great freedom of these mechanisms functioning provided for by manufacturers; however, the inconsistencies are higher than assumed. Results presented in the Table 2 are mean values from 10 measurements of RSSI and time required to obtain the RSSI value for distances with differences of a meter. Based on these, it is easy to observe that the maximum range often exceeds 20 m , which is twice the value of 10 m resulting

Table 3 Mean values of measurements for various distances in a room (A—BH-104 position A; B—BH-104 position B; C—BH-104 position C; D—BH-104 position A)

No.	A		B		C		D	
	RSSI*	Time*	RSSI*	Time*	RSSI*	Time*	RSSI*	Time*
1	6	274	5	312	no	no	6	274
2	0	330	0	292	no	no	0	330
3	0	189	0	207	no	no	0	189
4	0	232	0	218	no	no	0	232
5	0	248	0	192	0	408	0	248
6	0	263	no	no	0	301	0	263
7	0	282	no	no	0	248	0	282
8	–6	246	no	no	0	241	–6	246
9	–9	213	no	no	0	239	–9	213

RSSI*—RSSI reading
Time*—time of measurement [ms]

from the device class. This difference is probably a result of redundancy used by a manufacturer in order to guarantee the 10-m range in rooms or in the events when the aerial is covered by the user. This also confirms the conclusions described below the previous table concerning different RSSI values for different devices and the negative influence of mechanisms for adjusting the transmitted signal power. These two factors make it impossible to determine a range of distances for which RSSI stays equal to 0. The maximum and minimum values for separate devices are different as well.

Based on observations described above, it can be stated that it is impossible to develop a method which would explicitly relate RSSI value to a distance for various devices. The results from measurements in a room (Table 3) were obtained in relation to the same base station placed in different positions, i.e. A, B, and C (Fig. 3). The wall is constructed of bricks and it is 15 cm thick. The door opening is 1 metre wide. The measurements were performed along the line marked with; the measurement along the dotted line was omitted. The Table 3 presents mean values from 9 measurements.

Based on the obtained results, it can be observed that the RSSI value does not change virtually along the distance of up to 8 m . What is even more interesting, if there is an obstacle of a wall, the signal remains within the optimum area of received power for longer. This can be a result of radio signal reflections off side walls. Due to no changes in RSSI values in the room which is 8 m long, RSSI is useless in assessing distance as in open spaces. Based on conducted research it has been determined that the concept based on RSSI value cannot be used to measure distance and, as a result, to localize devices. However, it can be useful in combination with other methods to give approximated information on whether a device is located near to or far from the base station. It is worth considering combining RSSI with Link Quality (LQ) and

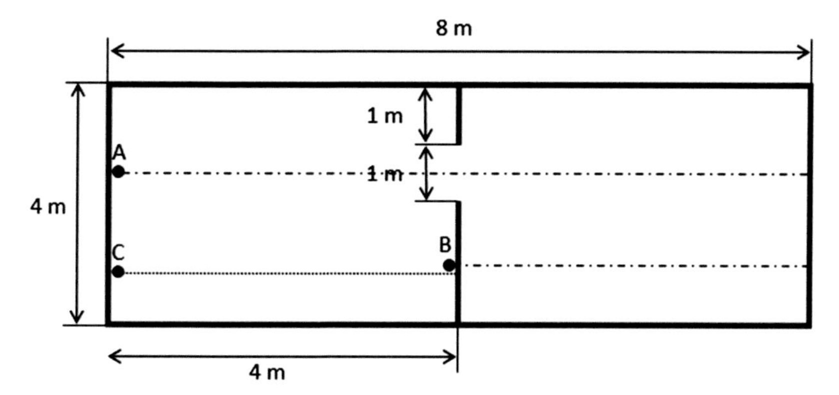

Fig. 3 Distribution of base stations in the room

Transmit Power Level (TPL). Unfortunately, these indexes are not available from the level of a pile operating on Windows Mobile; however, they are possible to obtain in Linux through the BlueZ file.

5 Alternative Solutions

In their theoretical considerations, the authors had taken into account three other methods which were rejected, having obtained results presented above.

5.1 Permanent Assignment of the Value Related to the Maximum Range of a Device

This method makes it possible to approximate distance. It is based on the point—point structure, which enables permanent assignment of the value related to the maximum range of a device resulting from the device class. This solution results in the least accurate measurements; the inconsistencies are equal to the device range. In the test, it was 10 m for the used class 3 module. The measurement itself comes down to checking whether the device which the distance is measured to is located within the activity range or outside it. The majority of present commercial solutions operate based on this method. Although it results in low accuracy of distance measurements, they are more precise when estimating the location up to 3 m. This is possible owing to the increased number of devices with the range over the area in question and to using the trilateration method. This method was omitted in the tests because of the simplicity of the solution and a great number of already existing systems based upon it.

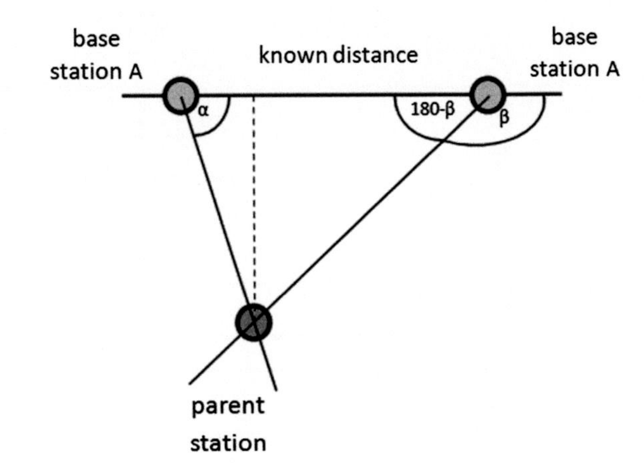

Fig. 4 Distribution of base stations in the room

5.2 Angle of Arrival (AoA) Measurement

This is a measurement method operating within the point—multipoint structure, based on the measurement of the angle of arrival of the signal (AoA). Apart from the measuring device and the base station, the third device is also required. In addition, the two base stations cannot change their positions and the distance between them has to be known beforehand. It is a virtual line based on which a triangle is constructed (Fig. 4). The last method of estimating distance is based on the TDOA technique measuring differences between arrival times of signals at two base stations.

5.3 Measurement of Time Difference of Arrival (TDoA)

In this method, the reference stations must also be stationary, with known distance between each other. Then, based on differences between the arrival times of a signal to the base stations, the distance between them and wave speed, it is possible to determine the distance from the measuring device to any station. In order to calculate time differences, the stations must be synchronised to make them start the counter at the same time and stop it at the moment when the signal from the emitter arrives. The next step is to send the readings of the counters to the emitting station which will calculate reception differences based on the difference between the readings and wave speed. Here, the specificity of Bluetooth based communication should be taken into account, as it is based on time slots and on the possibility of synchronisation only between the master and the slave devices and not between two lower stations. This problem may be solved by synchronising reference stations first and then starting the counters. Then the emitting device should be synchronised with each station. After that, messages should be sent in suitable time slots, stopping the counter and updating the reading on the counter of the second receiver with the time difference of sending the message. In practice, that would come down to the procedure developed in the first method, increasing the costs of labour and resources. It is also possible to use the reverse procedure, i.e. base stations transmit a signal at the same time or with a significant time shift and the third device receives the signals, measuring differences of time of arrival. This eliminates the necessity of exchanging counter readings between the stations; however, it still requires simultaneous synchronisation of 3 devices. As mentioned before, the solution does not eliminate the necessity to synchronise the devices; in addition, it requires simultaneous synchronisation of 3 stations. Although the time needed by the signal to travel the distance is not measured, differences between the arrival time of the signal to separate stations are measured. This requires the same precision in calculating time as in the first method, i.e. the minimum of 3.5 ns. These factors resulted in the rejection of this method due to the parameters of available devices.

6 Alternative Solutions

The concepts based on time of arrival (ToA) and time difference of arrival (TDoA) met an obstacle, i.e. no technological resources which would enable time measurement to an accuracy of a nanosecond. Synchronisation inconsistencies up to 10 μs, allowed by the Bluetooth standard, are also a problem. In future, when technological development leads to installation of oscillators with 1ns resolution, it will be possible to synchronise internal timers of Bluetooth devices more precisely. In such an event, the developed concepts should be successful and should provide the distance measurement accuracy of a metre. The concept using the received signal strength indication (RSSI), which seemed the most suitable for the performed tests, also proved unsuccessful. Although it involved quick data update and did not require a connection between the devices, it did not ensure repeatable results. Furthermore, strength adjustment mechanisms resulted in additional changes of results; therefore, it was not possible to relate RSSI values with the distance between the devices. The information on Link Quality (LQ) and Transmit Power Level (TPL) could help with relating RSSI to distance. However, in the event of the devices used in the tests, it was not possible to obtain values of these indices due to no mechanisms provided by the manufacturer which would enable exchange of information between the application and the device. In order to assess userfulness of additional indices, the platform should be changed to a different one supporting Linux or one should wait until manufacturers provide suitable interfaces to other systems. The concept based on the angle of arrival was rejected at the very beginning of the test due to the requirement of directional transmitting-receiving antennas. The only method which partially enabled meeting the requirements for the constructed system was the concept based on the division of the area into range fields. However, due to the fact that it is virtually useless for distance measurement, it was omitted in the conducted tests.

References

1. Boukerche, A., Oliveira, H., Nakamura, A., Loureiro, A.: Localization systems for Wireless Sensor Networks, Wirel. Commun. **14**(6), 6–12 (2007). IEEE, Ottawa
2. Cirdeiro, C., Agrawal, D.: Ad Hoc Sensor Networks Theory and Applications. World Scientific, Singapore (2006)
3. Ghaleb, I.: Bluetooth Connect Without Cables. Alexandria University, Alexandria (2002)
4. Iskander, M., Zhengqing, Y., Zhijun, Z.: Outdoor/indoor propagation modelling for wireless communication systems. In: Antennas and Propagation Society International Symposium no. 2 2001. IEEE, Utah (2001)
5. Miller, M.: Discovering Bluetooth. Sybex, Alameda (2001)
6. Patwari, N., Ash, J., Hero, A., Moses, R., Correal, N.: Locating the nodes: cooperative localization in wireless sensor networks. IEEE Signal Process. Mag. **22**(4), 54–69 (2005) Michigan
7. Yanying, G., Lo, A., Niemegeers, I.: A survey of indoor positioning systems for wireless personal networks. Commun. Surv. Tutor. IEEE **11**(1), 13–32 (2009) Delft

Self-configuration Networks

Rafał Wojciechowski, Artur Sierszeń and Łukasz Sturgulewski

Abstract The observation of the network behavior and the quantitative description of the network condition is one of the most important activities for network administrators and analysts. Networks are dynamically changing over time, while their structure stays fixed traffic or device load can rapidly change causing networking failures, inaccessibilities or delays due to nodes overload. The article is focused on the active monitoring and configuration of network nodes using the general common standards. The work concerns a design of a decision system based on predefined actions invoking network reconfiguration when observed parameters exceed acceptable limits. The general concept of the self-configuration networks system is presented; the architecture, requirements and basic activities are widely discussed.

Keywords Network monitoring · Network configuration · Decision systems

1 Introduction

A design of a network topology developed by network architects reflects a customer's needs in details. After the deployment and production start-up, the topology meets the real life traffic and is verified in terms of its correctness and efficiency. Frequently, the network needs to be customized because of the growth and development of new sites, security aspects or just to improve its performance. Hardware changes are applied in topologies very rarely because of economic costs or the risk related to the network compliance and functionality. The observation of the network behavior and

R. Wojciechowski (✉) · A. Sierszeń · Ł. Sturgulewski
Institute of Applied Computer Science, Lodz University of Technology,
ul. Stefanowskiego 18/22, 90-924 Łódź, Poland
e-mail: rafal.wojciechowski@p.lodz.pl

A. Sierszeń
e-mail: artur.sierszen@p.lodz.pl

Ł. Sturgulewski
e-mail: lukasz.sturgulewski@p.lodz.pl

R.S. Choraś (ed.), *Image Processing and Communications Challenges 7*,
Advances in Intelligent Systems and Computing 389,
DOI 10.1007/978-3-319-23814-2_34

the quantitative description of its condition is one of the most important activities for network administrators and analysts. Networks are dynamically changing over time the traffic or the device load can rapidly change causing networking failures, inaccessibilities or delays due to nodes overload. The network structure stays fixed but variable conditions or the unpredictable change of the network traffic and the device load may suggest that the network be reconfigured.

There exist some technologies of network automatic reconfiguration which cause changes of the logical topology, e.g. the dynamic routing or link aggregation protocols; however, they represent only a small part of the whole network functionality. Moreover, they are based on a close set of parameters or metrics normalizing the mechanism behavior, which frequently are not generally, logically confirmed. For example, if a node of the OSPF network providing a path to the external network is overloaded (CPU and MEM usage is at high level, the interfaces not participating at OSPF are congested), we can enforce the decrease of the path priority re-direct the traffic outgoing to the external network to another node. The metric based on a set of simple parameters, including CPU, MEM load and mean usage of all interfaces (not only OSPF interfaces), is a more adequate benchmark of the node condition.

2 Network Parametrization

The network activity and the current configuration are reflected by a set of parameters corresponding with interface settings, current load statistics, routing paths etc. This information is valuable for IT analysts and administrators because of the knowledge of the actual network condition, its weaknesses and optimization possibilities. A typical set of monitored parameters is based on the interface/path load; however, other information can be very useful for a complex network analysis, including node resource load or service statuses.

There are a lot of technologies for gathering information about specified networking parameters, e.g. dedicated applications monitoring protocols/systems like SNMP, NetFlow, Zabbix or Cacti [2, 5–10, 12]. For non-typical parameters handling, the low-level scripting using bash or TCL languages or syslog surveys can be used to access device configurations or runtime properties [1, 11]. Some of those technologies also offer the application of changes to the configuration at the runtime (Fig. 1).

3 Self-configuration Network Design

3.1 System Overview

The self-configuration networks design is a software solution intended for the active monitoring and configuration of network nodes using the general common standards.

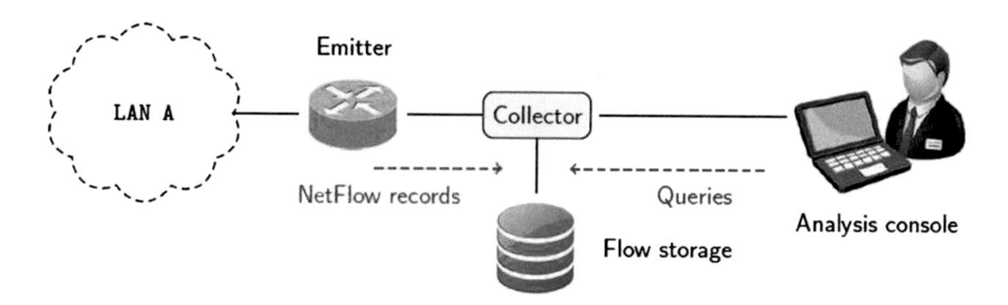

Fig. 1 Typical network monitoring system based on NetFlow

The network behavior is dynamically changing over time the network traffic or the device load can rapidly change causing the networking failures, inaccessibilities or delays due to the node overload. The reason is that the network structure stays static, while the observation of the network parameters suggests that it be reconfigured. The main aim of the project is to balance the networking resource load based on the rules defined by the system analyst involving any measurable and configurable parameter. The main features and requirements of the system are:

- active monitoring and configuration of network nodes,
- fully customizable monitoring and configuration of network nodes according to the user's needs,
- solution based on the common standards like SNMP, NetFlow, TCL, bash scripts, etc. [1, 6, 7, 10, 11]; due to that most systems are closed binary distributions or the software license does not allow one to modify the system structure,
- distributed system architecture limited only to the connectivity between system components,
- independence from the operating systems.

3.2 System Architecture

The architecture of self-configuration networks relies on the separated functional tiers model (Fig. 2). The purpose of the system forces the division into independent subsystems which differ in structure and requirements but have to cooperate with each other. The tiers logically group the functionalities of the system.

The knowledge repository is a database/storage system which enables the definition of the network parameters used in the monitoring process to determine the quantitative value of the parameters and the decision rules describing the reaction of the system to an exceeded limit defined with the parameter value. The monitoring subsystem is a standalone solution for the measurement and calculation of the parameters defined in the knowledge repository using the common standards like SNMP, NetFlow, etc. The current values of the parameters are stored in the repository

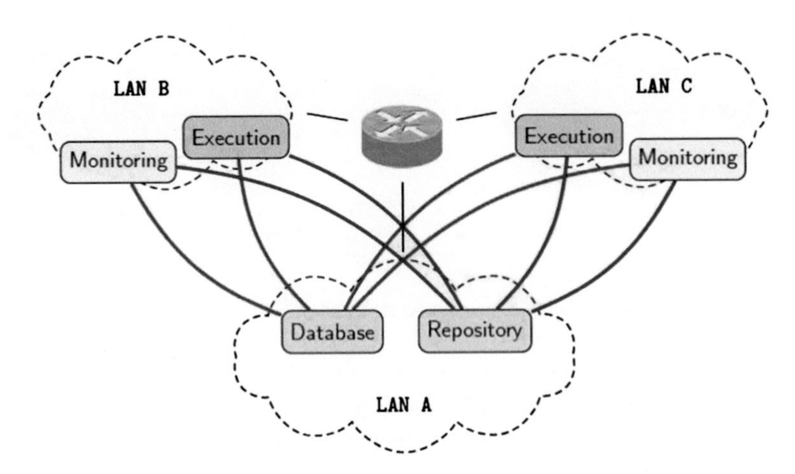

Fig. 2 The system architecture including the knowledge repository, monitoring and execution subsystems

subsystem. The execution subsystem is an application providing the reaction for an inappropriate parameter value. The reaction can be defined as a set of reconfiguration scripts to run on various network nodes.

3.3 Configuration Repository

The most important part of the system containing the main logic is the configuration repository. It consists of the rules which define network parameters and which should be observed, and the actions which determine the reactions on the parameter overlimit. The rules, actions and current parameter values are stored in the database system. The additional scripts for non-standard parameter handling, both for monitoring and reconfiguration purposes, are stored in the server filesystem in the scripts repository. The rules and scripts are defined by the system architect or analyst aware of the network structure and activity as well as of the impact of the topology change on the network behavior.

The example diagram of the database structure is presented in Fig. 3. The `t_parameter` dictionary table contains the definitions of parameters, including the name of a parameter, the IP address of the monitored device, the getting method (SNMP, bash, TCL) and the script name (TCL in case of bash) or OID (SNMP) [1, 8, 9, 11]. The parameters can be grouped together in a complex parameter, which is calculated in a function defined in the `t_complex_parameter` table (average, sum, ...) and related to the parameters by a group identifier and a special weight. It allows one, for example, to define a parameter of mean load of a node as an average of memory and CPU usage and specified network interfaces utilization. The cur-

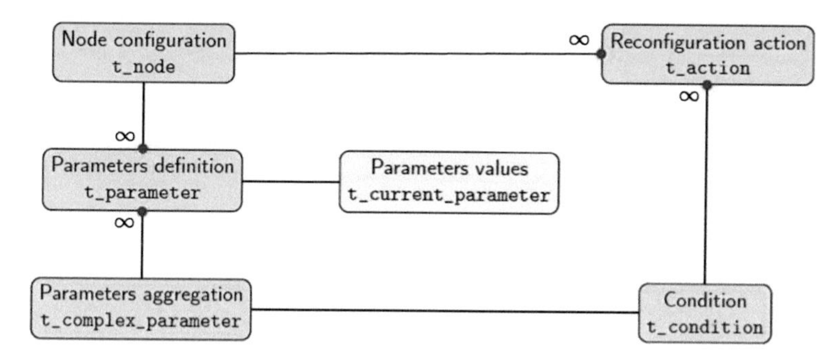

Fig. 3 The entities diagram of the database structure including dictionary and operational tables

rent value of the parameter is stored in the `t_current_parameter` operational table. The activity of the execution unit is controlled by a simple condition action model. The `t_condition` table consists of the list of limit parameters value, the exceeding of which causes reconfiguration. The execution unit calculates the current value of complex parameters and compares them with the defined conditions. The `t_action` table defines actions related to the specified condition like the parameters table, it consists of the IP address of the device, the reconfiguration method (SNMP, bash, TCL) and the script name.

Depending on the purpose and the target device (servers monitoring/configuration, Cisco/Juniper/other devices) [3, 4], scripts can be implemented as bash or TCL scripts which are remotely executed on the specified node.

3.4 Network Monitoring Subsystem

The monitoring subsystem is a standalone part of the self-configuration network solution which is responsible for collecting data from remote monitored devices in a specified network area. The module activity logic is based on the configuration stored in the database system and the script repository, which is loaded at the module startup (Fig. 4, pos. 1). As mentioned above, the configuration describes, in detail, the parameters to get, including the getting method. For each parameter, the module connects to the remote device and gets the parameter value according to the reading method—using SNMP protocol or running the bash or TCL script remotely (Fig. 4, pos. 2). Finally, all parameter values are written in the operational table stored on the database system (Fig. 4, pos. 3).

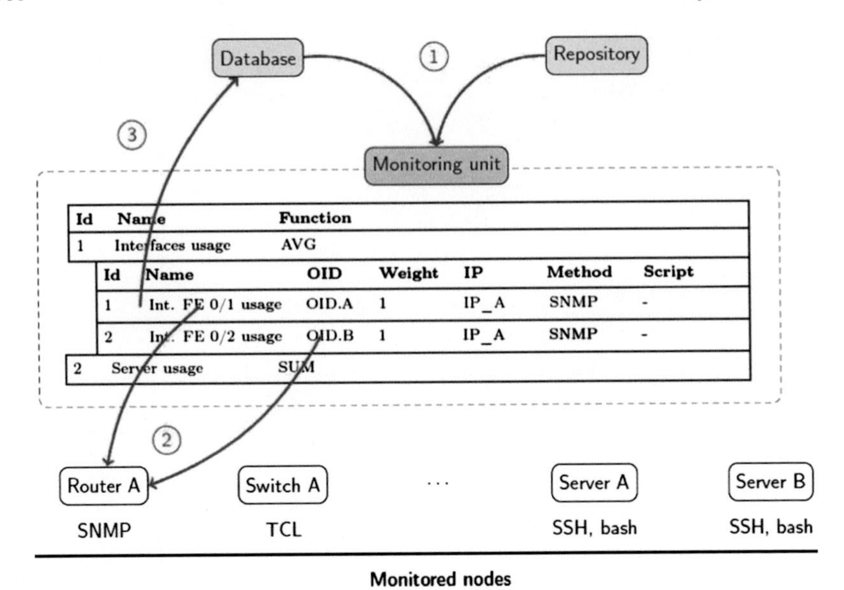

Fig. 4 Network monitoring subsystem

The monitoring subsystem provides information primarily for the reconfiguration subsystem but the data can also be used for the external monitoring visualization software as the data source. The monitoring subsystem is optional and can be replaced with any other software providing the desired and predefined information to the database system.

3.5 Network Reconfiguration/Execution Subsystem

The network reconfiguration subsystem is a decision system based on the predefined rules and actions describing the system reaction on the parameters value overlimit. The main aim of the subsystem is to reconfigure the network topology/mechanism as a result of the parameter's reaching a forbidden value. Like in the monitoring system, the definitions of the parameter limits and reconfiguration actions are loaded at the module startup. For all stored simple parameter values, the appropriate function calculates the aggregates of complex parameters based on mathematical functions—the result is the list of the current values of parameters, which are compared with the limit definition. If a parameter value exceeds its limit, the appropriate reconfiguration scripts are executed on specified network nodes causing a change of the topology. Finally, the database is notified that the topology change was performed.

The network reconfiguration subsystem is only an execution system based on predefined rules. The experience of the analyst or the administrator and the knowledge of the network behavior determine the efficiency of the solution.

4 Self-configuration Examples

4.1 Case 1: The Web Server Migration

The scenario assumes that in the corporation network exist two web server machines—the primary one and the mirror of the main server stored in the second one—the web server backup. The monitoring unit task is to observe the CPU utilization of the `httpd` process in the primary machine. If it exceeds the 70 %, the `httpd` process should be stopped in the first machine and start in the second one.

The configuration of parameters for monitoring unit is presented in Tables 1 and 2. The monitoring unit executes the script given in the parameter definition (Listing 34.1), current parameter value is stored in the operational part of the database system. The execution unit compares the current parameter value with the reference against the condition (Table 3). If the condition is fulfilled, the execution unit runs simple action scripts halting and running `httpd` service on the appropriate nodes (Table 4).

Listing 34.1 Script httpd_CPU.sh.

```
#!/bin/bash
ps axu -C | grep httpd | awk '{sum += $3} END {print sum}'
```

Table 1 Complex parameters definition

Id	Name	Function
1	httpd CPU usage	SUM

Table 2 Parameters definition for parent complex parameter httpd CPU usage

Id	Name	IP	Method	Script
1	CPU	192.168.1.1	bash	httpd_CPU.sh

Table 3 The aggregate view with condition and operational tables (condition, reference and current values of the complex httpd CPU usage parameter)

Id	Name	Condition	Value	Current value
1	httpd CPU usage	GREAT_EQ	70	50

Table 4 The actions defined for primary web server CPU overload

Id	Name	IP	Method	Script
1	httpd_stop_primary	192.168.1.1	bash	httpd_stop_pri.sh
2	httpd_start_backup	192.168.1.2	bash	httpd_start_bck.sh

5 Conclusions

The observation of the network behavior and reaction for any deviant situation is one of the most important duties of networking support staff. Frequently, there are strict restrictions concerning the network condition, especially in critical applications, where the core network and corresponding services should work 24/7 and the desired SLA is close to 100 %. The network usage is dynamically changing over time—the traffic or the devices load can rapidly change causing networking failures, inaccessibilities or delays due to nodes overload. The automatic network monitoring mechanisms give the administrator up-to-date quantitative information about the network performance and, coupled with the appropriate decision system, create the powerful system of the network guard improving the overall performance and condition.

The self-configuration network design is focused on the active monitoring and configuration of network nodes using the general common standards like SNMP, NetFlow, SSH, supported by most hardware manufacturers. It focuses on the design of a decision system based on predefined actions invoking network reconfiguration when the observed parameters exceed the acceptable limits. The main aim of the project is to balance the networking resources load based on the rules defined by the system administrator involving any measurable and configurable parameter. The rules and monitoring/configuration scripts are defined by a person aware of the network structure and activity as well as of the impact of the topology change on the network behavior. The solution allows one to define complex parameters involving many related logical parameters affecting the network functionality and to create script sets for the reconfiguration of networking nodes, including routers, switches, APs, firewalls or servers.

References

1. Bash project website: http://gnu.org/software/bash (2015)
2. Cacti—monitoring system: http://www.cacti.net (2015)
3. Cisco Networks: http://www.cisco.com (2015)
4. Juniper Networks: http://www.juniper.net (2015)
5. Przyłucki, S., Płachecki, K.: Self-similarity analysis of real network traffic inside service provider MAN. Polish J. Environ. Stud. **16**(4A), 248–251 (2007)
6. RFC 1155: Structure and Identification of Management Information for TCP/IP-based Internets (1990)
7. RFC 1157: Simple Network Management Protocol (1990)
8. RFC 1212: Concise MIB Definitions (1991)
9. RFC 1213: Management Information Base for Network Management of TCP/IP-based internets: MIB-II (1991)
10. RFC 3954: Cisco Systems NetFlow Services Export Version 9 (2004)
11. TCL project website: http://www.tcl.tk (2015)
12. Zabbix—enterprise-class monitoring system: http://www.zabbix.com (2015)

Author Index

R.S. Choraś (ed.), *Image Processing and Communications Challenges 7*,
Advances in Intelligent Systems and Computing 389,
DOI 10.1007/978-3-319-23814-2